The Theory and Operation of Spectral Analysis Using ROBFIT

Instrument and Measurement Science Series

Coldwell & Bamford: The Theory and Operation of Spectral Analysis Using ROBFIT

Rabinovich: Measurement Errors: Theory and Practice

The Theory and Operation of Spectral Analysis Using ROBFIT

Robert L. Coldwell
Gary J. Bamford

Instrument and Measurement Science Series

Library of Congress Cataloging-in-Publication Data

Coldwell, Robert L., 1941–
The theory and operation of spectral analysis using ROBFIT / Robert L. Coldwell, Gary J. Bamford.
p. cm.
Includes bibliographical references and index.
ISBN 0-88318-929-1. — ISBN 0-88318-941-0 (pbk.).
1. Spectrum analysis — Computer programs. 2. Spectrum analysis — Data processing. I. Bamford, Gary J., 1958– . II. Title. III. Title: ROBFIT.
QC452.C63 1991
535.8′4′02855369–dc20 91-23561
CIP

This book is dedicated to the researcher who is sure the effect is in the data and needs help proving it!

Contents

Chapter 5. Representation of the background

Chapter 6. Representation of peaks

Chapter 7. Treatment of errors

Chapter 8. Application of ROBFIT

Preface

This book contains the information needed to use the spectral fitting code ROBFIT. It explains the physical and mathematical foundations of the algorithms used in the code and describes how to apply the code to real spectra. ROBFIT is a computer-automated spectrum analysis package that tries to extract the maximum information from a histogram of intensity versus channel number. A nonlinear minimization routine using a generalization of the Newton-Raphson technique has been developed and tested at the University of Florida's Department of Physics and Institute for Astrophysics and Planetary Exploration. This algorithm, SMSQ, is the core of ROBFIT; it ensures that the code finds the minimal fit even with large numbers of nonlinear constants in the fit.

The intended user of the code is the person needing quick and accurate peak fitting with little user intervention. We have endeavored to make the code as "user acceptable" as possible. A user's guide, included as Appendix A, will enable the reader to start running the code immediately. Spectral analysis, however, is not quite that simple. To extract the maximum detail from the data, a great deal of information must be provided. This includes information on peak types and on how the peak widths of each type vary with channel number, and estimates of the deviations to be expected from these estimates for the various peak types. The code uses this information during the fitting process to produce a file of peak locations, widths, and strengths, plus an estimate of the errors in each of these. Chapters 3–6 explain the physics involved in the input parameters used by the code. Chapter 7 explains the method used to generate the error estimates and the limits of their validity. Chapter 8 contains analyzed sample spectra that should enable the reader to see the practical limitations of the code.

Although this book is primarily concerned with ROBFIT, it is more than just a user's manual for the code. It can also serve as a teaching aid for those who are new to the field of spectral analysis and as a guide for those with more experience. The book contains information on mathematical techniques such as spline fitting, robust estimation, and nonlinear regression, which is available in this form nowhere else. We illustrate how the minimization algorithm SMSQ can be used outside ROBFIT as a general nonlinear fitting routine (NLFIT). The book also addresses some of the more practical aspects of spectral fitting. We show how to avoid overfitting a spectrum while fitting peaks down to the noise level. This includes a discussion of how to choose the correct weighting for a particular application. The more experienced reader may also be interested in the way in which the code handles its error analysis. Although this error analysis is a straightforward application of accepted techniques, Equation 7.3, which allows easy calculation of the errors in any functional of the data, is almost impossible to find elsewhere. In addition, the discussion of Section 7.5, which details how to add imperfectly known information to a functional, can be found nowhere else!

A full listing of the code is given in Appendix B. We have included this for two reasons. First, it gives the reader an idea of what is involved in writing a code such as ROBFIT and shows the newcomer to computational methods a complex FORTRAN code. We cannot in all honesty say that the code was designed according to strict software rules because, in real research, codes such as this tend to evolve rather than be designed. In addition, much of the code has been compacted and optimized to make use of the last dregs of speed. This makes it a little difficult to read; however, we have supplied descriptions at the beginning of each routine, which we hope will guide the eye. Second, the listing contains many useful routines. These range from a simple routine called LOCATE, which finds the position in an array of a given value, to a minimization routine QMIN. Of course, unearthing these routines does require a little patience and a certain level of proficiency in FORTRAN programming. However, we hope they will be of use.

For those interested in running the code, a free copy of the most recent version can be obtained either by supplying the authors with a BITNET address to which the code can be sent, or by sending a diskette onto which the code will be copied and a stamped addressed envelope for return of the diskette. Send enquiries to the Institute for Astrophysics and Planetary Exploration, Department of Physics, University of Florida, Gainesville, FL 32611 or by electronic mail to ROBFIT@NERVM.NERDCNFL.EDU.

ACKNOWLEDGMENTS

To bring the ROBFIT suite of programs to their present state has taken well over 10 years. It would be impossible to mention all the people who have contributed to this effort. However, we would like to acknowledge certain individuals whose input has been invaluable.

First, we must blame Dr. Carl Rester for originally suggesting the writing of a spectral fitting code. At the time this suggestion seemed eminently reasonable; however, if we had known in advance that it would take 10 years to write, we may have been a little less enthusiastic in embracing the idea. In his defense, though, during the development Carl has been saddled with not only finding funding for the project but also actually having to use the infant code to analyze spectra. A number of other users of the early versions of the code also need to be mentioned, in particular Mr. Wayne Smart, Dr. Norm Delamater, and especially Dr. Penny

Haskins. The feedback provided by these desperate users has sometimes been painful but without a doubt has helped improve the code. It did not take long to realize that a purely mathematical approach to spectral analysis is only half of what is required to produce a usable code. We thus turned to actually studying nuclear physics at work; to this end, Drs. Buzz Piercey and Gene Dunnam were the nuclear physicists whose intuition we attempted to put into the code. Since the early work, considerable modifications have been made to the routines that make up ROBFIT. Significant contributions to the code have been made by Dr. Steve Alexander and Mr. Charles Coldwell, whose input has helped make the code "user acceptable"; we are still striving for friendly!

Producing a manuscript that describes all this work has been a not-inconsiderable task in itself. A number of people have contributed to converting our garbled scrawl into something that is readable. The typing and retyping of the various versions of the manuscript was performed by three people: Mrs. Susan Lupi, Mrs. Audrey Hunter, and Mrs. Cathy Lawson. We would also like to thank Dr. Amanda Bamford and Mrs. Joey Owen for reading through the many revisions of the manuscript and trying to bring a certain level of comprehension to the work.

We would also like to acknowledge the contributions of Dr. Stuart Jefferies and Mr. Ed Anderson for providing us with the solar physics data; Dr. Ron Schoenau and the entire Northeast Regional Data Center staff, especially Mr. David Nestle, who provided support and encouragement for the IBM mainframe versions of ROBFIT; Lt. Col. George Lasche and Dr. Ralph Alewine for support throughout the development of the code; Dr. Charles Hooper, who enabled us to give the first public presentation of a primitive version of the code; and, finally, Dr. Samuel Trickey, for not only helping in acquiring funding for the code's development, but also contributing significantly to whatever writing clarity is present here.

This work was supported by the Defense Advanced Research Projects Agency, Nuclear Monitoring Research Office, through grant N00014-87-G-1259, monitored by the Office of Naval Research.

The initial development of the ROBFIT user's guide and the development of an IBM mainframe vectorized version of ROBFIT were supported by an IBM grant to the University of Florida and the Northeast Regional Data Center.

To all the aforementioned people and institutions, we are extremely grateful for their involvement and patience in producing this work. Of course, all the errors, omissions, and other deficiencies of the book and code are our responsibility.

Finally, we wish the best of luck to those people who use the code in their research work.

CHAPTER 1

Introduction

This introductory chapter will help prospective users of the code quickly decide whether ROBFIT meets their needs.

The first section gives an overview of how the code operates, and the second section goes into this operation in more detail. This chapter introduces the user to the basics of running ROBFIT. The following chapters contain information on specific areas of the code. Once you have become familiar with the general operating characteristics of the code, you can skip to the chapter that you are most interested in. However, newcomers to spectral analysis are advised to read all chapters.

1.1 WHAT IS ROBFIT?

ROBFIT is a computerized spectral analysis package. Spectra are histograms of intensity versus channel number and contain peaks superimposed on an underlying background function. ROBFIT has been developed to aid the analysis of spectra containing many peaks residing on a complex background function. ROBFIT runs on various computer systems. At the present time we have versions that run on IBM and Macintosh personal computers and on VAX and IBM mainframe computers. The code is written in FORTRAN and is almost entirely self-contained. The major changes needed to adapt it to new operating systems can be found in a separate file. This file also contains the graphics commands for viewing the fitted data.

ROBFIT performs a least-squares fit to spectral data using user-chosen peak shapes and a background function composed of cubic splines. Cubic splines can be thought of as piecewise third-order polynomials. During analysis, the entire background is fitted to a single spline function. This technique allows the background fit to follow complex fluctuations in the data and to incorporate all possible information. This results in a more accurate representation of the background. We provide an introduction to this cubic spline fitting in Chapter 3.

An algorithm called SMSQ has been written to minimize, in a least-squares sense, the difference between the data and the cubic splines of the fit to the data. Full details of this minimization procedure are given in Chapter 4.

SMSQ is an enhanced minimization scheme, which means that ROBFIT can analyze spectra even with many constants in the fit. Quick fits for studying the gross properties of a spectrum can be completed in minutes on a personal computer. The code is configured so that it can run for long periods of time, successively analyzing spectrum 1, spectrum 2, and so on. This is necessary because on smaller personal computer systems a fit to the noise level of 100 peaks and a background composed of 20 constants requires three to four hours. In fact, we have used the code to analyze spectra that called for as many as 100 hours. After this period, the user has a representation of the data that is statistically correct and a fit that uses all the available data in its peak and background determinations.

Many spectral analysis packages require the spectrum to be broken into small sections, each of which is then fitted separately. Since these packages need substantial user intervention, analyzing a 100-peak spectrum would require much more time than that taken by ROBFIT. Also, because these analyses are done on a small scale, the resulting fits may misrepresent the background continuum in regions with multiple peaks or complex background functions.

A further advantage of ROBFIT is that the code can represent any peak shape. Common peak shapes such as Gaussian, Lorentzian, and Voigt profiles are supplied with the code, but the user can generate other shapes either by picking a clean peak from the data and generating a peak shape from that, or by fitting some previous calibration of the peak shape. A peak shape is built from back-to-back cubic splines, which the code then uses as a representation of that peak shape during the fitting phase. A maximum of five independent peak shapes can be used in a single fit.

1.2 ROBFIT OPERATION

In its operation, ROBFIT separates spectra into two functions: background and foreground. The background contains slowly varying features of the spectra, and the foreground contains the high-frequency content. Accurate separation of these functions allows the code to detect small peaks and decompose multiple-peak structures. ROBFIT iterates on background and foreground fitting to move smaller peaks from the background to the foreground.

The background is fitted over the entire spectrum as a set of cubic splines with adjustable knots. A knot is the place at which two cubic splines meet. This background fitting is explained further in Chapter 5. Fitting over the whole spectral range allows the background features to be continuously fitted with fewer constants, resulting in a more accurate representation than is possible when features are fitted in small sections with the peaks. Two algorithms make this possible. The first, data compression, uses a robust averaging technique to reduce contributions to the background from peaks and spurious high points. The second, the minimization algorithm SMSQ, minimizes chi-square (χ^2) with respect to the constants of the background and foreground. With the background represented as a smoothly varying function, peaks can be identified as regions of the spectra that lie above this background curve.

The foreground is the peak content of the spectra and may contain many peaks of various shapes. Each shape is represented by a collection of back-to-back cubic splines. These shapes are then fitted to the spectrum to determine where the peaks reside. Overlapping peaks, of various shapes, are fitted simultaneously. This process is explained further in Chapter 6. A fast residual locater for finding peak regions, a routine for systematically controlling the allowed widths, and SMSQ are the principal algorithms used in peak determination.

To control ROBFIT, the user chooses the number of constants to be fitted to the background and picks the peak shapes to be used. A typical first run of the code, illustrated in Figure 1.1, could proceed as follows:

1. Select a peak shape. A Gaussian could be chosen for the first run of the code. Although this may not be the correct shape, the code will still operate. The output can be checked for an incorrect peak shape once fitting has ended.
2. Run the main code using the default options and the shape just generated. A large cutoff value can be chosen for a first run. This has the effect of identifying only large peaks within the spectrum.

ROBFIT will fit the data and generate peak and background constant files and a graphical file for viewing the fit. Several cycles of fitting different peak shapes with various cutoffs (defined in Appendix A) may be needed to find the correct peak shapes in the data. During this cycling the user can also monitor the background function. By increasing the number of background constants in the fit, an optimum number can be decided upon. Once these preliminary quantities have been set, the cutoff can be lowered to enable detection of smaller peaks.

The code works well for detecting small peaks within a spectrum, where "small" means the same size as the local background fluctuations. It also performs well in regions with low or complex backgrounds. Under these conditions, the effects of imperfections in the representation of the peaks are negligible. Problems can arise when the data contain large peaks on a smooth background. Here the effects of peak shape variations may disrupt the fitting. Although the code can adjust for peak height and width variations, it cannot change the shape of the peak during fitting. It assumes that the peak shape stays the same across the entire spectral range. Any variation in this peak shape will lead to a bad fit. A bad fit to a large peak has a significant effect on the χ^2 minimization and will limit the code's ability to home in on the best fit. Generally, this can be quite readily seen as the code breaks the large peaks into many smaller ones, a situation we call braiding. When this happens, the user must examine closely the form of the peak shape being fitted. Shape variation can be accounted for, to some extent, by generating several peak shapes for various channel regions. As the code scans through the spectrum, it can choose a peak shape that is closer to the actual data peak shape.

Throughout the analysis, ROBFIT keeps track of the errors introduced into each of the constants. This ensures that peak positions, widths, and strengths are statistically correct when the program ends. This error treatment is explained further in Chapter 7.

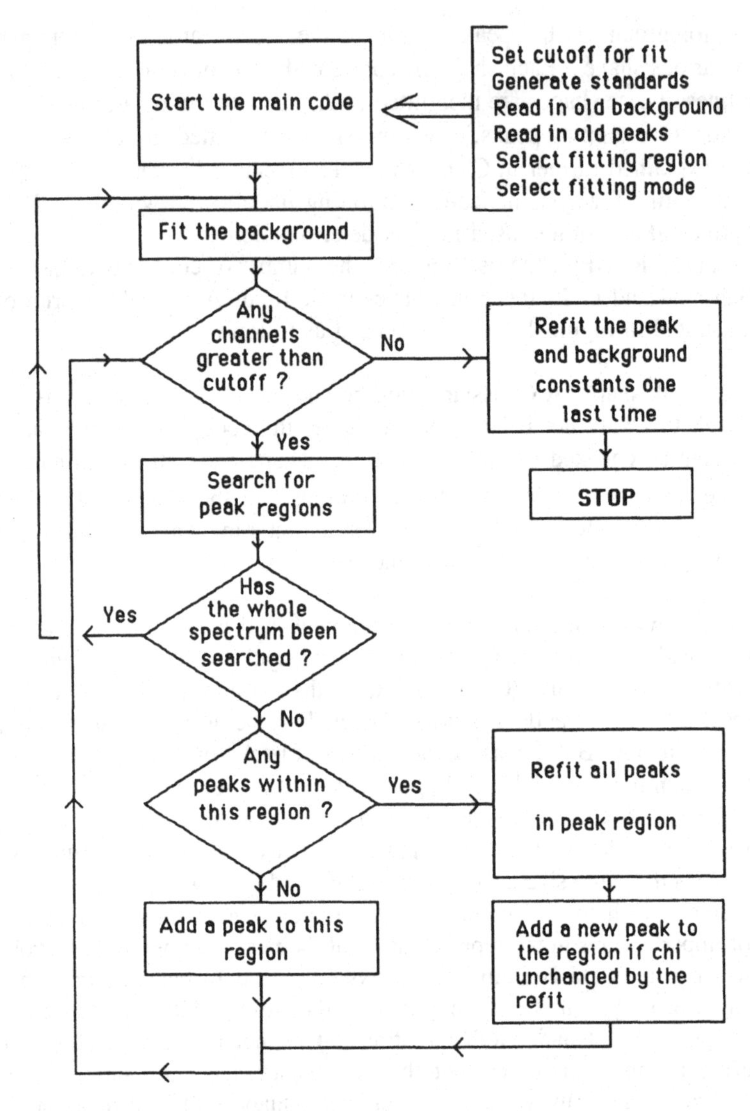

FIGURE 1.1
The general flow of ROBFIT.

Although the code has been mainly used in the study of gamma-ray spectra, we expect that it will be of use in many areas of spectral analysis. In Chapter 8 we highlight the flexibility of the code by fitting a set of data samples taken from computer-generated test cases, gamma-ray astrophysics, and solar seismology.

The preceding description has given a brief glimpse of the computer techniques used by ROBFIT in analyzing a spectrum. To introduce the reader to some popular methods used in other spectral analysis codes, we give a brief review of computerized spectral analysis in Chapter 2. In Chapter 9 we discuss when to consider using other analysis packages.

Appendix B contains a full listing of the code for those users who wish to study the actual coding of the algorithms used by ROBFIT. This listing will be used by those who wish to pirate certain parts of the code. The minimization routine SMSQ, for example, can be used outside ROBFIT as a general minimization algorithm. Appendix C contains a listing of a nonlinear least-squares fitting routine that uses SMSQ. We call this routine NLFIT, and it is described in Chapter 4. For those readers wishing to get their hands dirty straight away, Appendix A contains a user's guide to ROBFIT.

CHAPTER 2

Review of Spectral Fitting

This chapter gives an overview of computer methods used in spectral analysis. We concentrate on gamma-ray spectra throughout this book. This is partly because ROBFIT was developed on these data and partly because, by their very nature, they provide stringent tests for any analysis package. However, we hope the code can be used in other areas of spectral analysis.

We begin by discussing the requirements of any spectral analysis code, and we then show how these requirements are fulfilled in the more popular analysis packages. Finally, we illustrate how ROBFIT performs these standard procedures.

2.1 A TYPICAL SPECTRUM

ROBFIT has been used as a data reduction package in the Antarctic research program to study gamma rays produced by Supernova 1987A.[1] Figure 2.1 shows a typical gamma-ray spectrum from the supernova, illustrating the complexity of these spectra. Some regions contain well-separated peaks, and others have overlapping peaks.

The interesting features in these spectra are not the easily identified large peaks, but the small ones that are the same size as the noise. In the supernova case, the signals also fell in regions of overlapping peaks. Figure 2.2 shows an actual supernova signal identified by ROBFIT. Refer to Chapter 8 for the analysis of the supernova data. We mention this analysis here only to illustrate the level to which the fitting must be taken to extract these very small signals.

The goals of analysis of such spectra can be summarized as:

1. To correctly determine all single-peak parameters
2. To correctly separate peaks in overlapping peak regions
3. To provide accurate analysis in the presence of noise

The reader interested in learning more about gamma-ray spectra is referred to Knoll,[2] who gives a comprehensive review of the subject.

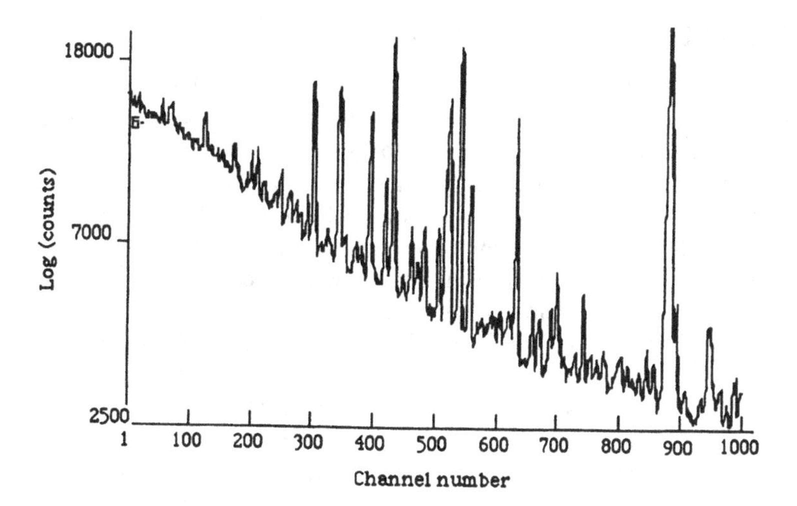

FIGURE 2.1
A typical gamma-ray spectrum from Supernova 1987A.

Although the supernova data are complicated, a limited analysis can be done manually either by graphically viewing the data and fitting functions by eye or by applying stripping techniques.[3] Stripping techniques are used to successively subtract individual components from the spectrum. These methods, however, are laborious and prone to error. A computer analysis provides a more accurate evaluation and also automates the peak finding process, easing the burden on the user. The following section gives a brief review of computer fitting techniques.

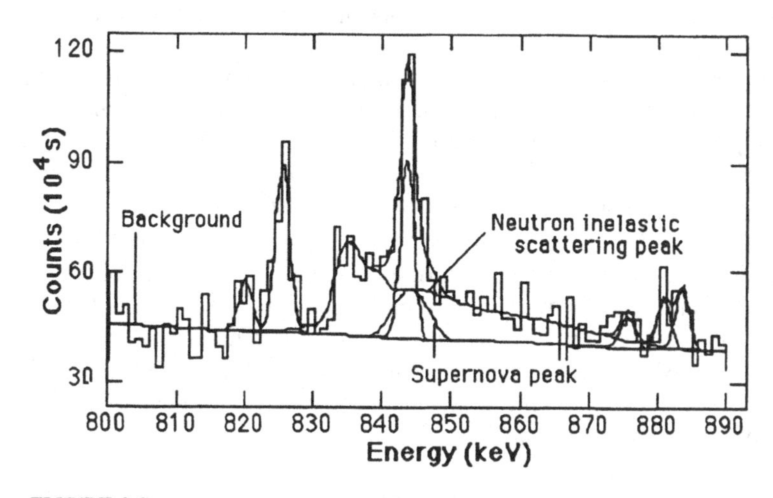

FIGURE 2.2
View of a small channel region surrounding the supernova signal.

2.2 COMPUTERIZED FITTING

The aim in fitting any spectrum is to identify "all" the peaks. Before any peaks can be found, the background must be estimated. Separating the peak and background functions, however, is a difficult task. To fit a peak, the background must be accurately determined, but in determining the background, all the peaks need to be identified. The various analysis routines break this cycle in many ways. In this section we do not intend to review each of these routines in detail, but rather to give an overview of the techniques used. We will concentrate only on the areas of peak and background identification.

Background determination is generally done in one of two ways:

1. By considering only small regions of the spectrum and assuming the background is a simple function that can be fitted with a low-order polynomial
2. By taking the whole spectrum and in some way filtering out the peaks to leave a smooth background function

The first method was used in the earliest computerized analysis programs. Its simple low-order polynomial fit makes it easy to encode. However, there are dangers with such an approach. Describing the true background may call for more than just the low-order polynomial. This is especially so when the region contains multiple peaks or a rapidly changing background function. This means that the background will interfere with the determination of peak parameters, such as widths, areas, and position. Even so, the more popular spectral analysis codes[4,5,6] still use this procedure.

The second method is the more desirable approach to background fitting. Here all the spectral information is available, providing a better representation of the background. Several techniques have been used to perform this automated background calculation. The two most popular methods are listed here.

1. *Averaging techniques* are used to smooth the background continuum. Points that lie beyond a specified number of standard deviations from the smoothed data are replaced by the smoothed value. This averaging effectively filters out the high-frequency content of the peaks. Iterating on this procedure progressively removes peaks and leaves a smoothed background function.
2. *Minimal search techniques* use the fact that peak-free regions of a spectrum contain fewer counts than neighboring peak regions. A search for background minima can be carried out by determining local minima over the entire spectral range on a channel-by-channel basis. A local minimum is identified as any channel in which the number of counts is less than the numbers in adjacent channels. The local minima are then connected using an interpolation scheme to give the background function.

A review of four of the more widely used background estimating codes is given in Burgess.[7]

Once the background has been calculated, a search for peaks can begin. Unlike the background, the peak shapes may be known a priori or, if not, then approximated by some standard function. It is generally enough simply to find the peak positions and then insert the peaks at those points. A least-squares fit can then be used to better determine peak widths, strengths, and positions. The critical task is the accurate determination of peak positions. These are generally chosen by one of two methods.

The first method looks for regions of high curvature within the data. Peaks can be detected in two ways:

1. *Studying the "first differences,"* which are calculated by subtracting a given channel value from the next highest channel value. Peaks appear here as a significant change in the sign of the difference.
2. *Studying the "second differences"* by measuring the degree of curvature at each channel position. Sharp photopeaks have high curvature, whereas the background, such as Compton continua, generally shows only mild curvature.[8]

The second method performs a correlation between the data and a given peak shape.[9] For efficient application, the width of the correlation function must be about the same as the width of a peak.

Whichever technique is used, the idea is to detect as many true peaks and as few spurious peaks as possible. Once a peak position has been found, the next stage is to accurately determine all peak positions (peak centroids), widths, and areas (number of counts under the peak). Most peak fitting procedures use an iterative nonlinear least-squares method for such determinations. Here the parameters that describe a peak shape are varied while the shape is being fitted to the data. The optimum values are determined by minimizing the χ^2 for the fit. These optimum values then describe the peak.[10] An introduction to least-squares fitting and the χ^2 function can be found in Bevington.[11]

This peak fitting methodology has troubles if the actual peak shape is unknown. Generally, gamma-ray peak shapes can be approximated by a Gaussian function, which is usually combined with some additional tailing criterion. However, other shapes are not so easily parameterized. One advantage to using ROBFIT is its ability to use the experimental data to accurately deal with odd peak shapes.

2.3 ROBFIT FITTING MODES

ROBFIT contains a number of modes of operation—three for curve fitting, three for display of the data and curve fits, and one for calibration. For fitting, we have codes to:

1. Generate peak shapes for use in fitting
2. Fit the background alone
3. Perform a full spectral fit

For graphically viewing the fits, we have codes to:

1. Display the raw data
2. Display the peak shapes
3. Display the full spectral fit

For calibration, we have a code to calibrate the x-axis and determine efficiency. Earlier versions of the code have been described in two papers by Coldwell.[12,13]

In Chapter 1 we gave a brief introduction to ROBFIT operation. We now give a more detailed account to illustrate the code's background and peak fitting algorithms.

Determination of the background is carried out by smoothing the spectra. This involves compressing the data by replacing every 16 channels by the average over that region. The average is a robust average, which helps reduce remaining peak contributions. This smoothing has two functions: it filters out large data fluctuations, and because it reduces the number of points in the fit, it speeds up the background fitting. After smoothing, the entire spectral channel range is used to calculate the shape of the background. The user then chooses the number of constants to be fitted to this smoothed background. ROBFIT adds splines to its representation of the background, refitting after each spline until the user-supplied number has been reached. This background calculation can be interspersed with peak fitting when a full spectral fit is performed. One must always be cautious not to overfit the background by specifying too many constants, as this will allow it too much flexibility. In practice, the background can be kept "stiff," with constants added slowly to "slacken" it as needed. This way the background fit will not rise into the peak regions of the spectrum.

In its peak fitting phase ROBFIT needs knowledge of the peak shapes it expects to find within the data. These shapes are termed standards. ROBFIT's standard-generating mode allows the user either to create a standard from the raw data or to use one of the three most common shapes—Lorentzian, Gaussian, or Voigt. Each standard is fitted to a polynomial plus a set of back-to-back cubic splines. The polynomial is optional. It is used to represent the background under a peak when a standard is generated from real data. The splines represent the standard peak. Complex-shaped standards can be generated using this technique.

ROBFIT uses the standards to search for peaks within the data. Each peak is determined by minimizing a weighted sum of the differences between the data and a background-plus-peaks function, with respect to the peak height, location, and width. After initial peak fitting has been carried out, the weights and the background-plus-peaks function are redetermined. The old peaks are refitted, and new peaks are added. This cycle is repeated until no peaks exceed a user-specified statistical significance. Before finishing, all the fitted parameters are reoptimized to ensure correct representation of peaks and background. This iterative procedure was outlined in Figure 1.1.

The full spectral fitting routines form the core of the fitting code. They perform a complete spectral analysis on the data. After the code has been cycled, the peak positions, widths, and strengths are fully determined. ROBFIT is configured so that it can start with or without information on the background and/or peak parameters. For example, if the background has been determined in a previous run, this information can be entered at the beginning of the next run. Then the code will not have to redetermine the background coefficients, and the fit will be speeded up. Alternatively, the positions or widths of certain peaks may be known beforehand. In such a case, this information can also be linked into the program, again speeding up the fit and possibly increasing the accuracy of the final peak parameters. Once all peak positions and areas have been determined, the final step is to calibrate the peak list so that the channel numbers and areas can be related to "real world" quantities. These would be energies and intensities in nuclear physics applications.

To summarize the main features of ROBFIT, we can say that:

1. Any peak shape can be accommodated.
2. By using the entire spectrum to determine the background, a better background representation is obtained.
3. Iterating on peak and background fitting means that background contamination of peaks and vice versa does not become a problem.
4. Reoptimizing all the splines after fitting gives a more accurate fit to the background and foreground functions and provides an accurate measure of the errors on these quantities.
5. The preceding features means that extremely small peaks can be picked out of a noisy spectrum.

CHAPTER 3

Why Use Cubic Splines?

The key to ROBFIT's ability to decompose spectral features lies in its use of spline functions. The idea is that all background and foreground features can be represented by smooth functions. Splines are the way in which these smooth functions are created. In this chapter we review the properties of cubic splines and illustrate how they can be used to build up the spectral features. We start with a look at polynomial fitting as an introduction to the techniques used in least-squares fitting. We will also illustrate when polynomial fitting breaks down. We then show how the use of cubic spline functions can eliminate some of these problems.

3.1 WHY NOT USE A SIMPLE POLYNOMIAL?

The usual way to fit an unknown function of data points y_i is to start by approximating the function by a polynomial:

$$y_{\mathrm{ap}}(x) = c_0 + c_1 x + c_2 x^2 + \cdots + c_M x^M$$

This approximation is then fitted to the data using a least-squares fit. In a least-squares fit, one tries to minimize the difference between the data and the approximating function by choosing the optimum c_n constants. This is done by minimizing a set of differences

$$\Delta_{\mathrm{diff}} = (y_i - y_{\mathrm{ap}}(x_i))^2$$

where y_i and x_i represent one data point. An illustration of polynomial fitting is given in Figure 3.1. We go into the problems involved in polynomial fitting in a little more detail in Chapter 4.

The optimum set of constants can be calculated by minimizing the sum of the differences from each of the data points:

$$\mathrm{Optm} = \min\left(\sum \Delta_{\mathrm{diff}}\right)$$

To take the illustration of polynomial fitting a stage further, let us consider a specific example of a two-constant (first-order) polynomial fit. Here

$$y_{\mathrm{ap}} = c_0 + c_1 x$$

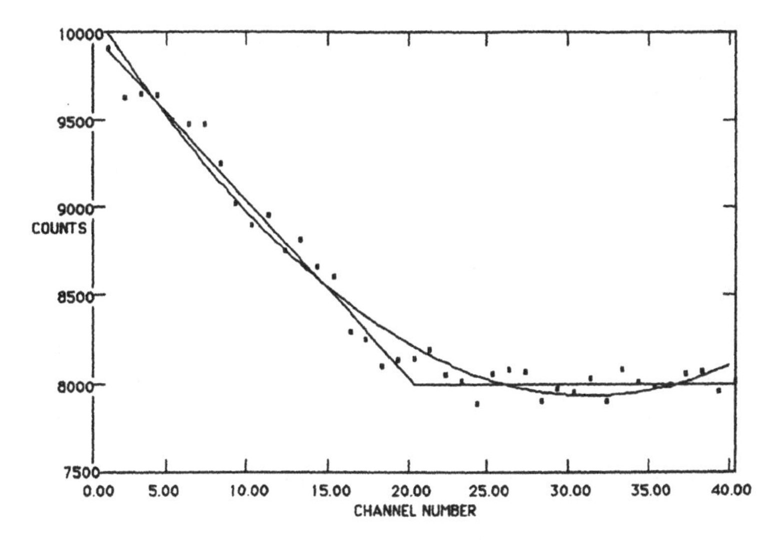

FIGURE 3.1
Illustration of polynomial fitting. The straight lines represent the function being fitted. The data points are randomly distributed about this function in a Gaussian manner. The curved polynomial fit is from a four-constant fit to the data points.

or, in the more usual format,

$$y_{\text{ap}} = a + bx$$

which is a straight-line fit to the data. We can now show some general features of polynomial fitting. Our minimizing function, χ^2, is

$$\chi^2 = \sum (y_i - a - bx_i)^2$$

which must be minimized with respect to the constants a and b. We can find the minimum for each of the constants by differentiating the minimizing function with respect to that constant. The minimum for a is given by

$$\frac{\partial \chi^2}{\partial a} = -2\sum (y_i - a - bx_i) = 0$$

and for b it is

$$\frac{\partial \chi^2}{\partial b} = -2\sum \{x_i((y_i - a - bx_i))\} = 0$$

This gives us two simultaneous equations to solve. In the general case of an mth coefficient fit, we end up with m simultaneous equations. This means that to fit the m coefficients, we need at least m data points. In our straight-line fit this is clearly illustrated by the fact that we cannot fit a straight line to one point, but we can to two points. Using more than two points allows us to include

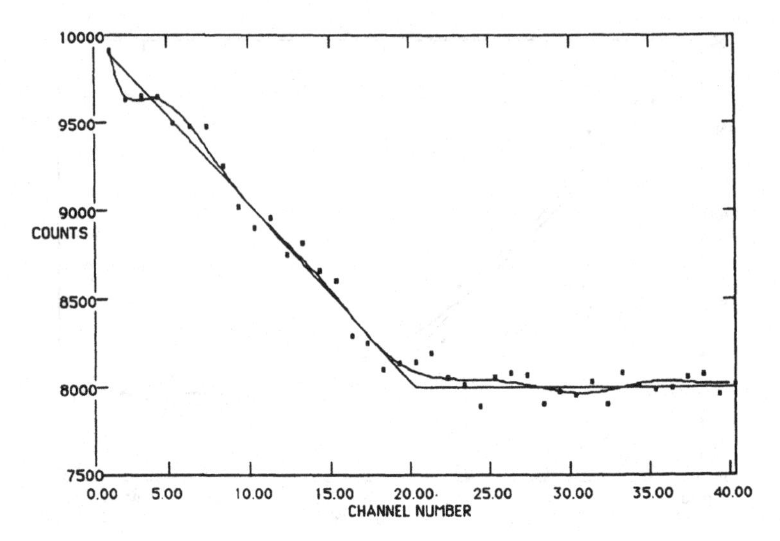

FIGURE 3.2
Illustration of "looping" caused by using too many constants in the fit.

additional information in our minimizing function, giving a better measurement of the constants. The simultaneous equations can be expressed in matrix form and solved (for the constants) by inverting the matrix. This matrix inversion can cause problems, however, when the value of m is large.

In considering smooth curves for m values of approximately 5, or for m much less than the number of data points, the preceding minimization scheme works well. However, for large m the matrix becomes nearly singular, calling for extremely high precision in the inversion. Although these problems can be overcome by using orthonormal polynomials, a basic problem associated with high m fitting still persists: the tendency for the fit to loop though the data points as it overfits the data. This "looping" is illustrated in Figure 3.2.

This phenomenon is a numerical representation of the familiar Gibbs phenomenon seen in oscilloscope patterns of sharp waveforms. It is caused by the fact that in the fitting function all derivatives are continuous. For example, in Figure 3.2 the large negative value of the first derivative before channel 20 continues for some distance past channel 20, and the resulting miss of the data is corrected by a positive first derivative to the right of channel 25. Normally, the problem occurs in derivatives higher than the first. The solution is to replace x^k by a spline of order L that has discontinuities in the Lth derivative.

3.2 CUBIC SPLINES AND THEIR PROPERTIES

The spline function has been likened by Schoenberg[14] to the draftsman's spline, which is a flexible strip used for drawing smooth curves. In a mathematical spline, the strip is replaced by a sequence of polynomial functions connected at points called knots. Any curve can then be portrayed by a sequence of these low-order polynomials. The advantage of using a sequence of low-order polynomials is that complicated nonlinear functions can be described, even in instances where they cannot be represented by a simple polynomial. In a cubic spline, the polynomial segments are of degree 3. At the knots, cubic splines have the property that they are continuous in themselves and their first and second derivatives.

In our usage, a spectrum consists of a sequence of abscissas, ordinates, and standard deviations:

$$\{x_i\}, \{y_i\}, \{e_i\} \quad (i = 1, \ldots N)$$

We assume this spectrum measures a series of peaks superimposed on a background. The object is to find properties of the peaks, such as location, area, and width; to estimate the form of the background; and to determine the uncertainties in these properties. We use cubic splines to represent the peak and background features.

For example, the cubic spline used to represent the background can be written as

$$S(x) = \sum_{i=0}^{3} c_i x^i + \sum_{j=1}^{M} d_j (k_j - x)_+^3$$

with

$$\begin{aligned}(x)_+ &= 0 \quad \text{for } x \leq 0\\ (x)_+ &= x \quad \text{for } x > 0\end{aligned}$$

where $\sum c_i x^i$ describes a third-order polynomial background, which is used as the basis for background fitting. The number of splines (M) to be included is set by the user. These add a contribution $\sum d_j(k_j - x)_+^3$ to the background representation. The terms c_i and d_j are the scaling constants for the polynomial and individual spline coefficients. The k_js are the knot positions. Figure 3.3 illustrates the effect of increasing the number of knots in a particular fit.

The background function derivatives are

$$S'(x) = \sum_{i=1}^{3} c_i i x^{i-1} - 3\sum_{j=1}^{M} d_j (k_j - x)_+^2$$

$$S''(x) = \sum_{i=2}^{3} c_i i(i-1)x^{i-2} + 6\sum_{j=1}^{M} d_j (k_j - x)_+$$

$$S'''(x) = 6c_3 + 6\sum_{j=1}^{M} d_j \theta(k_j - x)_+$$

where

$$\begin{aligned}\theta(x)_+ &= 1 \quad \text{for } x > 0\\ &= 0 \quad \text{for } x < 0\end{aligned}$$

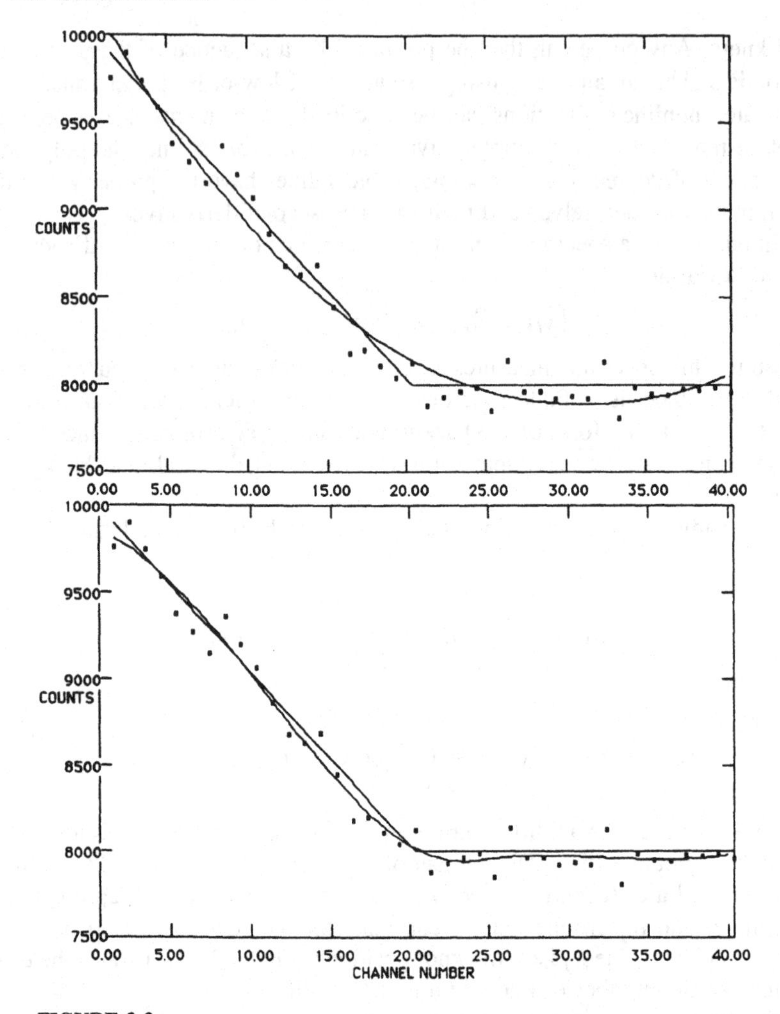

FIGURE 3.3
Effect of increasing the number of knots in a fit. The upper figure shows a two-knot fit, and the lower figure shows a four-knot fit to the same data.

These are plotted for $M = 4$, $c = \{1, -1, 1, -1\}$, $d = \{1, -1, 1, -1\}$, and

$$k = \{0.2, 0.4, 0.6, 0.8\}$$

in Figure 3.4. Note the fact that straight lines connect the points

$$S''(0, 0.2, 0.4, 0.6, 0.8, 1.0) = (-0.4, -1.6, -1.6, -2.8, -2.8, -4.0)$$

The theorem of interest to us was proved by Holladay[15] in 1957 and is restated by Ahlberg, Nilson, and Walsh[16] as Theorem (Holladay). Let $\Delta : a = x_0 < x_1 < \cdots < x_n = b$ and a set of real numbers $\{y_i\}$ $(i = 0, 1, \ldots, N)$ be given. Then of all functions $f(x)$ having a continuous second derivative on $[a, b]$

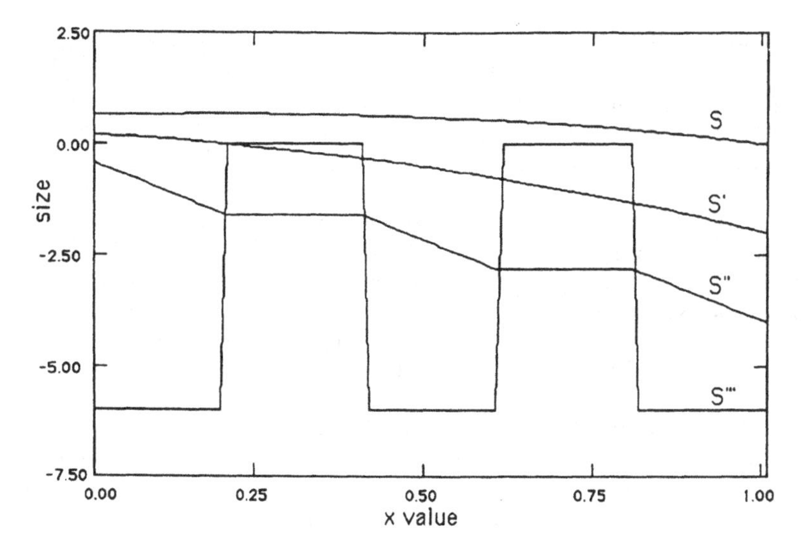

FIGURE 3.4
The background function S and its first, second, and third derivatives.

and such that $f(x_i) = y_i$ $(i = 0, 1, \ldots, N)$, the spline function $S_\Delta(f_j x)$ with junction points at the x_i and with $S''_\Delta(f_j a) = S''_\Delta(f_j b) = 0$ minimizes the integral

$$\int_a^b |f''(x)|^2 dx$$

In other words, the spline function is the function representing the data with the smoothest possible second derivatives. In essence, this is true because for splines, straight lines connect the relevant data points. These must be connected, but no others need be, so the straight lines minimize the integral.

Cubic splines have also been shown to converge in the interval (a, b) to all functions with continuous fourth derivatives such that

$$\| f - S \|_\infty \leq \frac{5}{384} \| f^{(4)} \|_\infty h^4$$

where f is the function, S is the cubic spline, $\| f^{(4)} \|_\infty$ is the largest value of the absolute value of the fourth derivative in (a, b), and h is the maximum spacing between knots. Also, they converge to the derivatives $f(n)$ with larger constants and lower powers h^{4-n}. This means that in any interval, enough knots guarantee convergence to the background. That is, if the data can be represented by an underlying function (in the mathematical sense) with continuous and defined fourth derivative, then it can be fitted, to within our statistical error limits, with a finite number of terms. This may not seem too valuable, but when all else is failing it can be somewhat comforting to know, and it is frequently all that one has.

The cubic splines used in this book differ quantitatively from the splines just discussed. About 4,000 data points will be fitted with splines containing between 10 and 50 knots. The smoothness of the fit will allow us to pick out sharp features, which will normally be fitted separately as peaks. The desire to find the smallest possible peaks led to de Boor's statements on varying knots:

> Use of a code for finding a (locally) best approximation from $S_{k,n}$ [the knot placement] is expensive. It is warranted only when precise placement of some knots is essential for the quality of the approximation (e.g., when the data exhibits some discontinuity) *and* an approximation with as few parameters as possible is wanted. Otherwise, an approximation with two or three times as many well-chosen knots is much cheaper to obtain and usually, just as effective.[17]

The first of de Boor's conditions is true. Finding each knot location takes ROBFIT about 50 times as long as finding the constant associated with the knot. The background, however, is a cubic polynomial between knots and, therefore, subject to the Gibbs phenomenon of oscillating through the data points shown in Figure 3.2. Extra knots imply regions in which there are constants not needed to fit the background.

To illustrate this point, in Figure 3.5 a single knot at 18 can be found by minimizing the χ^2 to about 36 with respect to the six constants in the fit. These constants include the location of the knot. A χ^2 of 45 can be found much faster by simply placing the knot at 24, as would seem to be indicated by the data. The fit responds by trying to keep the first derivative constant at 24, which artificially drives the curve low at 26, making the randomly high fluctuation at this value appear to be a spurious peak. To produce our goal of maximum information,

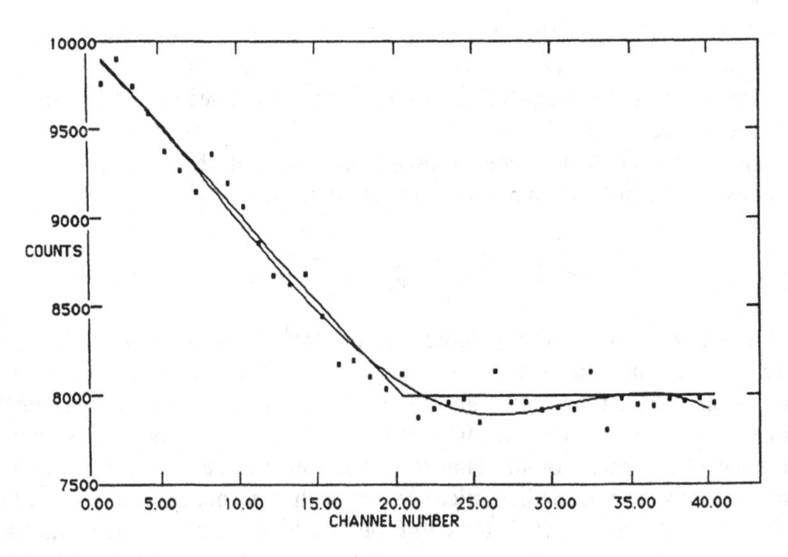

FIGURE 3.5
Spline fitting using a four-constant polynomial plus a single knot fit.

each constant in the background must be so involved in reproducing the structure of the background that it cannot cause trouble. Thus, it is usually essential that the background be fitted with the minimum number of splines. Extracting the maximum information in this manner consumes more computer CPU time than would otherwise be called for to fit the data. Once the knot locations have been determined, one can save time and gain accuracy, in some cases, by reusing them. This will be discussed further in Chapters 6 and 8.

One last comment about the splines used to fit the background is that frequently the background is positive definite. When this is the case, it can easily be incorporated by making the fitting function

$$f_A(x) = \exp(S(x))$$

which cannot become negative. While retaining the local properties of $S(x)$, $f_A(x)$ can vary by many orders of magnitude and is therefore a better approximation to positive definite data. The LOG option, which gives this in ROBFIT, should be used unless the desired fit should have negative regions.

CHAPTER 4

The Nonlinear Minimization Scheme

This chapter gives a review of weighted least-squares fitting and its applicability to spectral fitting. We introduce various methods for calculating channel weights and detail the minimization scheme used within ROBFIT. The final section illustrates how ROBFIT uses nonlinear Newton-Raphson minimization and accelerated extrapolation routines to find the least-squares minimum.

4.1 A REVIEW OF WEIGHTED LEAST-SQUARES FITTING

The basic approach to any type of curve fitting is to design a figure of merit that measures the agreement between the data and a chosen model. This value is usually designed to become smaller as the agreement between data and model gets better. The model parameters can then be adjusted to minimize the figure of merit until optimum agreement between data and model is achieved. This is the basis of the least-squares method of curve fitting.

Real data, however, are a bit more complicated. Each measurement is subject to random measuring errors, so a theoretically ideal model may never precisely describe the measurements. Fortunately, we can avoid the problem by performing a weighted fit. In a weighted fit, greater significance is given to the more accurately determined data points. All that is needed is some measure of the "goodness" of the fit. In least-squares fitting, the figure of merit and this "goodness" measurement are the same quantity, but this is not always so, as we demonstrate later in this section. In least-squares fitting, the quantity to be minimized is a function $S(c_n, x_i)$, with $n = 1, \ldots, M$ the number of constants in the fit and $i = 1, \ldots, N$ the number of data points, where

$$S = \sum_{i=1}^{N} \omega_i (f_i - f_{\text{th}}(\mathbf{c_n}, x_i))^2 \tag{4.1}$$

with

x_i = ith channel of the spectrum
$\mathbf{c_n}$ = Constants for the fit
f_i = Actual data value at channel i
f_{th} = Theoretical model of the data determined using the constants c_n
ω_i = Weight on the data point at channel i

In a weighted fit, the weighting for each individual channel must be specified. ROBFIT contains the following options for choosing the weights:

1. Weight = 1/Data value. This is generally used with gamma-ray spectra. Here, the assumption is that the deviation in a particular channel follows Poisson statistics.
2. Estimated from the spread in the data. The channel-to-channel fluctuations are used to give an estimate of the variance and, hence, the weight from the expression Weight = 1/Variance^2.
3. User-defined at each channel. The code has the capability to read a file containing the weights for each data point.

Each of these weighting schemes is discussed in greater detail in the following sections.

If the data deviations are distributed about the mean position in a Gaussian manner, then the quantity S is equivalent to the goodness-of-fit estimator. This means that minimizing S can be used to give the best parameters for the fit. For a more detailed account of least-squares fitting and an explanation of the χ^2 function, we refer the reader to Bevington.[11] We discuss the treatment of non-Gaussian statistics in Section 4.1.2 and in Chapter 8, but for now we simply indicate that Equation 4.1 is the quantity that is minimized. The goodness of the fit is also measured using Equation 4.1. For an ideal fit, the value of S should be equal to the number of data points (N) minus the number of constants (M) within a standard deviation of $\sqrt{2N}$. The quantity $N-M$ is called the number of degrees of freedom of the data. In statistical terminology, one says that the chi-square per number of degrees of freedom, $\chi^2/(N-M)$, equals unity for a good fit. Values significantly greater than unity indicate a bad fit, and values significantly less than unity indicate an overestimation of the errors. Once the minimizing function S has been correctly formulated, an optimization procedure can be applied to extract the "best fit" parameters. ROBFIT uses the extended Newton-Raphson technique, described in Section 4.3.

The following subsections describe how to choose the correct weights under various operating conditions.

4.1.1 The Standard Weighting Scheme

Frequently, the channel weights can be calculated directly from the individual data points. For example, in gamma-ray spectra, a data value corresponds to the number of measurements of a particular gamma-ray energy in a certain time interval. These measuring statistics follow a Poisson distribution. Each data point is a single measurement of the mean number of counts, f_i, for that interval. The standard deviation on this value is given by $\sigma_i = \sqrt{f_i}$. These deviations can be used to calculate the weights on individual channels from $\omega_i = 1/\sigma_i^2$.

Problems can arise in dealing with small data values, where "small" means less than five counts per channel. In this situation, the weights must be chosen with care.

4.1.2 Dealing with Low Numbers of Counts

When data containing count rates of about five or fewer per channel are analyzed, fluctuations in the data values can lead to large variations in the weights. In these regions we can still express the weights as

$$\omega_i = \frac{1}{\sigma_i^2}$$

where σ_i is the standard deviation in the counts at each channel i, as described in Section 4.1.1. For moderately good statistics we can get an estimate of the weights directly from the data by setting the variance (σ_i^2) as

$$\sigma_i^2 = f_i \qquad \text{giving} \qquad \omega_i = \frac{1}{f_i}$$

where f_i is the data value at channel i. As the f_i values become low, these two equations start to break down, showing that the single data point can no longer be used as a good estimator of the standard deviation.[18] This can be illustrated by considering the data values to be

$$f_i = \overline{f_i} + \Delta(f_i)$$

where $\overline{f_i}$ is the expected value of f_i as calculated from a large ensemble of measurements of f_i, and $\Delta(f_i)$ is the random statistical error in f_i. The correct variance would be

$$\sigma_i^2 = \overline{\Delta(f_i)^2} \qquad \text{with} \qquad \overline{\Delta(f_i)^2} = \overline{f_i}$$

for Poisson statistics. We do not have a good estimate of $\overline{f_i}$ with only a single measurement. Taking the data value f_i as $\overline{f_i}$ introduces large fluctuations in the weights. If the weighting criteria are not changed and a peak is fitted to the limited data using the weighted least-squares method, the fit will systematically underestimate the area of the peak. This can be explained by considering the positive and negative fluctuations of $\Delta(f_i)$. For positive fluctuations f_i is larger than $\overline{f_i}$, and this will underweight these positive-going values. For negative fluctuations f_i is smaller than $\overline{f_i}$, leading to overweighting of the negative-going values. The size of the fitted function for small counts is seen to be systematically biased to lower peak heights. This bias in the weighting is also true for the larger data values, but there the $1/f_i$ weighting is not as sensitive to small changes.

Several changes to the method of least squares have been tested and found to give unbiased weights in the low statistics region.[19] The most successful methods use a smoothing algorithm to replace the actual data value with its smoothed estimate. This helps reduce the data fluctuations, enabling them again to be used in the weighting determination. ROBFIT uses a similar method; it replaces the data value by the fitted value or 1, whichever is larger. The fitted value is calculated using the entire spectrum and provides a better estimate of the underlying function. As the code adds peaks and changes the background, the weights are continually updated. At the outset of peak fitting there is a tendency to misrepresent the weights under a peak. However, iterating on peak and background fitting and redetermining the weights ensures that the correct weights are eventually found. This process is illustrated in Figure 4.1.

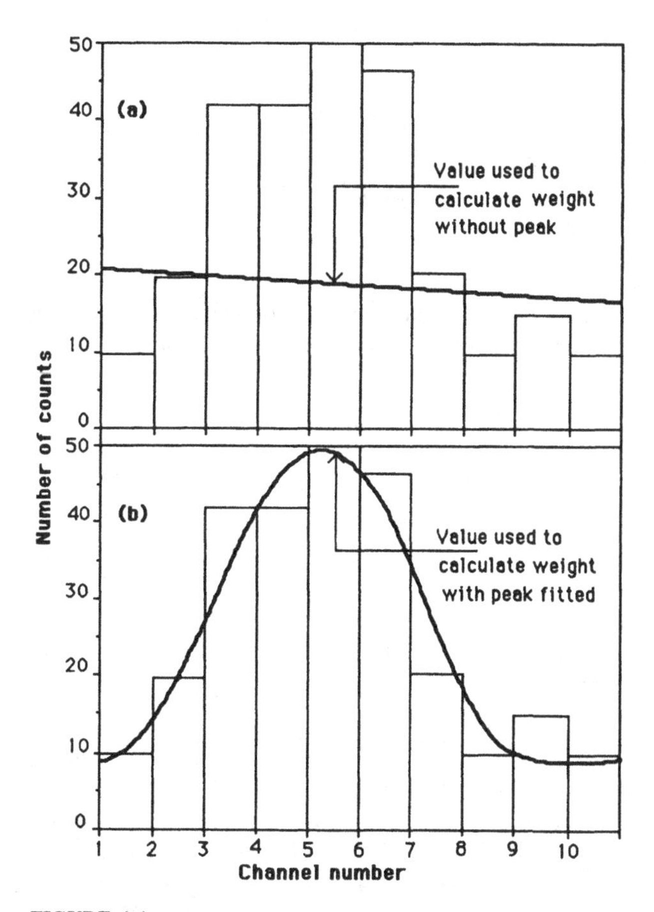

FIGURE 4.1
(*a*) A fit consisting of a background function alone. The weight of channel 5 is calculated from the fitted value indicated. (*b*) The effect of adding in the peak alters the fit so that now the weight is calculated correctly.

4.1.3 What to Do When the Weights Are Unknown

With gamma-ray spectra the individual channel weights are simple to calculate. For spectra in which the statistical nature of the data fluctuations are unknown, the channel weights can be estimated from the channel-to-channel variations in the data. If the background is stored as a function $f_b(x_i)$ and the foreground, or peaks, as $p(x_i)$, the fitted function can be represented as

$$f_{\mathrm{th}}(x_i) = f_b(x_i) + p(x_i)$$

The data fluctuations around the fitted curve are given by

$$d_i = f_i - f_b(x_i) - p(x_i)$$

with d_i being the difference between the data and the fit, as shown in Figure 4.2. The usual definition of a weight for the ith channel is given by

$$\omega_i = \frac{1}{\sigma_i^2}$$

where σ_i^2 is an estimate of the variance at the ith channel. Generally, the variance of a set of measurements (m_j) is given by

$$\sigma_i^2 = <(m_j - \mu)^2>$$

where μ is the mean of the parent distribution and $<>$ represents an expected value. The expected value of a function $f(x)$ is defined as the weighted average value of the function averaged over all possible values of the variable x, so that

$$< f(x) > = \lim_{N\to\infty} \left[\frac{1}{N} \sum_i f(x_i) \right]$$

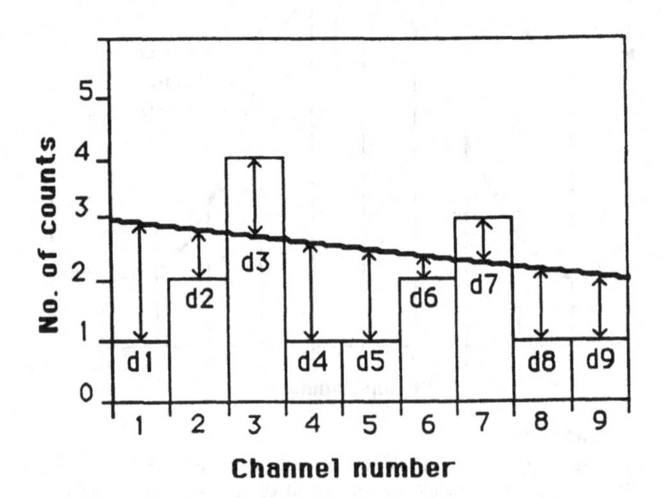

FIGURE 4.2
Data fluctuations around the fitted curve.

Using the curve fit, at channel i, $f_b(x_i) + p(x_i)$ represents the mean, μ. The variance can therefore be expressed as

$$\sigma_i^2 = < d_i^2 >$$

Once the fit has been established, the channel weights can be calculated using this equation. At the outset of fitting, however, the background and foreground are not well defined, and d_i must be calculated by an alternative method. An estimate of d_i, d'_i, can be made by considering the differences on either side of the ith channel. Figure 4.3 shows that this estimate is given by

$$d'_i = 2f_i - f_{i+1} - f_{i-1}$$

and so

$$< d_i'^2 > = < (2f_{\text{th}} + 2d_i - (f_{\text{th}} + d_{i+1} + f_{\text{th}} + d_{i-1}))^2 >$$

If the differences d_i, d_{i+1}, and d_{i-1} are randomly distributed about the mean, f_{th}, which is assumed constant across the three channels, then the expectation values for the cross products will average out to zero, leaving

$$< d_i'^2 > = < 4d_i^2 + d_{i+1}^2 + d_{i-1}^2 >$$

from which we estimate

$$< d_i^2 > = \frac{1}{6} < d_i'^2 > \qquad (4.2)$$

A better estimate of the variance uses the fitted f_{th}, which is not quite constant across the three channels, to form

$$\sigma_i^2 = \frac{1}{6} < 4(f_i - f_{\text{th}})^2 + (f_{i+1} - f_{\text{th}})^2 + (f_{i-1} - f_{\text{th}})^2 > \qquad (4.3)$$

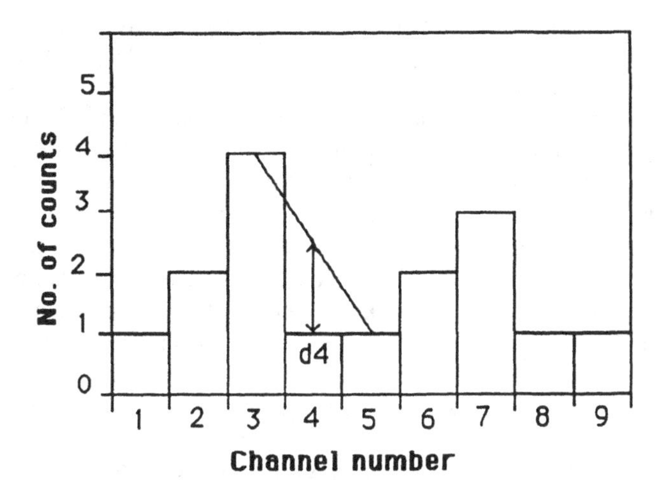

FIGURE 4.3
Estimating the difference at channel 4 at the outset of fitting.

and the weights can be estimated in the usual manner using

$$\omega_i = \frac{1}{\sigma_i^2}$$

At the outset of fitting, the weights are determined using Equation 4.2. They are then continually updated during fitting using Equation 4.3. This combination ensures the best possible weighting at each stage of fitting.

Before this weighting scheme is implemented, a further correction is made to the weights. Within the range of a peak, where range is defined as twice the peak width (wp), there is a tendency for the fitted function to follow the data. This is because the 2wp region is fitted with a three-constant function (position, width, and strength of the standard peak), which results in overfitting. This is more pronounced for small peak widths. To correct for this, the weights are scaled by a factor

$$\frac{\text{Number of channels}}{\text{Number of channels} - 3}$$

where the number of channels is 2wp, giving the scaling factor

$$\frac{\text{wp}}{(\text{wp} - 1.5)}$$

This correction compensates for the reduced degrees of freedom for these small regions.

4.1.4 Weighting Subtracted Spectra

ROBFIT gives the user the ability to subtract one spectrum from another. Subtraction is useful for comparing two spectra that differ by a small signal that is present in only one of the spectra. An example of the use of this technique is the analysis of the Antarctic supernova data.[1] Here, two spectra were effectively taken: the first "on" supernova, with the instrument pointing directly at the supernova, and the second "off" supernova, with it pointing away. All nonsupernova signals are then, ideally, contained within the "off" spectrum while the tiny supernova signal is picked up in the "on" spectrum. The "off" spectrum contains the background to the supernova signal. This is not to be confused with the background continuum that resides under the peaks in both spectra. The procedure is to subtract the "off" background from the "on" spectrum, leaving the residual supernova signal.

Subtraction of spectra can be used to enhance a small signal, but one must be cautious concerning its accuracy. After a subtraction, the random fluctuations from channel to channel in the residual spectrum will have increased. This can be shown by considering the subtraction of two numbers x_1 and x_2 and their random errors $\Delta(x_1)$ and $\Delta(x_2)$:

$$(x_1 \pm \Delta x_1) - (x_2 \pm \Delta x_2) = (x_1 - x_2) \pm \sqrt{\Delta x_1^2 + \Delta x_2^2}$$

Here we see that the error in the difference is larger than in the individual numbers. This is precisely what happens when two spectra are subtracted. This may mask the effects of a small peak. If the spectra contain only a simple underlying background continuum, generally there is no gain from using the subtraction method. Small peaks can be identified by fitting the background continuum with few constants and then removing its contribution from the spectrum. For a complex background, a subtraction helps only if the "off" spectrum contains an adequate number of counts. In doing an experiment, it is usual to spend most of the valuable observing time taking data for the "on" spectrum. However, if a subtraction is to be the way of analysis, then a reasonable amount of time must also be spent building up the "off" spectrum. Otherwise, the poorly defined "off" spectrum will dominate the errors of the subtraction. Therefore, there is a trade-off between the number of constants used to fit the background continuum and the number of counts recorded in the "off" spectrum. Of course, this is dependent on the actual experiment being carried out.

ROBFIT can subtract spectra of differing normalizations. Two scaling factors are entered, one for each spectrum. The second spectrum is then rescaled by a normalization value equal to Factor 1/Factor 2. If these are unequal, then an additional error proportional to (Factor 1/Factor 2)^2 is introduced. After subtracting the two spectra, the code stores the subtracted data and the sum of the spectra in two separate files. The summed data contain the individual channel weights for the subtracted data. These weights can be read by the main curve-fitting code and applied to each channel in fitting the subtracted data.

Refer to Chapter 3 of Knoll[2] for a review of the statistical considerations involved in gamma-ray spectroscopy.

4.2 SMSQ, THE MINIMIZATION ALGORITHM

This section details the minimization algorithm, SMSQ, used to optimize the fit to the data. The object is to minimize χ^2, defined in Section 4.1, with respect to the peak and background constants. The required inputs into SMSQ are χ_0^2 (the initial χ^2), $\partial\chi_0^2/\partial c_l$, and $\partial^2\chi_0^2/\partial c_l \partial c_m$, where the cs are the constants for the fit. Other inputs are Fr, the fractional reduction in χ^2 asked for in this minimization step, and $\mathbf{c}_0$, the starting values for the constants. In ROBFIT, the number of constants defining a fit can be anything from 1 to several hundred. To increase the fitting accuracy, knots are continually added to the background and new splines are added to represent peaks. Each background knot adds two constants (position and height), and each peak adds three (height, location, and width). The time needed to fit the spectrum rises as the number of constants increases, so it is important to have a fast minimization scheme. ROBFIT uses a nonlinear Newton-Raphson technique, which is described as follows.

To recap, the quantity minimized is

$$\chi_0^2 = \sum_{i=1}^{N} \omega_i (f_i - f_{\text{th}}(\mathbf{c_0}, x_i))^2$$

where χ_0^2 is the initial value of χ calculated using $\mathbf{c_0}$. The procedure is to modify the constants $\mathbf{c}$ to reduce χ^2.

One minimization scheme would be to map out the χ^2 space for all possible values of the constants. For large numbers of constants, this method becomes impractical. Consequently, ROBFIT uses an alternative method based on a nonlinear minimization technique. The algorithms used are collectively called SMSQ. To find the optimum fit, ROBFIT performs the following sequence of events.

1. At the start of minimization, SMSQ asks for a fractional reduction (Fr) in χ^2 equal to Fr$*\chi_0^2 (= \chi_p^2)$. To find this reduced value, SMSQ needs exact values of χ_0^2 and $\partial\chi_0^2/\partial c_l$ along with an approximate $\partial^2\chi_0^2/\partial c_l \partial c_m$ for the M constants being fitted. These are used in the expansion

$$\chi_p^2 = \chi_0^2 + \sum_{l=1}^{M} \frac{\partial \chi_0^2}{\partial c_l}(c_{\text{Fr}l} - c_{0l}) + \frac{1}{2}\sum_{\substack{l=1\\m=1}}^{M} \frac{\partial^2 \chi_0^2}{\partial c_l \partial c_m}(c_{\text{Fr}l} - c_{0l})(c_{\text{Fr}m} - c_{0m})$$

2. Using nonlinear Newton-Raphson minimization and accelerated extrapolation techniques, SMSQ predicts the change of the constants from $\mathbf{c_0}$ to $\mathbf{c_{Fr}}$ needed to bring about the fractional reduction of step 1. $\chi_p^2(\mathbf{c_{Fr}})$ is the predicted value of χ^2 for these constants.
3. Using the predicted constants $\mathbf{c_{Fr}}$, ROBFIT makes a new fit to the data and calculates $\chi_N^2(\mathbf{c_{Fr}})$, the actual χ^2 for the $\mathbf{c_{Fr}}$ constants.
   a. If $\chi_N - \chi_p$ is less than some tolerance level, the prediction is following the actual data, and the fractional reduction can be made larger. Set $\chi_0 = \chi_N$, $\mathbf{c_0} = \mathbf{c_{Fr}}$, and Fr $=$ Fr $*$ Fr.
   b. If $\chi_0 < \chi_N$, then the step went too far. Leave χ_0 and $\mathbf{c_0}$ unchanged, and set Fr $= 0.75 + 0.25 *$ Fr.
   c. If $\chi_p < \chi_N < \chi_0$, then the step is converging too slowly. Set $\chi_0 = \chi_N$, $\mathbf{c_0} = \mathbf{c_{Fr}}$, and Fr $= 0.5 + 0.5 *$ Fr. SMSQ repeats the preceding steps, from step 1, until either $\chi_p = \chi_0$ (i.e., we can predict no lower value) or $(\chi_p - \chi_0)/\chi_0$ is less than some tolerance level. Additionally, a limit is placed on the total number of loops the code can make in this minimization.
4. Once the minimum has been found, additional constants can be added to the fit and the whole process repeated from step 1.

This sequence of events is shown in Figure 4.4.

The speed at which the code finds a minimum is directly related to the nonlinear minimization step 2, the step that calculates $\chi_p^2(\mathbf{c_{Fr}})$. At the heart of the

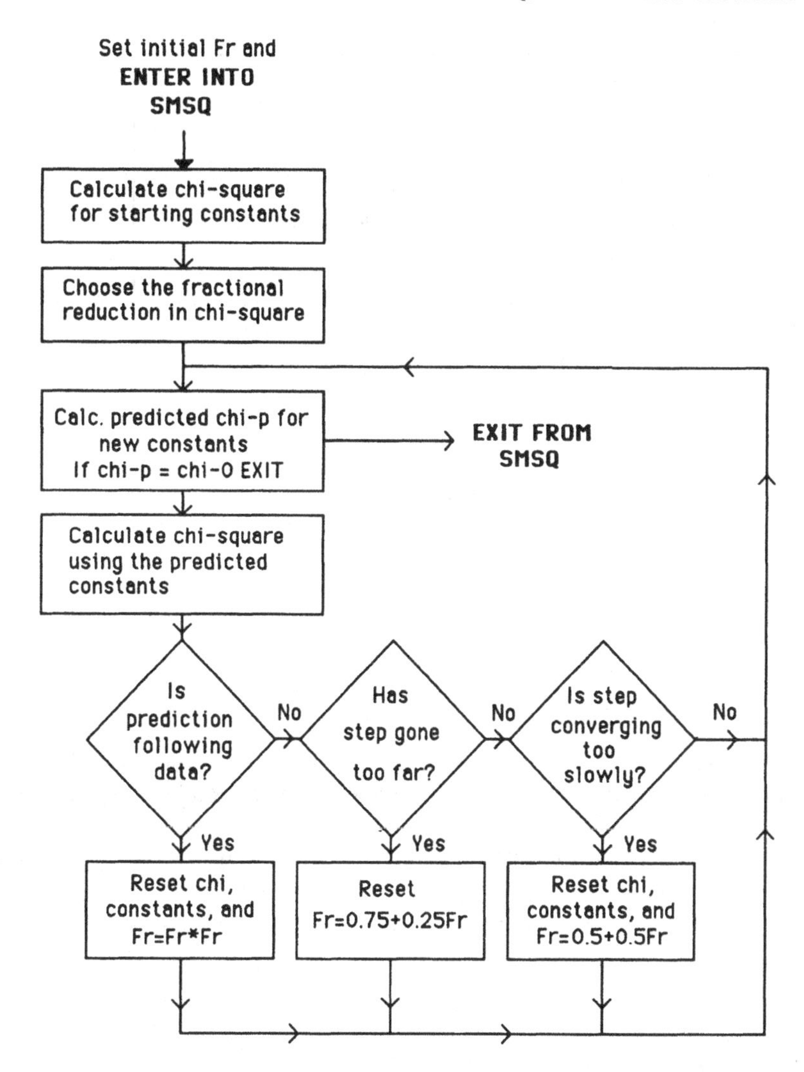

FIGURE 4.4
Flowchart of the minimization algorithm SMSQ.

minimization scheme is a parameter λ, somewhat akin to the Marquardt parameter, that allows the predictor to rapidly converge on the $\chi_0^2 * \mathrm{Fr}$ reduction. A schematic of χ_p^2/χ_0^2 against $\log(\lambda)$, shown in Figure 4.5, illustrates how convergence is effected. For large λ, the change in the constants is small, so χ_p^2 approximately equals χ_0^2. For small λ, there can be large changes in the predicted χ. The object is to calculate the predicted χ_p^2 at the Fr level. To home in on the turn in the curve, SMSQ decreases or increases λ until two points (A and B), lying on either side of Fr are found. Accelerated extrapolation is then used to quickly calculate the $\chi_p^2(\mathbf{c}_{\mathrm{Fr}})$ value.

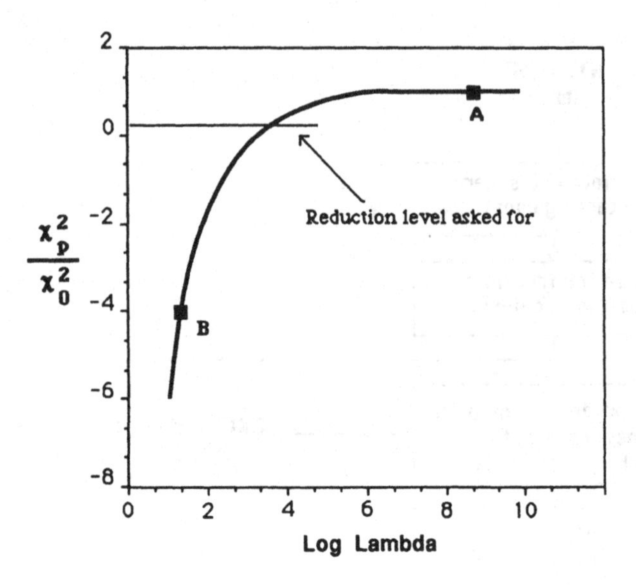

FIGURE 4.5
Plot of log λ against the predicted χ^2.

4.3 THE χ^2 PREDICTOR

A brief outline of the SMSQ χ^2 predictor was given in Section 4.2. Here we describe the mathematical basis of this nonlinear Newton-Raphson minimization scheme.

There are two stages to finding $\chi_p^2(\mathbf{c}_{\mathrm{Fr}})$, the predicted reduced χ_0^2. The first involves homing in on the $\chi_p^2(\mathbf{c}_{\mathrm{Fr}})$ region, which means finding values of χ_p^2 that lie above ($\chi_p^2 A$) and below ($\chi_p^2 B$) the $\chi_p^2(\mathbf{c}_{\mathrm{Fr}})$ value. The second stage involves extrapolation over the $\chi_p^2 A$ to $\chi_p^2 B$ range to find $\chi_p^2(\mathbf{c}_{\mathrm{Fr}})$. We will call stage 1 the extended Newton-Raphson minimization and stage 2 the accelerated extrapolation.

4.3.1 Extended Newton-Raphson Minimization

Before we proceed to a discussion of the SMSQ minimization routine, it is instructive to review the standard Newton-Raphson method. In Figure 4.6*a* the task is to find the point at which the function crosses the *x*-axis, which gives the root of the equation. At point x_1 the function value $f(x_1)$ and its derivative $f'(x_1)$ are used to predict the next point, x_2, closer to the root of the function at x_r. At x_2, the function value $f(x_2)$ and its derivative $f'(x_2)$ are used to predict a further point, x_3, closer still to x_r. This procedure is repeated until the predicted point x_p is close to x_r. This method stems from the Taylor expansion of the function around a given point:

$$f(x+\Delta) = f(x) + f'(x)\Delta + \frac{f''(x)\Delta^2}{2} + \ldots$$

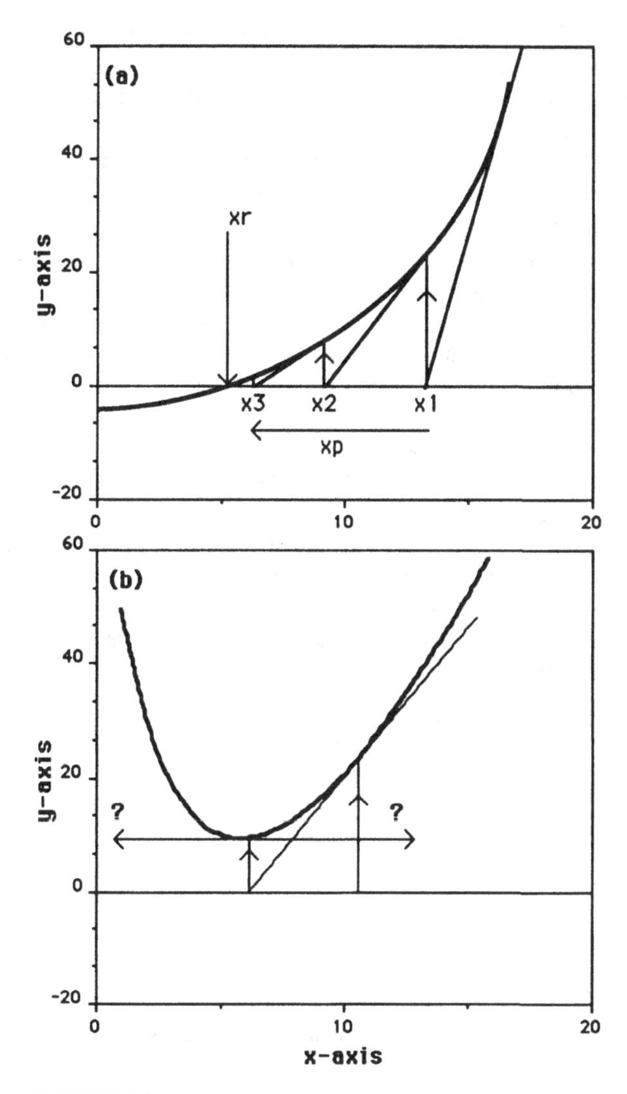

FIGURE 4.6
(*a*) The standard Newton-Raphson technique. (*b*) Problems that can occur if the curve contains a minimum.

For small values of Δ, terms of order Δ^2 or higher can be ignored. Around the x_r point, $f(x + \Delta)$ is approximately zero, giving

$$\Delta = -\frac{f(x)}{f'(x)}$$

The method then steps along the x-axis by the Δ amounts. This procedure, however, has problems when the function contains maxima and minima. Figure 4.6*b* illustrates this problem. If the predicted x_p happens to be close to either the maximum or minimum of the function, the next predicted step size can be extremely large. This makes the method unstable when dealing with complex functions containing many secondary maxima and minima.

To avoid this problem, ROBFIT uses an extension of the Newton-Raphson method. This method overcomes the large step size predicted by reducing the step size and trying again. The function to be zeroed is the derivative of χ^2, because we are looking for a minimum in the χ^2 function. If χ_p^2 at the new point is predicted to be larger than the starting value of χ_0^2, then the step size is reduced. This reevaluation of step size and calculation of χ_p^2 continues until either χ_p^2 becomes less than χ_0^2 or the step size is so small it becomes indistinguishable from the starting point. The step size is controlled by the parameter λ, which can be increased or decreased to decrease or increase the step size. If χ_p^2 is less than χ_0^2, then a larger step size can be used. The code continues to increase the step size to optimize the size of the final step. Varying the step size in this manner means that the procedure can rapidly move through flat regions and accurately step through complex regions in the χ^2 parameter space. This sequence stops once $\chi_p^2 A$ and $\chi_p^2 B$ in Figure 4.5 have been found. A flowchart of this sequence of events is shown in Figure 4.7.

The following provides the mathematics of the χ^2 iteration sequence. Consider the standard Newton-Raphson determination of χ_p^2:

$$\chi_p^2 = \chi_0^2 + \sum_i \frac{\partial \chi_0^2}{\partial c_i} \Delta_i + \frac{1}{2} \sum_{ij} \frac{\partial^2 \chi_0^2}{\partial c_i \partial c_j} \Delta_i \Delta_j$$

where

$$\frac{\partial \chi_0^2}{\partial c_k} = -2 \sum_i \frac{(f_i - f_{\text{th}}(\mathbf{c}_0, x_i))}{f_i} \frac{\partial f_{\text{th}}(\mathbf{c}_0, x_i)}{\partial c_k}$$

and

$$\frac{\partial^2 \chi_0^2}{\partial c_l \partial c_k} = 2 \sum_i \frac{1}{f_i} \frac{\partial f_{\text{th}}(\mathbf{c}_0, x_i)}{\partial c_l} \frac{\partial f_{\text{th}}(\mathbf{c}_0, x_i)}{\partial c_k}$$

The parameter λ is then introduced, so that the quantity to be minimized is

$$\chi_{\min}^2 = \chi_p^2 + \lambda \sum_i \Delta_i^2$$

The λ parameter corrects for the wild fluctuations that occur in the standard Newton-Raphson method. Differentiating, we find

$$\frac{\partial \chi_{\min}^2}{\partial \Delta_l} = \frac{\partial \chi_0^2}{\partial c_l} + \sum_i \frac{\partial^2 \chi_0^2 \Delta_i}{\partial c_i \partial c_l} + 2\lambda \Delta_l$$

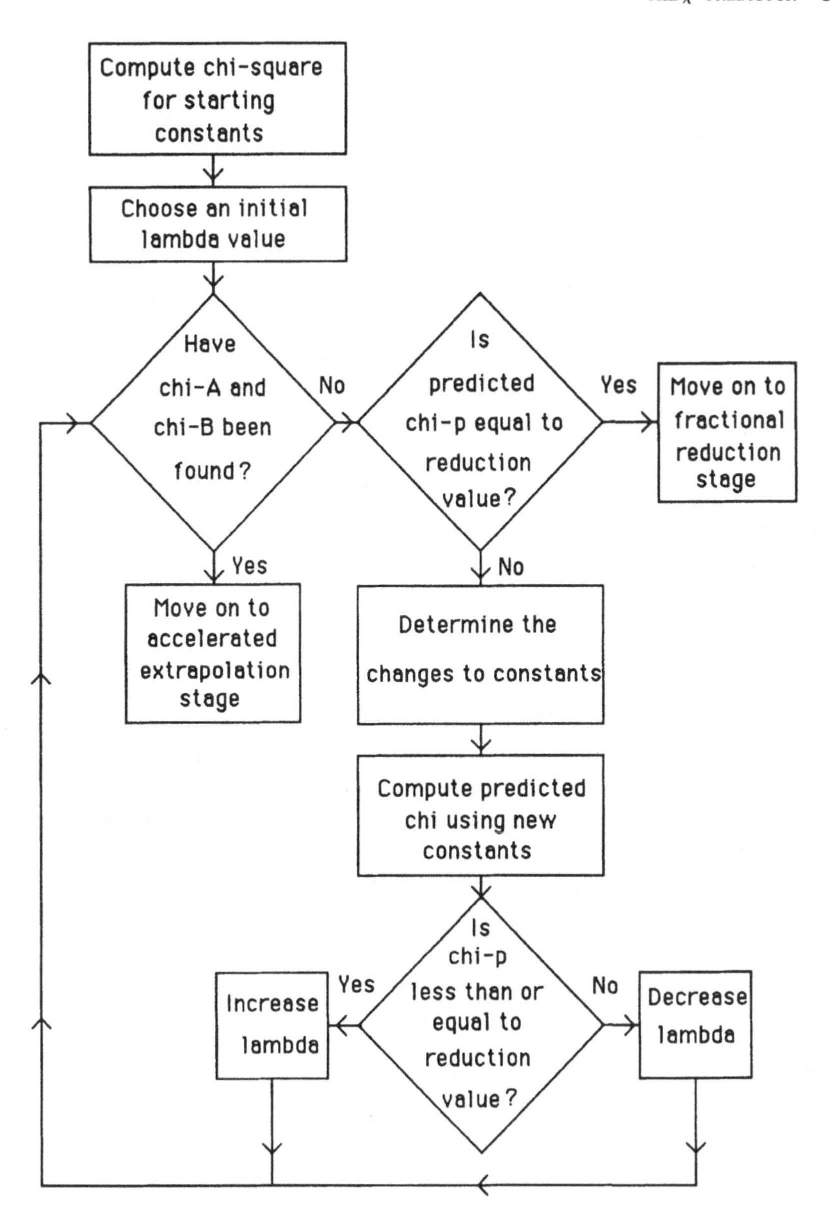

FIGURE 4.7
Flowchart of the procedures for calculating χ_A and χ_B.

so that

$$\frac{\partial \chi_0^2}{\partial c_l} + \sum_i \left(\frac{\partial^2 \chi_0^2}{\partial c_i \partial c_l} + 2\lambda\delta_{il} \right) \Delta_i = 0 \quad \text{at the minimum}$$

where δ_{il} is the Kroenecker delta function.

From this equation Δ_i can be calculated using

$$\Delta_i = \frac{-\frac{\partial \chi_0^2}{\partial c_l}}{2\sum_i \left(\frac{\partial^2 \chi_0^2}{\partial c_i \partial c_l} + \lambda\delta_{il} \right)}$$

which are used to update the fitted constants,

$$\mathbf{c}_0 = \mathbf{c}_0 + \mathbf{\Delta}$$

until the minimum has been found. The λ parameter can be adjusted, as previously described, to change the Δ_i values during fitting. Once $\chi_p^2 A$ and $\chi_p^2 B$ have been found, SMSQ moves to the accelerated extrapolation stage to find the value of χ_p^2 at the $\chi_0^2 * \text{Fr}$ level.

4.3.2 Accelerated Extrapolation

It is possible to continue the nonlinear Newton-Raphson minimization, detailed in Section 4.3.1, until the specified reduction in χ_0^2 has been achieved by using very small changes in λ. Such a minimization scheme would however be expensive in computer time. ROBFIT uses an accelerated extrapolation to speed up the calculation once $\chi_p^2 A$ and $\chi_p^2 B$ have been found.

The accelerated extrapolation method is a combination of linear interpolation and Aitken's delta-squared process, after Aitken, who first suggested it in 1926. The principle of the method is to solve an equation

$$f(x) = 0$$

by replacing it with an equivalent equation

$$x = g(x)$$

and computing a sequence using $x_{n+1} = g(x_n)$

$$x_1, x_2, x_3, \ldots, x_n, \ldots, x_r$$

that converges on the root of $f(x)$.

Suppose α is the root of $f(x)$; then, provided $x_n \neq \alpha$,

$$x_{n+1} - \alpha = g(x_n) - g(\alpha) = (x_n - \alpha)g'(\epsilon_n)$$

with ϵ_n being some point below x_n and α. Now

$$\lim_{n\to\infty} \frac{x_{n+1} - \alpha}{x_n - \alpha} = \frac{\partial g(x)}{\partial x}\Big|_\alpha = g'(\alpha)$$

or, equivalently,

$$\frac{x_{n+1} - \alpha}{x_n - \alpha} = \frac{x_n - \alpha}{x_{n-1} - \alpha}$$

provided the value of $g'(x)$ is approximately constant over the two steps. This gives a method for quickly calculating the root. For a function g that does not have a constant g', we may still write

$$\frac{x_{n+1} - x_p}{x_n - x_p} = \frac{x_n - x_p}{x_{n-1} - x_p}$$

and expect the value of x_p will be close to α. The number x_p is calculated from the expression

$$x_p = \frac{x_{n+1}x_{n-1} - x_n^2}{x_{n+1} - 2x_n + x_{n-1}} \tag{4.4}$$

Figure 4.8 illustrates how this accelerated extrapolation works in ROBFIT.

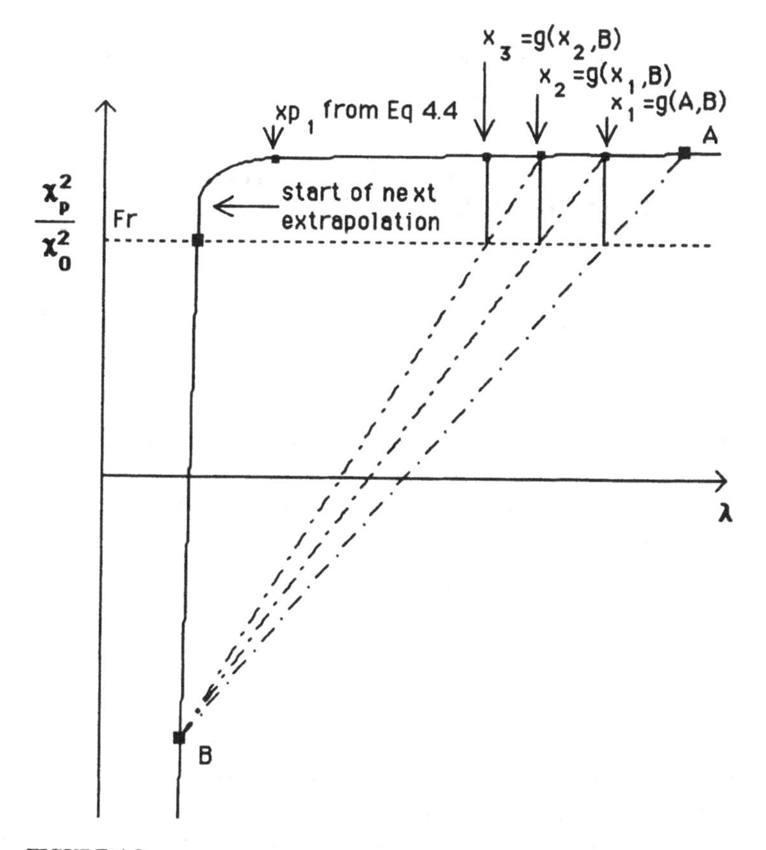

FIGURE 4.8
Iteration sequence of the accelerated extrapolation.

For a sequence of x_ns with $\chi_p^2(x_n)$ slightly above $\mathrm{Fr} * \chi_0^2$,

$$\chi_{n+1} = g(x_n) = \lambda_B + (X_n - \lambda_B)\frac{(\mathrm{Fr}\chi_0^2 - \chi_p^2(\lambda_B))}{(\chi_p^2(x_n) - \chi_p^2(\lambda_B))}$$

which is simply linear interpolation between the values closest on either side of $\mathrm{Fr} * \chi_0^2$ as shown in Figure 4.8. When $\chi^2(x_n)$ is not of the form shown in Figure 4.8, this sequence rapidly converges without Aitken's help. A usual and common sequence is, however, the one shown. In this case Aitken's extrapolation speeds up the convergence by jumping the sequence from x_3 to x_p as shown. The x_p position now forms the starting point for the next interpolation sequence. This process is carried on until the Fr level is reached.

4.4 USING SMSQ OUTSIDE ROBFIT

The SMSQ routines can be used as a general minimization package provided the operating parameters described in the previous sections are defined. Here we try to pull together the various aspects of the minimization procedure by showing how SMSQ can be used in a least-squares fit to artificially generated data samples. The following also introduces some of the more important points that need to be addressed in fitting histogram spectra.

4.4.1 Description of a Nonlinear Least-Squares Fit Routine that Uses SMSQ

As usual, the quantity to be minimized is

$$\chi^2(\mathbf{c}) = \sum_i \left(\frac{f_i - f_{\mathrm{th}}(\mathbf{c}, \mathbf{x_i})}{\epsilon_i}\right)^2 \tag{4.5}$$

where f_{th} is the function to be fitted to the f_i data points normally distributed about $f_{\mathrm{th}}(\mathbf{c}, \mathbf{x_i})$ with standard deviations ϵ_i. A routine called NLFIT is now used to perform the nonlinear minimization. A complete listing of this routine is given in Appendix C. It assumes the data are in a file with one data point per line and written in free format with the form

$$\begin{array}{ccccc} x_{11} & x_{21} & x_{31} & f_1 & \epsilon_1 \\ x_{12} & x_{22} & x_{32} & f_2 & \epsilon_2 \\ \vdots & \vdots & \vdots & \vdots & \vdots \\ x_{1N} & x_{2N} & x_{3N} & f_N & \epsilon_N \end{array}$$

where the dimension of the data (here three), the number of constants in the fit, and initial guesses at the constants, along with their relative nonlinearities, are read from another file that has the following form, where the first column of numbers lists initial guesses for the constants, the second lists their relative nonlinearities, and the third (not used on input) has the standard deviation of each constant.

```
ADAT.DAT,     NAME OF INPUT DATA FILE
 1,     NUMBER OF DIMENSIONS TO DATA X,F(X) IS ONE; X,Y,Z,F(X, YZ) IS 3
16,     NUMBER OF CONSTANTS BEING FITTED
nlfit.dir
  8776.88227475          1.00000000000    ± 395
  1704.60713974          1.00000000000    ± 597
  -906.526057399         1.00000000000    ± 318
  219.659192191          1.00000000000    ± 84.0
  -30.0692438396         1.00000000000    ± 12.7
  2.52538578582          1.00000000000    ± 1.18
  -0.135230639569        1.00000000000    ± 0.696E-01
  0.461951197757E-02     1.00000000000    ± 0.260E-02
  -0.955385800160E-04    1.00000000000    ± 0.584E-04
  0.984502023374E-06     1.00000000000    ± 0.645E-06
  -0.100122518353E-09    1.000000000      ± 0.103E-14
  -0.762275403168E-10    1.0000000        ± 0.573E-10
  0.519858199369E-14     1.00000          ± 0.103E-14
  0.554323697394E-14     100.0000         ± 0.463E-14
  0,10000
  0,1000000
```

We call this file the direction file. The first derivatives of $\chi^2(c)$ are given by

$$\frac{\partial \chi^2}{\partial c_l} = -2\sum_i \left(\frac{f_i - f_{\text{th}}(\mathbf{c}, \mathbf{x}_\mathbf{i})}{\epsilon_i}\right)\frac{1}{\epsilon_i}\frac{\partial f_{\text{th}}(\mathbf{c}, \mathbf{x}_\mathbf{i})}{\partial c_l} \tag{4.6}$$

and the second derivatives are adequately approximated by

$$\frac{\partial^2 \chi^2}{\partial c_l \partial c_m} = 2\sum_i \frac{1}{\epsilon_i^2}\frac{\partial f_{\text{th}}(\mathbf{c}, \mathbf{x}_\mathbf{i})}{\partial c_l}\frac{\partial f_{\text{th}}(\mathbf{c}, \mathbf{x}_\mathbf{i})}{\partial c_m} \tag{4.7}$$

The term dropped in the second derivative is

$$-2\sum_i \left(\frac{f_i - f_{\text{th}}(\mathbf{c}, \mathbf{x}_\mathbf{i})}{\epsilon_i}\right)\frac{1}{\epsilon_i}\frac{\partial^2 f_{\text{th}}(\mathbf{c}, \mathbf{x}_\mathbf{i})}{\partial c_l \partial c_m}$$

The second derivative of fth is hard to calculate accurately and is multiplied by a term that should average to zero. As long as this second term is small compared to Equation 4.7, ignoring it will do no more than slightly slow the rate of convergence.

The exact form of fth enters only in Equations 4.5, 4.6, and 4.7. These are input in the form of

```
SUBROUTINE POLY(X,P,NV,ND,CONS,FA)
```

in which X(ND) is the array of $\mathbf{x_i}$s, CONS(NV) is the current set of constants, FA is $f_{th}(\mathbf{c}, \mathbf{x_i})$, and P(NV) is $\partial f_{th}(\mathbf{c}, \mathbf{x_i})/\partial c_l$. In the event that the x_i is one-dimensional and $f_{th}(\mathbf{c}, \mathbf{x_i}) = \sum_l c_l x^{l-1}$, this routine consists of

```
P(1) = 1
FA = CONS(1)
DO 10I = 2,NV
P(I) = X(1)*P(I-1)
10 FA = FA + CONS(I)*P(I)
```

With a one-dimensional $\mathbf{x_i}$ and f_{th} a spline of the form

$$f_{th}(\mathbf{c}, \mathbf{x_i}) = \sum_{L=1}^{4} c_L x^{L-1} + \sum_{L=5,7,...}^{NV} c_L(c_{L+1} - x_i)_+^3$$

where

$$\begin{aligned}(x)_+ &= x \quad \text{for x} > 0\\ &= 0 \quad \text{for x} < 0\end{aligned}$$

the routine POLY consists of the preceding, together with

```
   DO 20 L = 5,NV,2
   P(L) = 0
   P(L+1) = 0
   XM = C(L+1) - X(1)
   IF(XM.GT.0) THEN
    P(L) = XM**3
    P(L+1) = -3*CONS(L)*XM**2
    FA = FA + CONS(L)*P(L)
   ENDIF
 20 CONTINUE
```

The spline knots, $C(L+1)$ for $L > 5$, are nonlinear. In the direction file their nonlinearities need to be made approximately 10^6 compared to the 1 appropriate for the other constants. They also need to be specified reasonably close to their final values, or else the routine will find an undesired local minimum rather than the desired minimum in χ^2.

The rest of NLFIT consists of rewinding the data file, calculating χ_0^2 and its partials from the values given by POLY, and calling SMSQ to find the new constants until the equivalence of χ_0^2 and χ_p^2 indicates a local minimum. The error treatment described in Chapter 7 is then used to convert a directly calculated inverse matrix into standard deviation estimates for the constants and the values of $f_{th}(\mathbf{c}, \mathbf{x_i})$. The routine writes a new direction file containing the constants and their standard deviations along with a file containing

$$
\begin{array}{cccccc}
x_{11} & x_{21} & x_{31} & f_1 & f_{th}(\mathbf{c}, \mathbf{x_1}) & \sigma_1 \\
x_{12} & x_{22} & x_{32} & f_2 & f_{th}(\mathbf{c}, \mathbf{x_2}) & \sigma_2 \\
\vdots & \vdots & \vdots & \vdots & \vdots & \vdots \\
x_{1N} & x_{2N} & x_{3N} & f_N & f_{th}(\mathbf{c}, \mathbf{x_N}) & \sigma_N
\end{array}
$$

where σ_i is the standard deviation in the estimated value of the function.

4.4.2 Sample Fits Using NLFIT

Two sets of data were generated by randomly selecting points normally distributed about the straight lines extending from (0, 10,000) to (20, 8,000) to (40, 8,000), as shown in the following figures. The standard deviations of the random points from the lines were made equal to the square root of the value at the line, as would be expected in typical spectra. The straight lines in this case are the actual function. The simulation depicted is similar to the build-up toward the Compton edge on the low-energy side of a large gamma-ray peak. The object is to be able to resolve small peaks that include approximately five data points in a row, the sum of which is two to three standard deviations above the background shown, without introducing an excessive number of spurious peaks.

The area of a small peak at channel N is given by

$$A = \sum_{i=N-2}^{N+2} f_i - \sum_{i=N-2}^{N+2} f_B(x_i)$$

The standard deviation of the first sum is given by

$$\sigma_s^2 = \sum_{i=N-2}^{N+2} \epsilon_i^2 = \sum_{i=N-2}^{N+2} f_i = S$$

and that of the second is approximately given by ϵ_i^2 at each point times the maximum of $\chi^2/(N - M))$ and 1 divided by N/M, the number of constants per region:

$$\sigma_B^2 = \sum_{i=N-2}^{N+2} \epsilon_i^2 * \frac{M}{N} * \max(1, \frac{\chi^2}{N - M})$$

Thus, the total error in A is given by

$$\sigma_A^2 = S\left(1 + \frac{M}{N} \max\left(1, \frac{\chi^2}{N - M}\right)\right)$$

where χ^2 is given by

$$\chi^2 = \sum_{i=1}^{N} \left(\frac{f_i - f_{\text{th}}(x_i)}{\epsilon_i} \right)^2$$

and is a measure of the quality of our fit. Note that because of the maximum function, there is no reason for finding $\chi^2 < N - M$.

The sensitivity to small peaks is given by the ratio

$$R = \frac{A}{\sqrt{S} \sqrt{\left(1 + \frac{M}{N} \max\left(1, \frac{\chi^2}{N - M}\right)\right)}} \tag{4.8}$$

When R for the number of channels, equal to the full width at half-maximum for a given peak type, becomes less than the user-defined cutoff, ROBFIT assumes that all detectable peaks have been found. Note that for $M << N$, which is usually the case, the second square root can be set to 1.

In the preceding analysis, it has been assumed that $f_{\text{th}}(x)$ is randomly displaced from the true function. Fits to our artificial sets of data using cubic polynomials with four adjustable constants are shown in Figure 4.9. Both of these fits have a χ^2 of approximately 65 for 40 data points, which makes the second square root equal to

$$\sqrt{1 + \frac{4}{40} \frac{65}{36}} = 1.09$$

which barely changes R. Both fits are too high by approximately 2σ in the region near channel 20 and too low by about 1σ in the regions near channels 10 and 30. This is despite the fact that we would expect the error in the fit to be

$$\sigma_B = \sigma \sqrt{\frac{4}{40}} * \sqrt{\frac{65}{36}} = 0.42\sigma$$

The extra error occurs because the polynomial is incapable of fitting the true function. The group of points near channel 10 in Figure 4.9*a* has an R given by

$$R = \frac{6\sigma}{\sqrt{5}\sigma} = 2.68$$

which means that it will be introduced as a spurious peak above the 2σ confidence level. It will also reappear in the independent data of Figure 4.9*b*. In addition, real peaks will need to reach almost to the 5σ level before they will be detected near the data bend at channel 20. A better representation of the background is needed!

The most obvious solution is simply to add more constants, as shown in Figure 4.10, where the data have now been fitted to fifth-order polynomials with

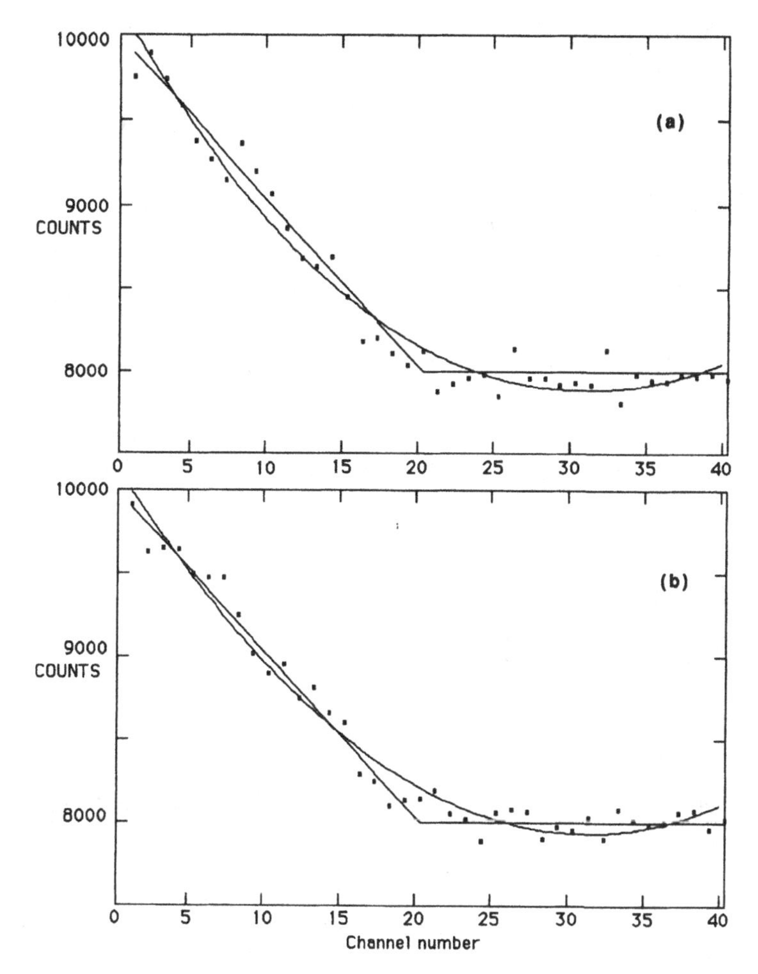

FIGURE 4.9
Two 4-constant polynomial fits. The straight line is the original function, the points are the randomized channel counts, and the curved line is the fitted function.

six constants. This gives a χ^2 of approximately 42. The spurious peak at channel 10 in Figure 4.9*a* moves to about channel 15 in Figure 4.10*a*, with an R given by

$$R = \frac{4\sigma}{\sqrt{5}\sigma} = 1.79$$

which means that it will not normally be detected. The point at channel 26 in Figure 4.10*b* will also have about the same value for R and should not be detected. Both fits, however, round the corner at channel 20, and the χ^2 of 42 is definitely above the 34 possible. The question is, will more constants help? Fits to ninth-

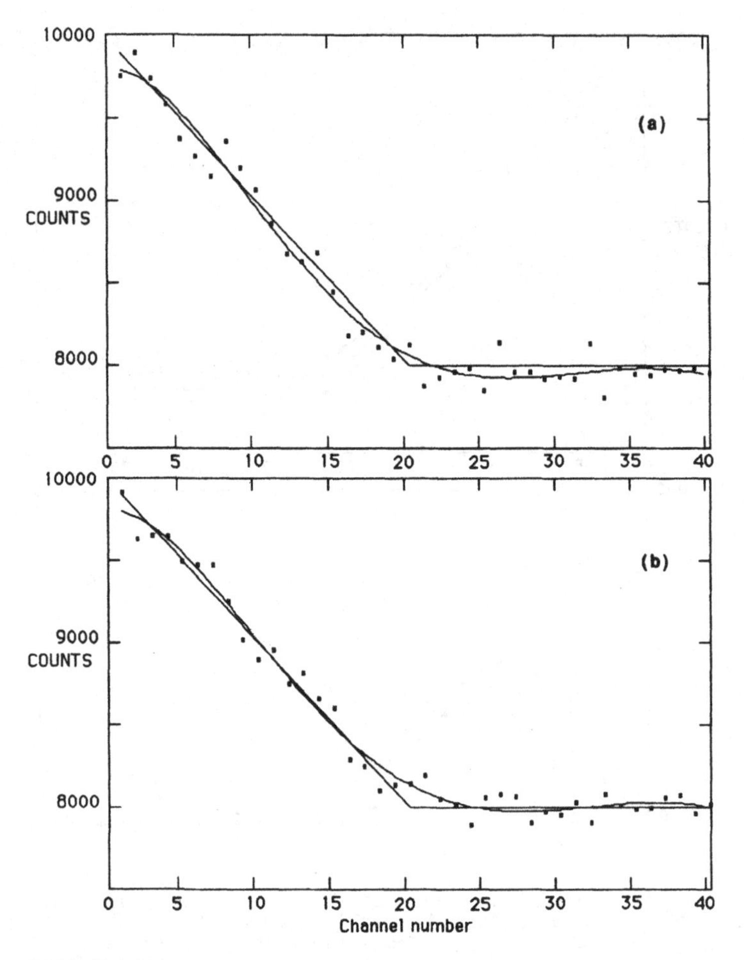

FIGURE 4.10
Two 6-constant polynomial fits.

order polynomials using 10 constants are required to make $\chi^2 = 30$ (40 points − 10 constants), which is the theoretical limit. These are shown in Figure 4.11.

Additional constants do not introduce spurious peaks, but they do exhibit the well-known phenomenon of looping through the data. This could hide peaks by reducing R in the peak neighborhood.

Adding constants to the fit can be carried to an extreme, as shown in Figure 4.12, where 13th-order polynomials are fitted to the data. Each constant now tends to reduce χ^2 by 1, but not continuously. Note that in Figure 4.12*b* the fit now shows the possibility of a peak even at channel 22. The fit is over-following the data, but that is not all bad.

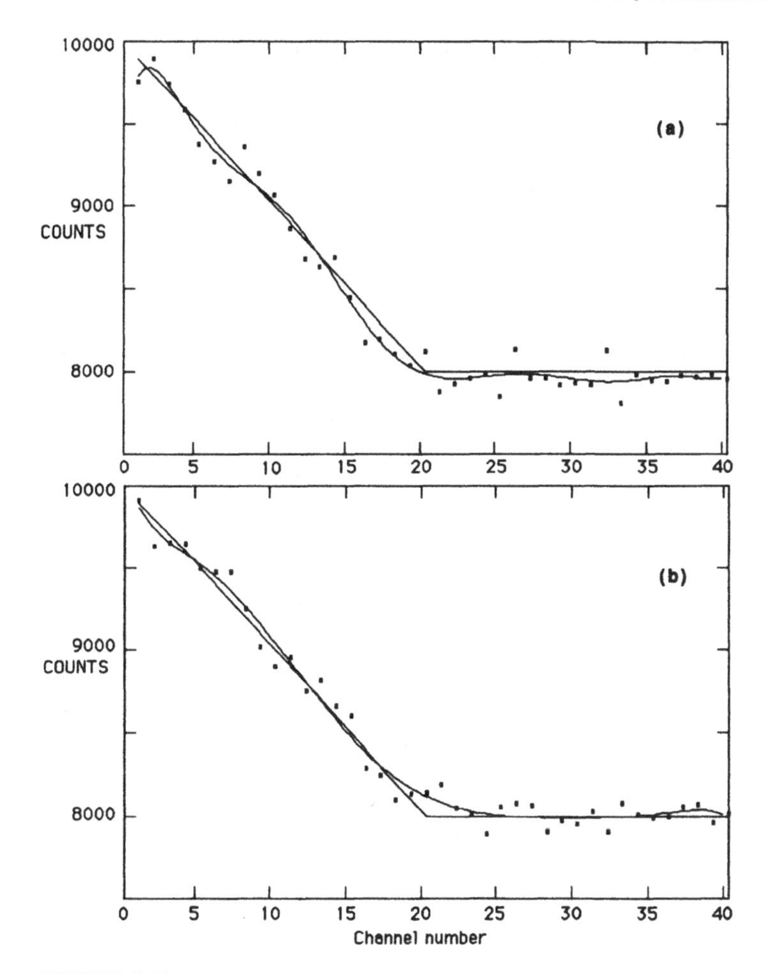

FIGURE 4.11
Ten-constant polynomial fits to the two data sets.

The looping tendency is caused in part by trying to keep all derivatives continuous. The solution is to use splines in the fit. In Figure 4.13 the data have been fitted using fixed knot splines

$$f_{\text{th}}(x) = \sum_{i=0}^{3} c_i x^i + \sum_{i=1}^{6} d_i(\xi_i - x)_+^3$$

where

$$\xi_i = i * 6\frac{2}{3}$$

and the 10 cs and ds were determined by NLFIT.

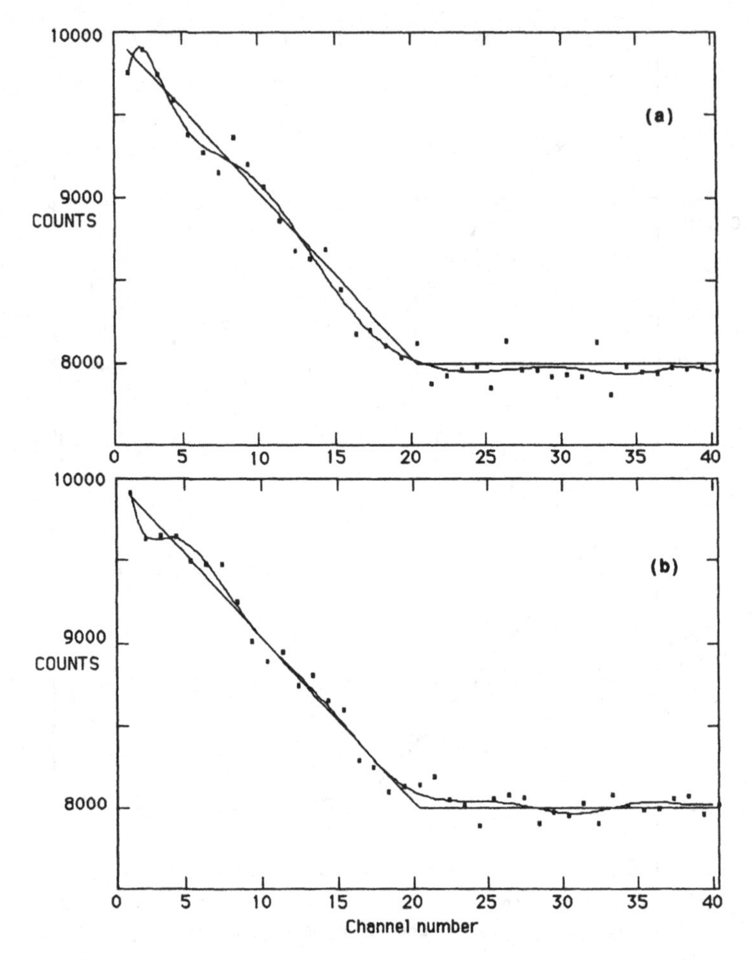

FIGURE 4.12
The limit of polynomial fitting for this function. These fits have been performed with 14 constants.

In comparison with the 10-constant polynomial, the match to the data is better, but it is still not perfect. It is worth noting that NLFIT handles constants in this form much better than in the polynomial form. The fit with 10 knots (14 constants) is shown in Figure 4.14, where again the data are followed too closely. In Figure 4.14*a* $\chi^2 = 26$, which is high for a fit containing 16 constants; in Figure 4.14*b* $\chi^2 = 21$, which is low. The expected standard deviation in χ^2 is $\sqrt{2N}$, or approximately 9 for 40 data points, indicating that these fits are consistent with one another. The fixed-knot splines are a little easier to work with than the polynomials and allow us to use more constants to fit the data, but they do not appear quantitatively better.

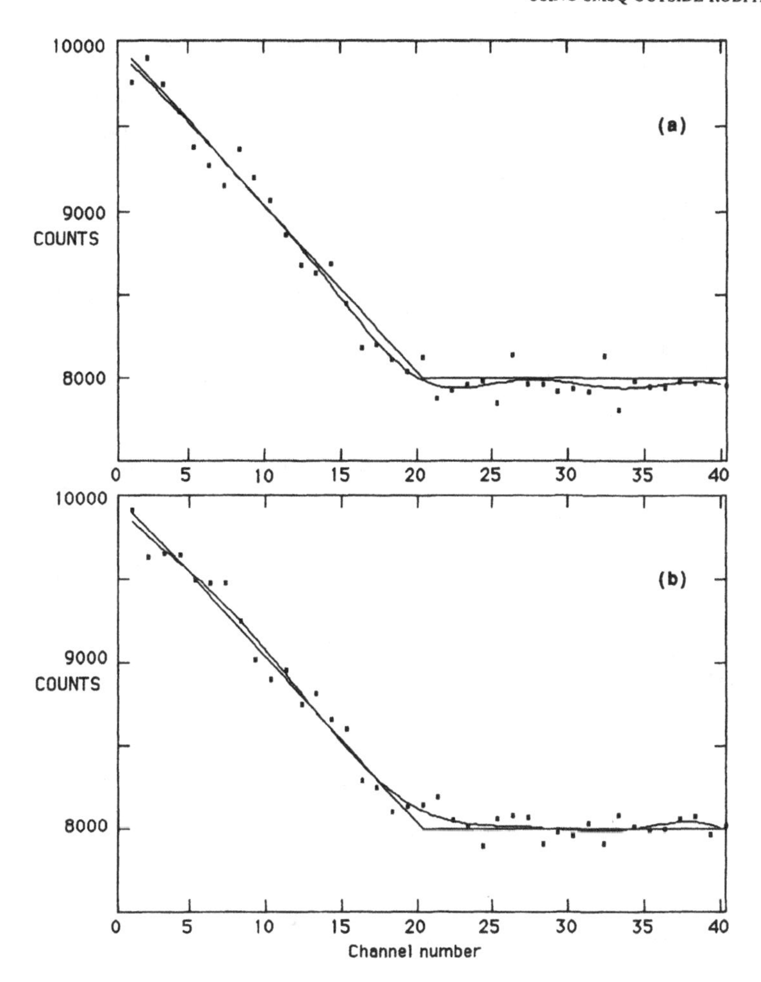

FIGURE 4.13
Ten-constant, fixed-knot spline fits to the two data sets.

The code NLFIT is capable of using splines with variable knots, which makes the fitting function nonlinear:

$$f_{\text{th}}(x) = \sum_{i=0}^{3} c_i x^i + \sum_{j \text{ by } 2\text{s}} c_j (c_{j+1} - x)_+^3$$

This dramatically slows NLFIT since it now takes tens to hundreds of steps to optimize with respect to the knot positions. In addition, there are now local minima, so that one does not always find the lowest possible χ^2. When knots are added, the new knot is introduced close to, and on the higher channel number side of, the

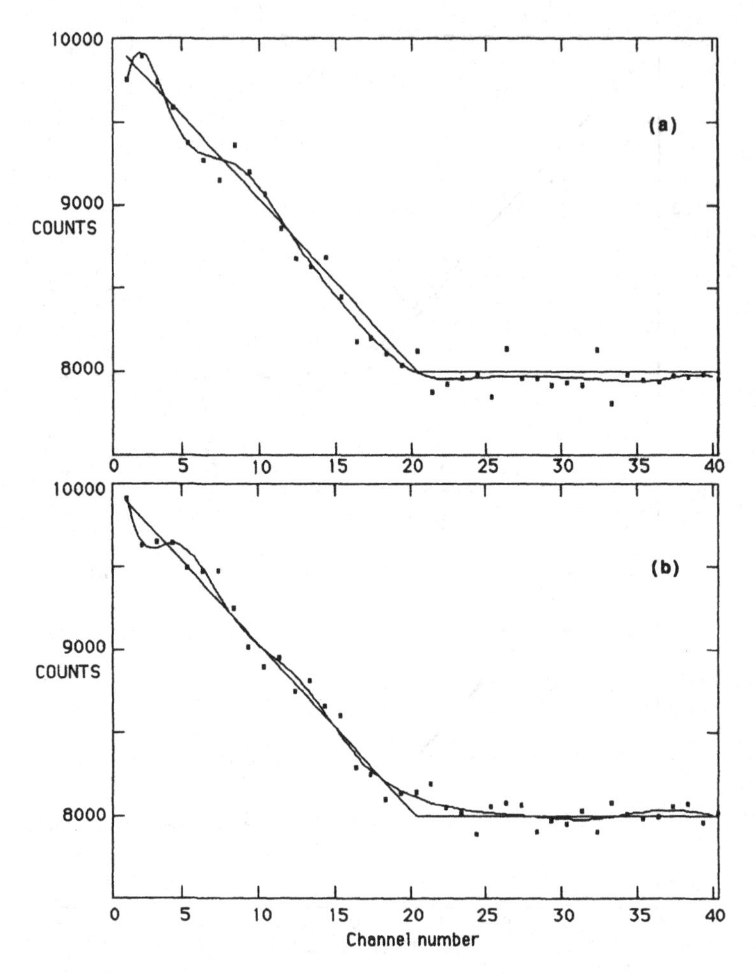

FIGURE 4.14
Fourteen-constant, fixed-knot spline fits to the two data sets.

largest residual. The knots were added one at a time to the best fit obtained without the knot. The result using 10 coefficients (three adjustable knots) is shown in Figure 4.15. Figure 4.15*c* illustrates the $+1\sigma$ and -1σ confidence limits for the fitted function of Figure 4.15*a*. These limits have been calculated using the error treatment discussed in Chapter 7. The knots are now involved in rounding the corner at channel 20, the spline with 14 parameters (five adjustable knots) is still reasonably smooth, as shown in Figure 4.16, compared to Figure 4.12 for polynomials and Figure 4.14 for fixed-knot splines. Use of too many constants means that the data are always followed too closely, as shown by the 16-parameter, six-adjustable-knot fits of Figure 4.17.

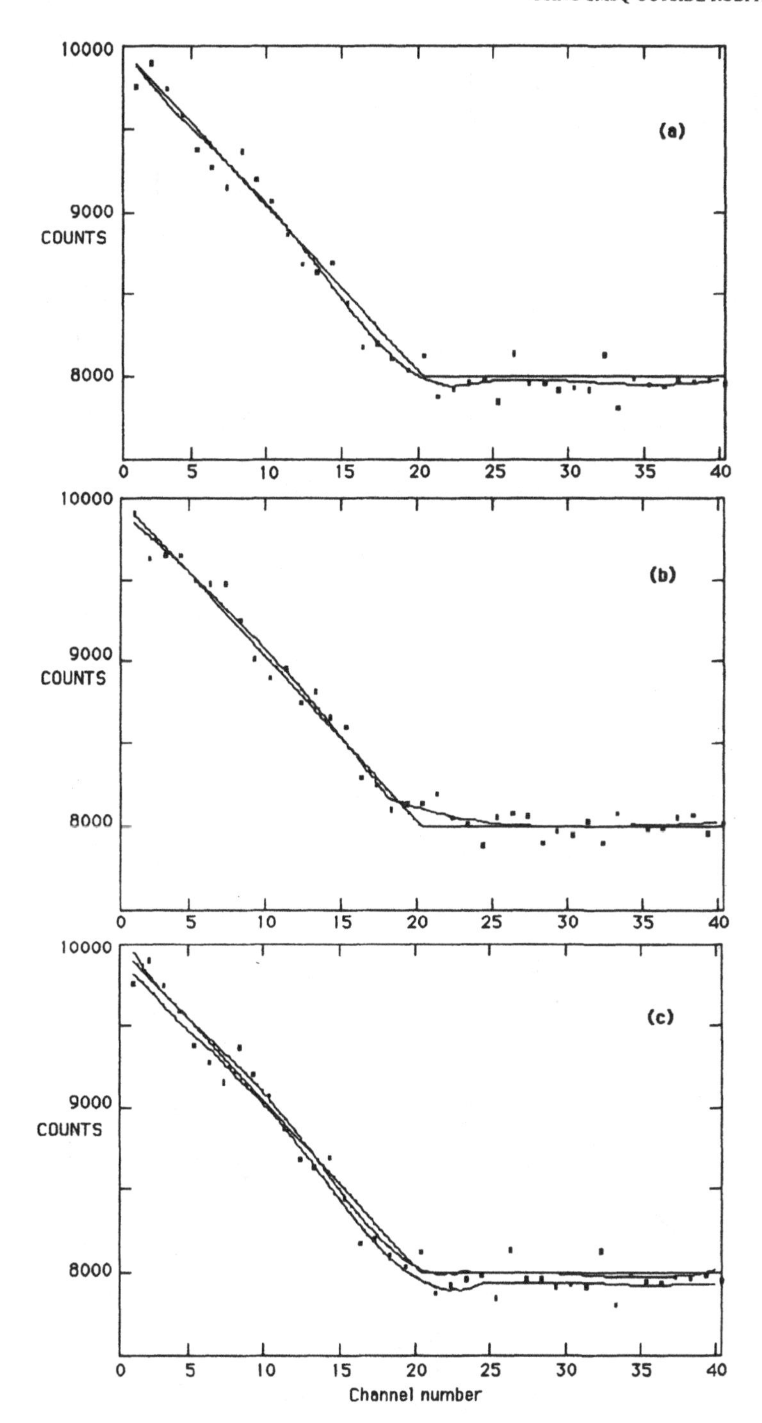

FIGURE 4.15
(*a*), (*b*) Ten-constant, variable-knot spline fits to the two data sets. (*c*) The region of error on the spline fit in (*a*).

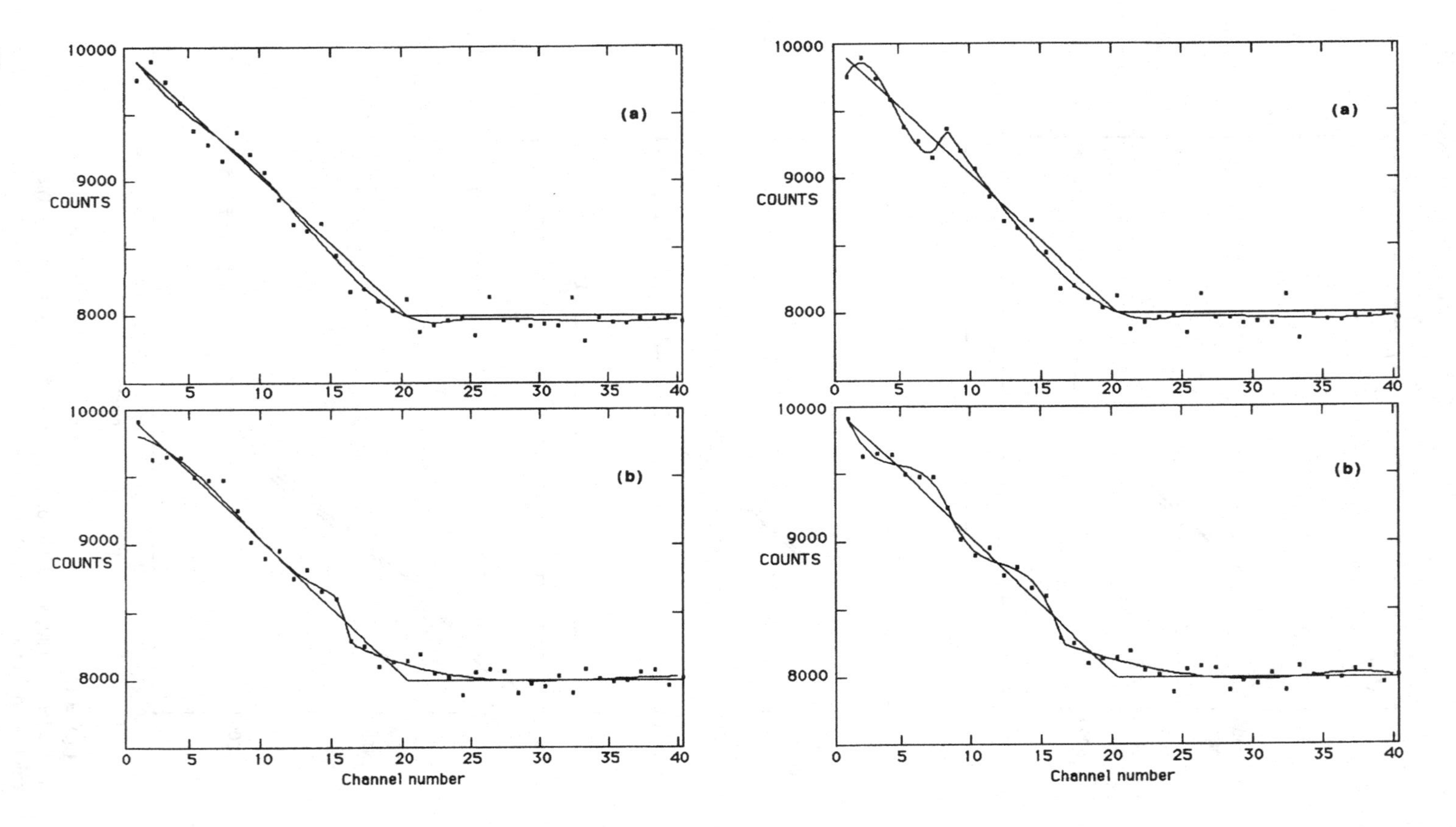

FIGURE 4.16
Fourteen-constant, variable-knot, spline fits to the two data sets.

FIGURE 4.17
The effect of overfitting the data with 16-constant, variable-knot spline fits.

The conclusion is that use of too few constants to fit the background can introduce regions containing spurious peaks. However, using too many constants will cause the background to approach the peaks too closely and thereby mask some of them. Polynomials are inherently limited to short ranges of data, and numerical precision stops us from going beyond the 13th order in these tests. Fixed-knot splines are fitted as quickly as polynomials, and, in addition, work for a larger number of constants. Variable-knot splines come closest to reproducing the underlying curve, as seen in Figure 4.15, and also have the most tolerance for excess constants. The routine ROBFIT uses variable-knot splines when CONT is specified in the startup menu and fixed-knot splines when FIXK is specified. In each case the knots are added two at a time; one at the largest residual and the other N/M higher in channel number.

CHAPTER 5

Representation of the Background

In this chapter we introduce the reader to the techniques used in fitting the background. The first section describes how robust estimation smooths the raw data and speeds up the background fitting process. The second section shows how this smoothed background is fitted with splines. The final section contains the details of how the knots are moved in the fitting of these splines.

5.1 ROBUST FITTING OF THE BACKGROUND

This section shows how ROBFIT implements robust fitting; in the background determination. Various forms of robustness have been defined in relation to statistical estimation (see Press[20]). The general technique is to make a particular analysis insensitive to points that are not accurately defined. Consider Figure 5.1, for example, which shows how a linear least-squares fit to a straight line can be disrupted by spurious outlying points. Robust fitting gets around the problem by weighting down the points with large errors.

We use a similar weighting technique within ROBFIT to reduce the contribution from peaks during fitting of the background. To a certain level, until all peaks have been found, the background will contain some peak contamination. The object of the robust fitting is to enable a background fit to be performed even in the presence of this peak contamination. Within the fitting, we can bring this about by biasing against high-lying points while biasing for low-lying points. The code effectively seeks out the low regions within a spectrum, as illustrated in Figure 5.2.

Problems can occur if the data contain channels with zero or unusually low counts. Possible causes of these low values are data dropout or data corruption during the experiment. These low-count regions are usually not related to the background function and will pull the background fit down if they are not corrected. ROBFIT contains an option for recreation of data channels. The code linearly interpolates across the problem regions between two user-chosen channels and successively replaces the data values with interpolated values. All anomalous

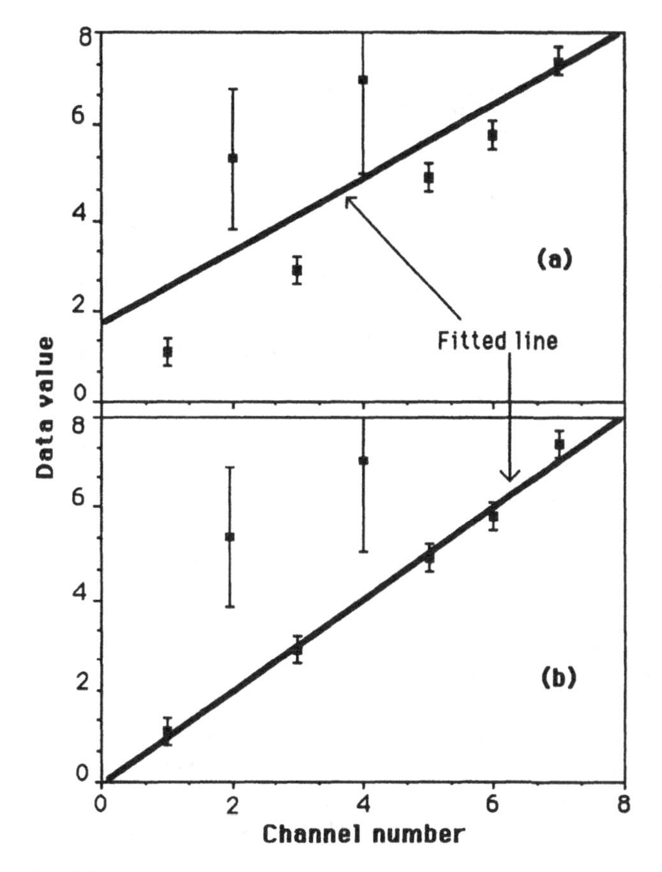

FIGURE 5.1
(a) An unweighted fit to the data. (b) A weighted fit to the same data.

low-count regions must be corrected before any fitting is done. The mechanism of choosing this option of the code is described further in the Appendix A user's guide.

Robust fitting is performed in ROBFIT by filtering the data over a 16-channel region. Before background fitting begins, the data are averaged in a robust manner over successive 16-channel regions in which the weights of low-lying points are increased by a factor

$$A\left(1 + \alpha(x - a)^2\right) \tag{5.1}$$

and the weights of high-lying points are decreased by a factor

$$\frac{B}{(1 + \alpha(x - a)^2)} \tag{5.2}$$

where a is the average over the region considered. A, B, and α are adjusted so that the χ^2 minimum occurs when $a = 0$ in the presence of 1 spurious point in

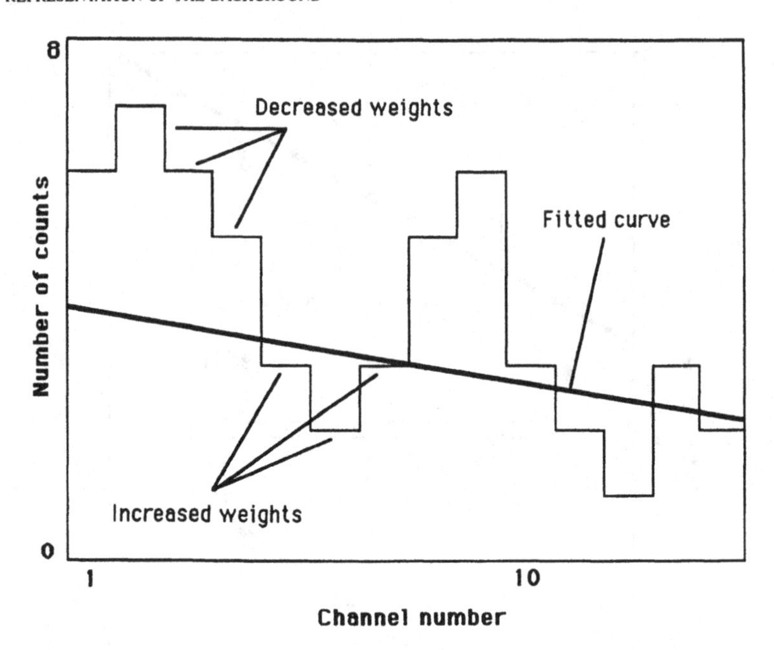

FIGURE 5.2
Illustration of the biasing toward low-lying points.

every 16. These functions have been chosen to yield a χ^2 of 16 when summed over the 15 points that are distributed randomly in a Gaussian manner, together with the 1 spurious point. They also ensure that the average of the 16-point region is not disrupted by the spurious point. The method of determining A, B, and α is given in Section 5.1.1.

The robust averaging has two advantages. The first, as just described, is that it allows the background function to be estimated in the presence of peaks. The second is the increase in the operating speed of the code. All fitting to the background is carried out on a reduced data set, resulting in a speed increase of over 16 times as compared with the raw data.

5.1.1 Determination of the Robust Fitting Parameters

In this section we describe how the robust fitting parameters A, B, and α are determined during fitting. By using Equations 5.1 and 5.2 as the weighting factors for points that lie respectively on the high and low sides of a, the average over a 16-point region, the code can be made insensitive to the presence of one spurious high-lying point in the region. The averaging starts by choosing the 16-channel regions. Channel averaging starts at channel 1 and includes the first 16 points, and then moves on to channel 17 for the next 16-channel block. This procedure continues until the entire channel range has been averaged.

Within any 16-channel region χ^2 can be defined as

$$\chi^2 = 15\left(A\frac{\int_0^\infty (x-a)^2(1+\alpha(x-a)^2)e^{-x^2/2}dx}{\int_0^\infty e - x^2/2} + B\frac{\int_0^\infty \frac{(x-a)^2e^{-x^2/2}dx}{1+\alpha(x-a)^2}}{\int_0^\infty e - x^2/2} \right)$$

plus the contribution from the one spurious point:

$$+B\frac{(\delta-a)^2}{1+\alpha(\delta-a)^2}$$

This minimizes at $a = 0$ when

$$\frac{\delta\chi^2}{\delta a}\bigg|_{a=0} = 0$$

$$= 15\sqrt{\frac{2}{\pi}}\left(-A\int_{-\infty}^{0}(2x+4\alpha x^3)e^{-x^2/2}dx\right.$$

$$\left. +B\int_0^\infty\left[-\frac{2x}{(1+\alpha x^2)}+\frac{2x^3\alpha}{(1+\alpha x^2)^2}\right]e^{-x^2/2}dx\right)$$

$$+B\left[-\frac{2\delta}{(1+\alpha\delta^2)}+\frac{2\delta^3\alpha}{(1+\alpha\delta^2)^2}\right]$$

Changing variables to $z = x/\sqrt{2}$ in the first integral and evaluating the second gives

$$0 = 30\sqrt{\frac{2}{\pi}}A(1+2\alpha)$$

for the first term,

$$-15\sqrt{\frac{2}{\pi}}4B\int_0^\infty \frac{ze^{-z^2}}{(1+2\alpha z^2)}\left(1-\frac{2\alpha z^2}{(1+2\alpha z^2)}\right)dz$$

for the second term, and

$$+B\left[-\frac{2\delta}{(1+\alpha\delta^2)}+\frac{2\delta^3\alpha}{(1+\alpha\delta^2)^2}\right]$$

for the third term.

The equation will be solved by a Newton-Raphson technique to give α once A and B are known. A and B can be determined by requiring for A

$$1 = A\frac{\int_0^\infty x^2(1+\alpha x^2)e^{-x^2/2}dx}{\int_0^\infty e^{-x^2/2}} = A(1+3\alpha)$$

and likewise for B

$$1 = B\frac{\int_0^\infty \frac{x^2 e^{-x^2/2}dx}{(1+\alpha x^2)}}{\int_0^\infty e^{-x^2/2}}$$

which gives

$$B = \frac{1}{\sqrt{\frac{1}{\pi}}4\int_0^\infty \frac{ze^{-z^2}}{(1+2\alpha z^2)}dz}$$

At this point A and B are still undetermined; however, by taking an initial value for α equal to 1×10^{-5}, the calculated A and B can be used to predict a new value of α. Iterating on A, B, and α, calculation quickly finds the minimal values.

5.2 REPRESENTING THE BACKGROUND WITH SPLINES

A gamma-ray spectrum of 4,096 channels can be expected to contain 200 or more peaks and to need about 100 coefficients to represent the background. Faced with such a complex situation, we must devise a method that will allow us to correctly describe each of these contributions. ROBFIT divides a spectrum into foreground and background. The foreground contains the peak contributions to the spectrum and is designated here by a function $p(x)$. During foreground fitting, constituent peaks in $p(x)$ are each allowed to vary in overlapping groups in order to determine a new $p(x)$. The background contains all the large structure contributions to the spectrum and is represented by a function $f_b(x)$. This background function is fitted as a sequence of movable cubic splines. The power of ROBFIT lies in its ability to vary both the foreground and background functions to optimize the fit to the data. It iterates on background fitting, then foreground fitting. Following a background fit, a peak search is performed, and new peaks are added. At the same time, all old peaks are recalculated regarding their position, height, and width. This iterative feature ensures that $f_b(x)$ remains a true representation of the background. The data, which consist of N measurements f_i at equally spaced points x_i, can be written in terms of the background and foreground as

$$f_i = f_b(x_i) + p(x_i) \pm \sqrt{\frac{1}{\omega_i}}$$

where ω_i is the channel weight assigned to the data value at channel i. Assuming x_i, f_i, and ω_i are known, the first step is to determine the background function $f_b(x)$. If the code is started with no background and no peak information, it takes the background as a third-order polynomial and $p(x)$ as zero. It proceeds to fit this function to the data using the fitting methods outlined in Chapter 4. An improvement in both operation speed and accuracy can be effected by compressing the data by a factor of 16 before fitting, as described in Section 5.1. This filters out high-frequency components from the background and helps reduce any undetected peak contribution in the fit to $f_b(x)$.

Background fitting is accomplished by making $f_b(x)$ a weighted least-squares fit to a modification of the difference $f_i - p(x_i)$. The modified $f_i - p(x_i)$ results from the fact that, until all the peaks have been found, this term contains peaks in addition to background. A schematic of how a fit is achieved is shown in Figure 5.3. The figure shows 10 channels containing an artificial peak with a width of 2 channels. Figure 5.3*a* illustrates the background fit with the peak contribution present. At this point, the background fit is simply a weighted average over the 10 channels. The peak region can be seen to stand out in the residuals shown on top of the figure. Figure 5.3*b* shows the improvements in the background determination once the peak contribution has been removed.

After the first round of peak fitting, which is described further in Chapter 6, f_b is redetermined. At this stage, knots are added to the background. Two knots are added at each background refit until the user-specified maximum number has been reached. Reoptimizing the background and peak parameters is an important feature of ROBFIT, as it correctly accounts for the correlations between them. Using splines ensures that f_b is a smooth, continuous function, with a structure that has been determined using the entire data stream.

Within the startup sequence of the code, the user chooses the number of knots to add to the background. The number of knots chosen must be consistent with the quality of the data. After fitting, the user must view the fit and decide whether the background has been correctly determined. A "better" background fit can always be achieved by adding more constants. In curve fitting of this nature, there is a trade-off between background and small peak detectability. If, for example, a large number of knots are used in the background fit, f_b will tend to "follow" rises and dips within the data. Consequently, small peaks may be washed out. The technique we use to get around this problem is to start with a "stiff" background, containing only a few knots, and progressively "slacken" it until the required flexibility has been achieved. This point can usually be defined by the size of the added background splines, a list of which is kept in the background constants file (see Appendix A). Once the optimum number has been reached, additional splines

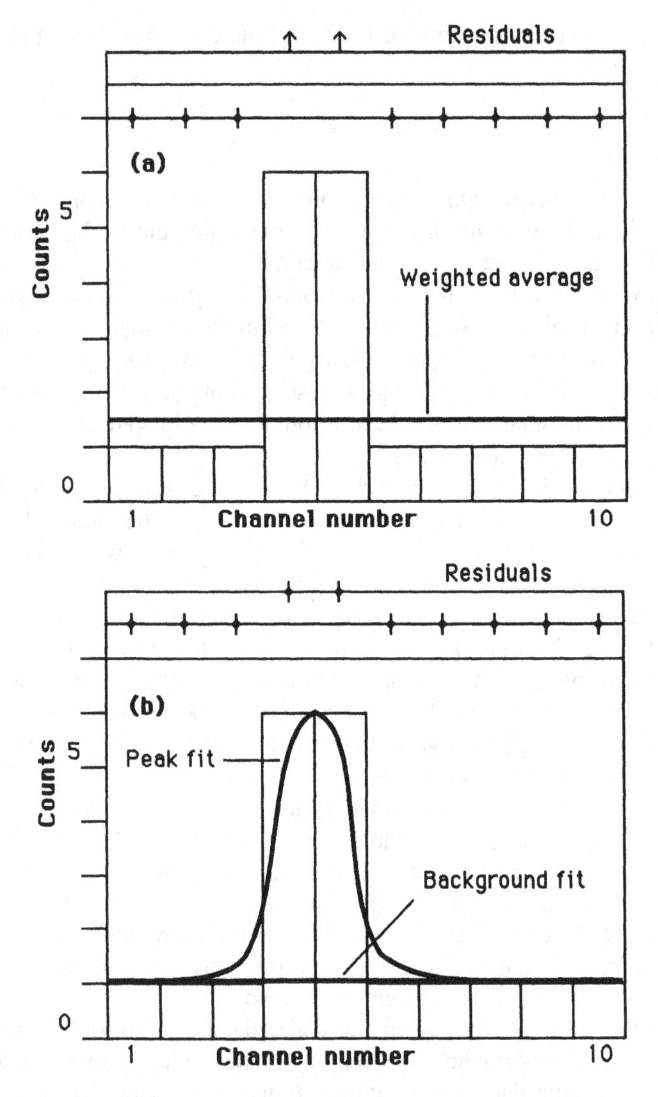

FIGURE 5.3
(*a*) A background fit to a 10-channel region containing a peak. (*b*) The improvement in the background fit once the peak has been found. Note that the peak has dropped with the background and would be raised again in the next iteration.

do not contribute significantly to the fit. This can be seen by simply inspecting the constants file and monitoring the size of the constants just added.

This background fitting technique is similar to the methods described by Morton.[21] The position of the knots is not so critical because the code will minimize χ^2 with respect to the knot positions. A bad knot placement will be forgiven by the minimizing routine, as ROBFIT will move it to the best position during the χ^2 reduction. ROBFIT adds background knots, two at a time, to follow the natural rise and fall of the background. It has been found satisfactory to have the code initially place the first new knot at the middle of the region with the largest "robust" error and the second a reasonable distance above it. Knot positioning is described in detail in Section 5.3. The minimization routine SMSQ then moves these and all other knots to their optimum positions.

ROBFIT contains options for selecting various background fitting scenarios. If a previous background fit has been done, the set of background coefficients determined in that run can be fed into the new fit. There is no need to recalculate the background because the code has been supplied with a starting point for fitting. Additional constants can be added to this starting file if a more detailed fit is called for. This is carried out by selecting the CONTINUE ADDING KNOTS option. Other operations may need either no background fitting or fitting of a fixed background function. Choosing the NO NEW KNOT option tells the code to hold the background fixed at the starting configuration; the NO BACKGROUND FIT option performs only a peak analysis.

In comparing similar spectra, it may be desirable to fix the position of the background knots from one fit to the next. This can be accomplished by fitting successive spectra with the spline constants fixed at positions determined in a previous fit and variable only in size. This cuts the number of movable coefficients in half and preserves the quality of the background from one fit to the next. It also allows errors in the differences between spectra to be estimated with fewer degrees of freedom. The FIXED KNOT option in ROBFIT allows for this kind of fitting. As a bonus, it is also much faster than the normal fitting with variable knots. These options are explained further in Appendix A.

5.2.1 Linear and Exponential Background Fitting

ROBFIT has two modes for fitting the background; exponential and linear. In an exponential fit, the background is expected to vary as an exponential function. Selecting an exponential background fit helps the fitting function quickly rise at lower channel numbers, a common feature of many background functions. In linear fitting, no transformation is used, and a straightforward linear fit is performed on the spectrum.

The background is positive definite in most spectra. Choosing an exponential fit makes use of this information and helps control the shape of the background. It must be remembered, however, that *negative backgrounds cannot be fitted with the exponential option switched on*. With subtracted spectra, or spectra in which the background goes negative, a linear fit must be carried out. In linear fitting, the background is expected to be a relatively smooth function that can be fitted with a few constants. We recommend the use of the exponential fitting function for most spectra, even though this fit takes slightly longer than the linear fit.

5.3 KNOT PLACEMENT

To summarize, the general technique used in fitting the background is to successively increase the number of knots in the representation of the background function. ROBFIT must therefore have some mechanism for choosing where to place the extra knots. The actual positioning is not critical, as the minimization routine can move the knots to their optimum positions. However, a bad positioning can take a considerable amount of computing time to correct. It is better to give the code a good starting point, thus limiting the range over which the minimization routine has to vary the knots.

The knots are actually added two at a time so that rises or dips in the background can be closely followed. One spline follows the upward trend while the other follows the downward trend. Adding knots two at a time means that only small changes in the residuals occur. This means that the minimization procedure, which relies on small variations from one fit to the next, can operate correctly.

The first knot is positioned at the largest residual in the compressed data. The second knot is placed a number of channels above this position, with the channel offset chosen as the maximum of 5 or $NN/(2*NV)$, where NN is the number of compressed channels and NV is the number of knots already present. This has the effect of selecting a smaller fitting region as the number of knots increases. This helps the fitting speed and accuracy. A review of knot positioning in least-squares fitting using cubic splines is given by Morton.[21]

CHAPTER 6

Representation of Peaks

In this chapter we show how ROBFIT handles the peak component of a spectrum. We describe how the peak regions are located and fitted with a given peak shape. We also show how the code decomposes multiple-peak regions and how peak width, area, and position parameters are varied to optimize the separation of background and foreground functions.

6.1 SPLINE REPRESENTATION OF PEAKS

You may expect to find a variety of peak shapes within a spectrum. A standard must be created for each of these shapes before ROBFIT is run. Each standard is used by ROBFIT to determine where peaks of that shape reside. The mechanics of creating a standard are explained in Appendix A. Here we concentrate on the mathematical representation of these standards. The standard peak is expected to be a relatively small set of N data values f_i, with weights ω_i at positions x_i. The user can either create a standard from a peak within the data or choose one of the peak-generating functions supplied with the code (Lorentzian, Gaussian, or Voigt). In the case of a fit to a peak within the data, the chosen peak is expected to be well separated from other spectral features. A fit to the peak region can then be made by assuming only a simple underlying background function.

The fitting function contains a user-specified number of polynomial coefficients to represent the background and a user-specified number of back-to-back cubic splines to represent the peak. The back-to-back cubic splines are given by

$$\Gamma_s(x) = ((1 + x)_+(1 - x)_+)^3$$

where

$$\begin{aligned}(x)_+ &= 0 \quad \text{for } x \leq 0 \\ &= x \quad \text{for } x > 0\end{aligned}$$

Figure 6.1 shows the typical shape of a back-to-back cubic spline. The splines are ideal for representing peaks because they have components that naturally follow the rise and fall of the peak shape.

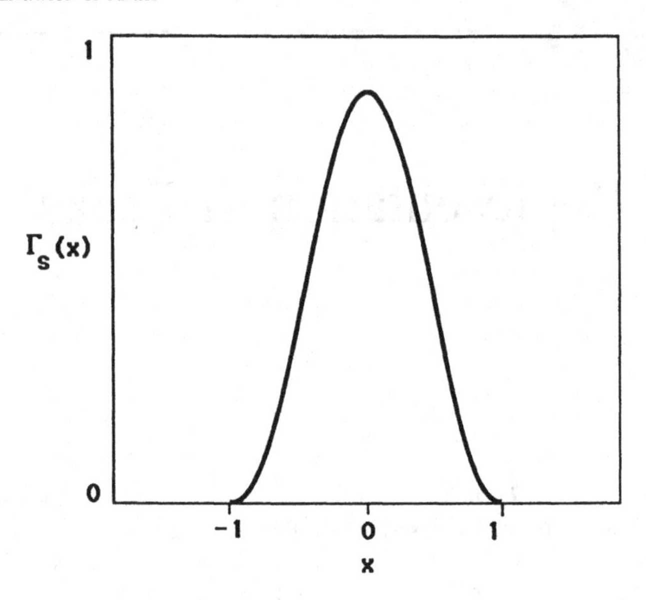

FIGURE 6.1
The shape of a back-to-back cubic spline.

Specifically, the N data points f_i are fitted to the form

$$g(x) = \mathrm{BG} + \mathrm{PK}$$

where BG is the background contribution

$$\mathrm{BG} = \sum_{k=1}^{K} a_k x^k$$

where K is the user-specified polynomial coefficient. Since this fit is over a limited region selected by the user, the background variation is not expected to be significant. This makes the polynomial representation adequate. For generating one of the standards supplied with ROBFIT, K is set to zero because, in this case, we have a peak with no background.

PK is the peak contribution

$$\mathrm{PK} = \sum_{l=1}^{L} b_l'^2 \Gamma_s \left(\frac{(x - p_l')}{w_l'} \right)$$

where L is the user-specified number of back-to-back cubic splines and p_l' and w_l' are the starting values of the positions and widths, respectively, of the splines. These are calculated by the code so that

$$g(x) = \sum_k a_k x^k + \sum_l b_l'^2 \Gamma_s \frac{(x - p_l')}{w_l'}$$

The optimum fit is then calculated by minimizing χ^2,

$$\chi^2 = \sum_i \omega_i (f_i - g(x_i))^2$$

with respect to a_k, b'_l, p'_l, and w'_l. In practice, an additional term is added so that the fitted peak width is kept larger than four channels. This gives

$$\chi^2 = \sum_i \omega_i (f_i - g(x_i))^2 + \sum_i (4 - w'_i)^3_+$$

Nonlinear minimization is carried out using the method described in Chapter 4. The code starts with a single Γ_s of initial width $w'_1 = 2$ and position $p'_1 = x_k$, where f_k is the largest data point and b'^2_1 is equal to the difference between f_k and the curve fit at that channel. At the outset of fitting, the curve fit is a linear interpolation between the endpoints of the selected region. This starting configuration is illustrated in Figure 6.2*a*. Once the optimum values of a_i, b'_1, p'_1 and w'_1 have been found, the contribution from this spline can be removed and the residuals searched again. A second Γ_s is added with $w'_2 = 2$, $p'_2 = x_k$, the position of the largest residual, and b'^2_2 equal to the residual at the kth channel. See Figure 6.2*b*. The function χ^2 is again minimized with respect to both old and new parameters, and the entire process repeated until the appropriate number of splines have been included.

The splines are then normalized so that

$$\Gamma(x) = \sum_l B_l^2 \Gamma_s \left(\frac{(x - P_l)}{W_l} \right) \tag{6.1}$$

represents the standard peak. B_l^2, P_l, and W_l have been scaled to make the height and width of the peak equal to unity.

The use of the positive definite $\Gamma_s(x)$ makes the form chosen for fitting the standard peak positive definite. This provides a rather large amount of intuition about the nature of the spectrum being fitted. It helps clarify what should be the background and what should be the peak. However, it will be a nuisance if one is trying to generate a standard for some purpose that requires a negative portion.

The normalized standards will be used in fitting the spectrum. In its peak fitting phase, ROBFIT utilizes a parameterized function,

$$p(x) = \sum_l b_l^2 \Gamma \left(\frac{(x - p_l)}{w_l} \right) \tag{6.2}$$

whose component parts are the standard peaks of Equation 6.1, to represent the peaks. This enables the code to rescale the standard to any height, width, or position it at any location. Note that $p(x)$ is also positive definite.

6.1.1 Gaussian and Lorentzian Standards

Both Gaussian and Lorentzian standards can be generated automatically using ROBFIT. The shape is chosen by selecting either GAUS or LORE on the standard-

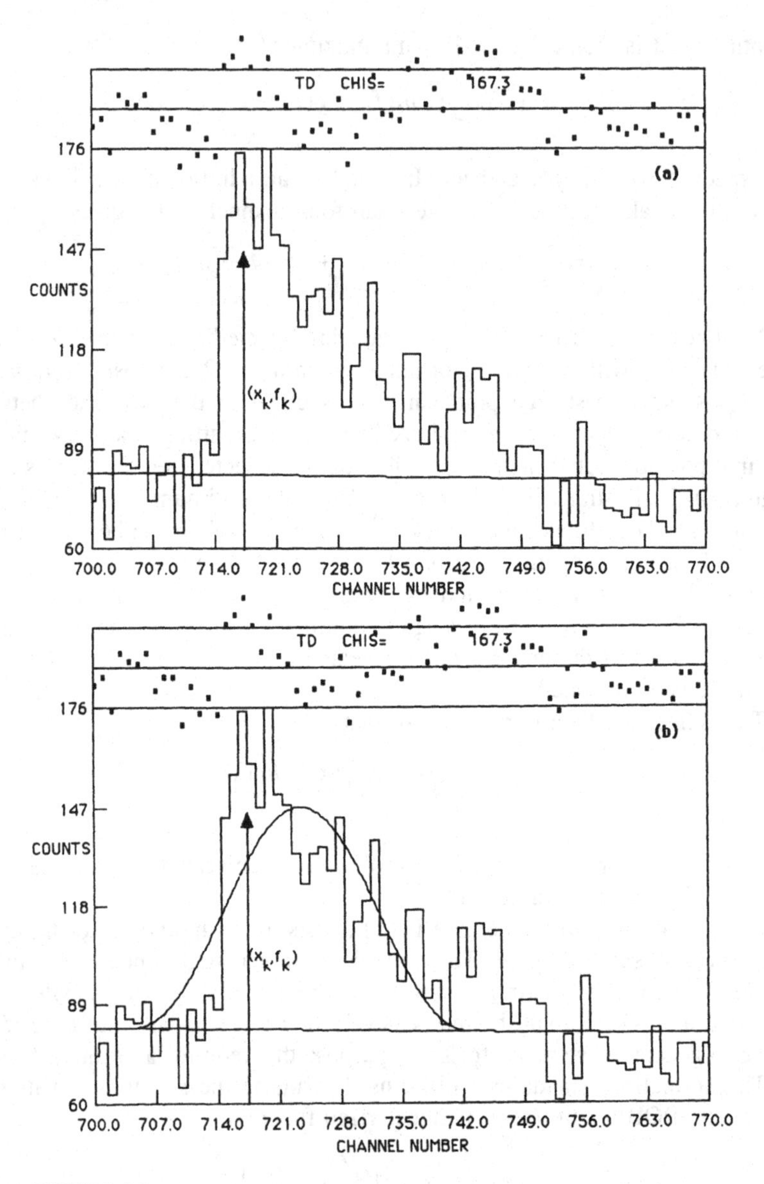

FIGURE 6.2
(*a*) The starting point of a fit to a neutron peak region of the supernova data.
(*b*) The first back-to-back cubic spline is added. See Figure 6.5 for the situation after a second back-to-back spline is added.

generating menu pages described in Appendix A. In fitting a pure peak shape—for example, a Gaussian—there is no background under the peak, and the number of background coefficients must be set to zero before generation begins.

Gaussian is generated by using

$$\text{Gaus} = \frac{1{,}000}{\sigma\sqrt{2\pi}} \exp\left[-\frac{1}{2}\left(\frac{(x-\mu)^2}{\sigma}\right)\right]$$

which gives a peak height of 1,000. The peak is positioned at channel zero ($\mu = 0$), and its width is set to FWHM = 2.

Figure 6.3 illustrates how the splines can be used to increase the accuracy of the fit. The residuals between the spline curve fit and the data values are shown

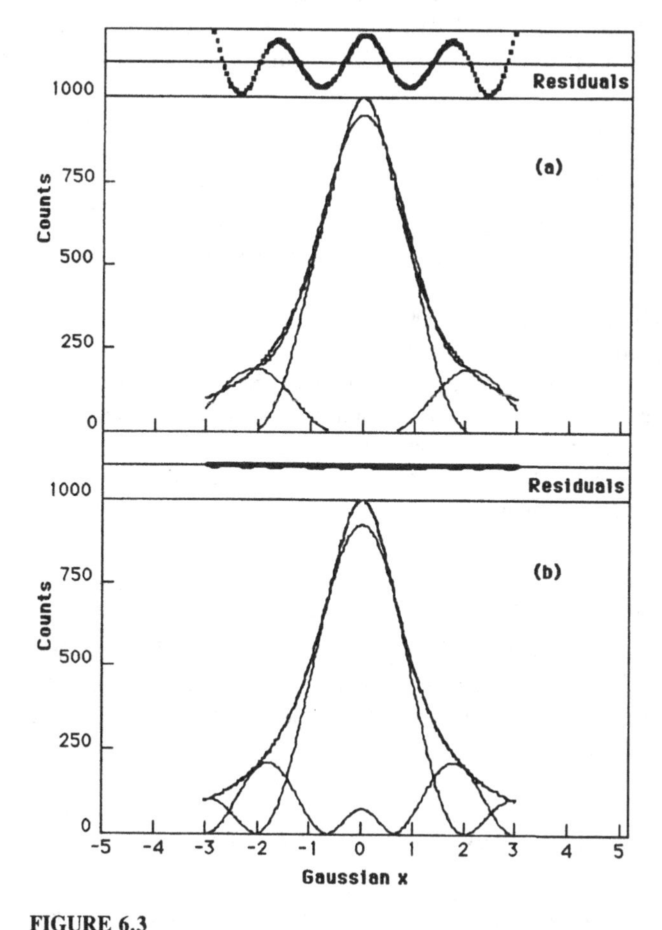

FIGURE 6.3
(*a*) A three-spline fit to a Gaussian. (*b*) The increase in accuracy using seven splines.

above each part of the figure. The center line is the zero residual level, and the top and bottom lines are the $+2\sigma$ and -2σ levels, respectively, where σ is calculated from $\sqrt{\text{max data value}}$. Figure 6.3*a* is a fit to a Gaussian using three back-to-back cubic splines. It shows that there is still structure in the residuals, indicating that the fit could be improved. By the time seven back-to-back splines have been added, the accuracy of the fit has been significantly improved (Figure 6.3*b*).

To detect small peaks within the data, all structure from the standard must be reduced to a level that is small compared with the peak size. For example, if the standard in Figure 6.3*a* is used to fit a peak in a full spectral fit, ROBFIT will attempt to fit the residual fluctuations caused by the mismatch between the standard and the peak shapes with additional small peaks. This leads to a situation in which each peak of the spectrum is decomposed into many secondary peaks; this effect is called braiding. Once the Gaussian has been fitted, the splines are normalized to give heights and widths equal to unity, as described in Section 6.1.

Lorentzian standards can be generated in a similar manner. These are created by the function

$$\text{Lore} = \frac{1{,}000}{\pi} \frac{\text{FWHM}/2}{(x-\mu)^2 + (\text{FWHM}/2)^2}$$

which gives a peak height of 1,000 counts. FWHM is set to two channels and is positioned at the origin ($\mu = 0$).

The user may wish to generate a peak shape that can be expressed analytically but that is not one of the standard shapes. It is easy to change the shape-generating function by simply including a function that calculates the required shape in subroutine BLI in the STGEN routine. Both STGEN and BLI are shown in full in Appendix B.

6.1.2 Voigt Standards

Another useful peak shape to have on hand is the Voigt. This shape is composed of both Gaussian and Lorentzian components. For example, the intensity distribution in a spectral line broadened by two independent effects is expressed by the equation

$$f(x) = \int_{-\infty}^{\infty} G(y)L(x-y)dy$$

where G and L are the line profiles that would result if the individual broadening functions were present alone. In most cases, $G(x)$ is a Gaussian distribution and $L(x)$ is a Lorentzian distribution.

One prescription used by Finn and Muggestone[22] helps to illustrate the mixing parameter η that ROBFIT uses to generate Voigt profiles:

$$f(x) = \frac{a}{\pi}\int_{-\infty}^{\infty}\frac{e^{-y^2}}{(x-y)^2+a^2}dy$$

where

$$a = \frac{\Gamma_L}{4\pi\Gamma_G} = \frac{1}{\eta}$$

with Γ_L = full width at half-maximum of the Lorentzian and Γ_G = full width at half-maximum of the Gaussian.

The η parameter effectively determines the mix between Gaussian ($\eta << 1$) and Lorentzian ($\eta >> 1$). The Voigt profile is normalized to a height and full width at half maximum of 1. All the user has to do to generate a Voigt profile is to define the mixing parameter η. Figure 6.4 illustrates the variation of η over a range $\eta = 0.1$, Lorentzian, to $\eta = 10$, Gaussian.

6.1.3 Selecting a Standard from the Data

Certain spectra contain peak shapes that are neither Gaussian, Lorentzian, nor Voigt. The spectrum in Figure 6.5 is taken from the supernova data and illustrates how a neutron-inelastic scattering standard can be generated. Chapter 8 describes how this standard is used in fitting the data. The problem with the neutron peaks

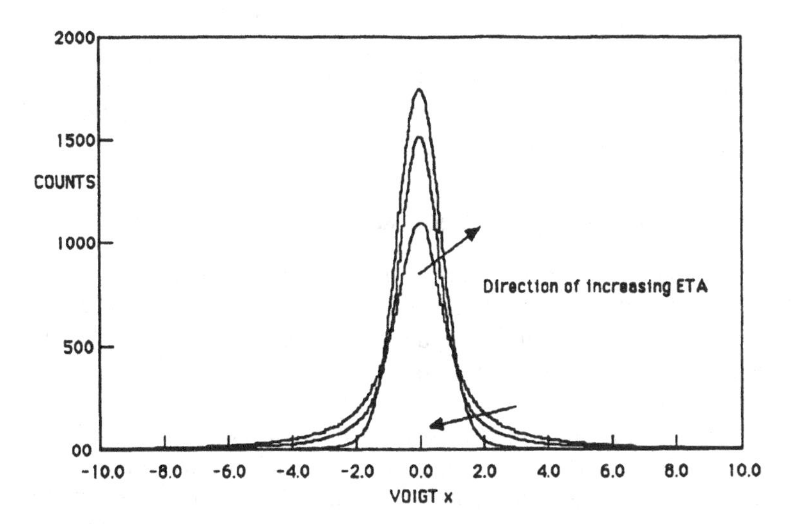

FIGURE 6.4
Variation in peak shape with changes in the Voigt η parameter.

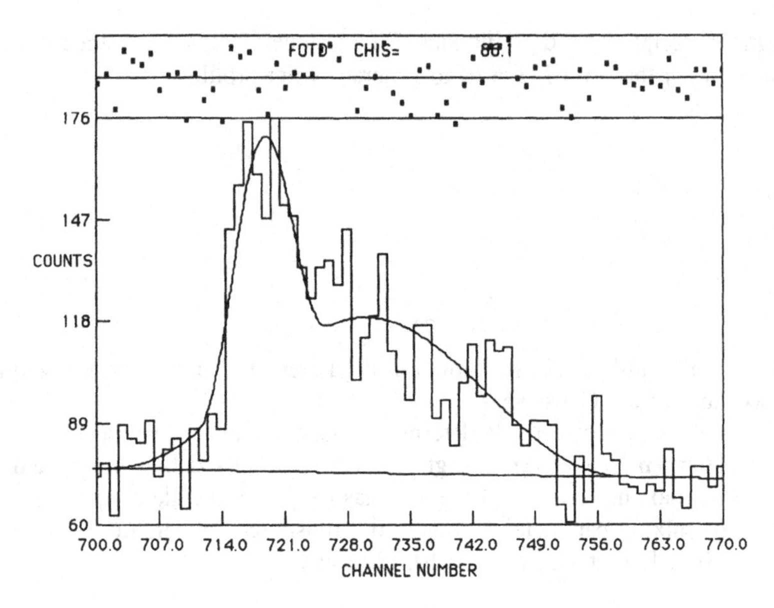

FIGURE 6.5
The supernova neutron peak region fitted with a two-constant background and two back-to-back cubic splines.

is that their shapes are not well defined. Therefore, we must generate this standard from the raw data. The peak indicated in the figure can be seen to be well separated from other spectral features and resides on a simple background. Under these conditions we can create a standard using the procedures detailed in Appendix A. A region surrounding the peak is chosen and fitted using a polynomial fit to the background and the back-to-back cubic spline fit to the peak. Figure 6.5 illustrates the result of a two-constant fit to the background and a peak fitted with two back-to-back cubic splines. The figure shows little fluctuation in the residuals, indicating that the peak has been fitted well.

6.1.4 Selecting Multiple Peak Shapes

When a real spectrum is fitted, usually no one standard will represent all peak shapes. Consequently, ROBFIT has the capability of using up to five standards in any one fit. At runtime, the code decides which of the peak shapes is more appropriate to fit a given region of the spectrum. This selection procedure is described further in Section 6.3.

Each of the peak shapes to be used in the fit is generated as described in the previous sections. These standards files are then read into the full spectral fit at the start of the fitting sequence. ROBFIT will fit individual peaks and decompose overlapping peak regions, taking into account all shapes present in the fit. Figure 6.6 shows a fit to the neutron peak region of the supernova spectrum using three different standards.

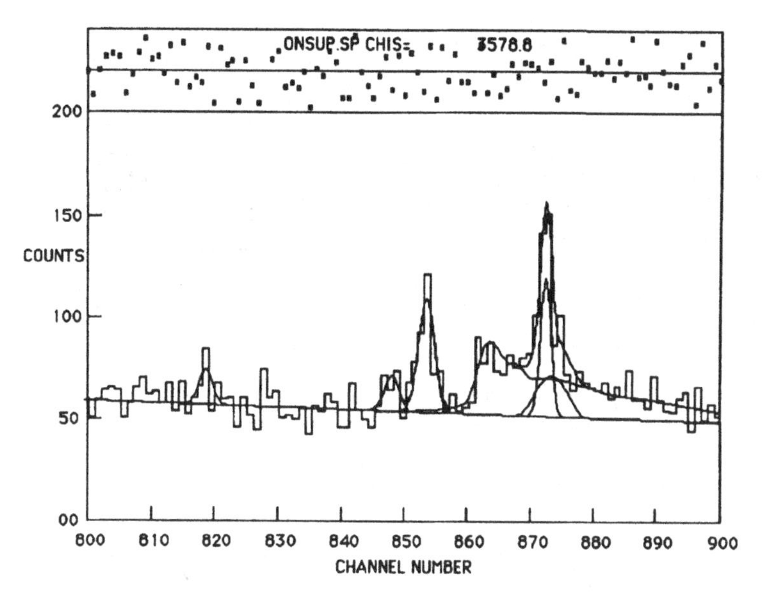

FIGURE 6.6
A fit to the neutron peak region of the supernova spectrum.

6.2 INITIAL SINGLET DETECTION

Before we get into the mathematics of peak detection, it is useful to describe the flow of the peak finding process. Central to this process is the setting of a cutoff level. This level is user-selectable and defines when the code will stop searching for peaks. The code will find peaks until there are no residuals greater than this cutoff level. The residual for a potential peak at x_i is equal to the excess counts in the data in a region equal in size to the full width at half-maximum of the potential peak, divided by the error in the integral of the fit across this same region. Having determined that there are residuals greater than the cutoff, the code proceeds to find the channel position of the largest residual. This is the channel at which the code will attempt to put the peak. A small region surrounding the largest residual is then selected. The χ^2 over this region is then minimized with respect to the peak parameters. At this point the background has not been recalculated. Due to the addition of this peak, the background may have been disturbed within a region surrounding the peak. Consequently, on the search for the next largest residual, the code does not position peaks within this region. Both the background and the peak constants are reoptimized following peak fitting. This ensures that both background and peak functions correctly describe the data and are not biased by fitting the peaks in small regions. Figure 6.7 illustrates a flow diagram for the peak fitting sequence.

The code has been designed so that a number of successively smaller cutoff runs can be executed consecutively. ROBFIT produces the fitted parameter file

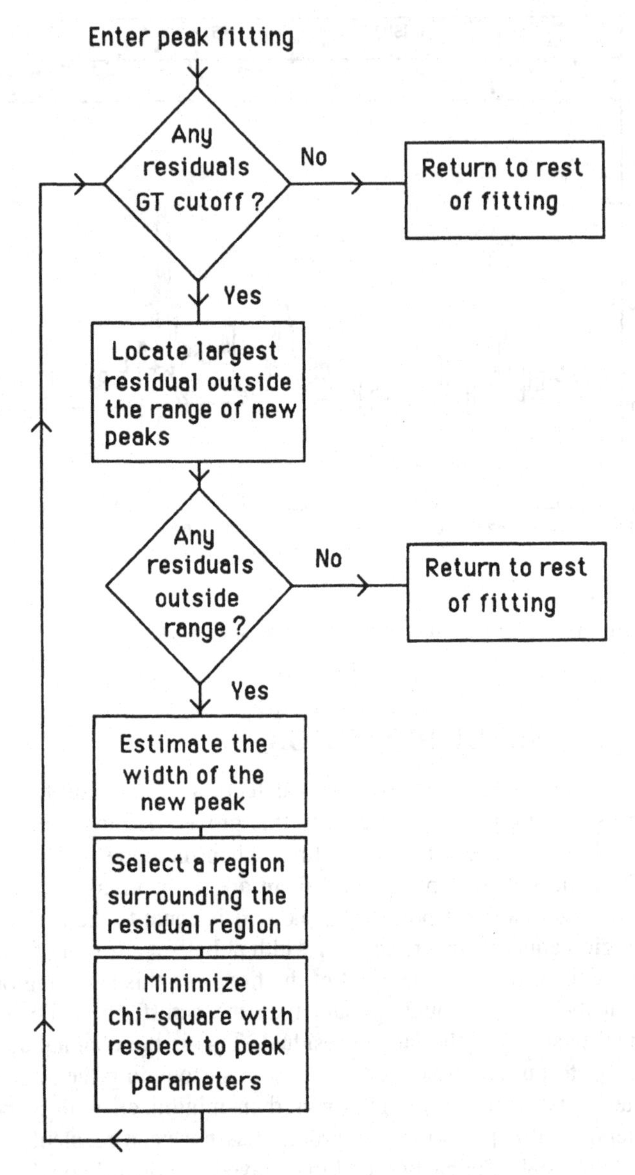

FIGURE 6.7
Flowchart of the peak fitting sequence.

after each cutoff run. This stage-by-stage reduction allows the user to step through the fitting sequence and monitor the progress of the fit.

We are now ready to begin the mathematics of peak fitting. To recap, the fitting function representation begins with N measurements f_i at equally spaced points x_i. It is assumed that we have found a relatively slow-varying function

$f_b(x)$ so that when it is added to the function $p(x)$, representing the peaks, the measurements are given by

$$f_i = f_b(x_i) + p(x_i) \pm \frac{1}{\sqrt{\omega_i}}$$

where all quantities are as defined in Section 5.2. Composite back-to-back cubic spline standards are assumed to represent the peaks. All different peak shapes within the spectrum are given a different standard. These standards have been described in Section 6.1.

Each standard is normalized to make the standard peak have height and full width at half-maximum of unity. Equation 6.2 defines the parameterized peak function $p(x)$. Two important features of the composite splines, $\Gamma(x)$, are that they are identically zero for most values of x and their derivatives can be found exactly.

To locate peaks, a set of smoothed residuals,

$$R_i^2 = \frac{1}{\text{FWHM}} \sum_{j=i-2}^{j=i+2} \omega_i (f_i - f_{\text{th}}(j))^2 \tag{6.3}$$

is searched for the largest residual R_L, where f_{th} is the theoretical estimate of the fitting function and is equal to the sum of the background and peak functions. FWHM is the full width at half-maximum, determined from the user-supplied width range, and ω_i is the weight at channel i. The value x_1 of the first peak standard, given by equation 6.2, is then initialized to x_L, the position of the largest residual. The width of the standard w_1 is set to an interpolated value between the user-supplied high-channel and low-channel estimates, which are read into ROBFIT at the outset of fitting. A minimization using the method described in Chapter 4 is then carried out on the quantity

$$\chi^2 = \sum_j \omega_j (f_j - p(x_j) - f_b(x_j))^2$$

with respect to the parameters in $p(x)$. For a single standard this would be a minimization of the position x_1, width w_1, and scaling factor b_1 of the standard. The optimization is performed over a limited region of the spectrum such that

$$x_l - 3w_l \leq x_j \leq x_l + 2w_l$$

The limited range of x_j reduces the computational time, and the short-range nature of the peak keeps it from sacrificing accuracy. After minimization, the matrix of second derivatives of χ^2 is inverted, without smoothing coefficients, to provide error estimates for the peak parameters. The effect of this peak can now be added to our theoretical representation of the function f_{th}.

The residuals are now recalculated, and large spectra can be searched again for other peaks. The fact that the peak just found was present in the determination of the background function f_b indicates that the background will be inaccurate in the vicinity of this peak. Thus, the code is not allowed to introduce another peak in this region until the background has been recalculated. Peaks are not added within a channel range of $6w_p$ on either side of the new peak, where w_p is the width of the peak just added.

When there are no more well-separated residuals greater than the user-supplied cutoff, the background is recalculated. Following the background refit, all channels are again made eligible for largest residual status. Additional peaks can come from two sources:

1. The residual was in the range of another peak on a previous peak fitting iteration.
2. The background has been lowered significantly, allowing the residual locator to find a new peak.

If there are peaks within range of the new residual, where again the range is determined by the user-supplied width values, ROBFIT first reminimizes χ^2 with respect to the old peak parameters. This is done because the background may have changed significantly since the old peaks were first added. After the re-minimization, if χ^2 does not change significantly, the new peak is added and χ^2 minimized with respect to the parameters of all peaks. This process holds for any number of peaks, up to a maximum of 10 within any one region. This sequence of events is shown in Figure 6.8.

Once ROBFIT determines that there are no residuals above the cutoff, the code refits all peaks one last time before ending that particular cutoff run.

An important part of ROBFIT is its allowing peak positions, widths, and strengths to vary. These variations allow the code to explore the "best fit" parameter space and thus estimate the errors on the parameters more accurately. There are three ways in which ROBFIT can manipulate the peak parameters.

The first method is variation during fitting. This has been discussed in the preceding peak parameter evaluation. ROBFIT continually fits and refits all peak parameters during its cycling so that all constants and their associated errors have been optimally determined.

The second method concerns the case when the peak widths are known. ROBFIT contains a mode of operation in which the peak widths and locations are held constant. Running the code in this manner greatly speeds up its operation, as the number of fitted constants drops to one-third.

The third method is that of modifying individual peak parameters prior to fitting. ROBFIT has the ability to read in peak constants from a previous fit. This gives the user the choice of modifying the input file. Peak constants can be changed, new peaks added, or old peaks deleted, as required. During fitting, added peaks that do not significantly reduce χ^2 will be removed, or if modified parameters are outside the limits of the best fit, they will be recalculated. This provides a powerful tool for analyzing spectra. The user can experiment with the contributions to the peak list and see the effects on the quality of the fit.

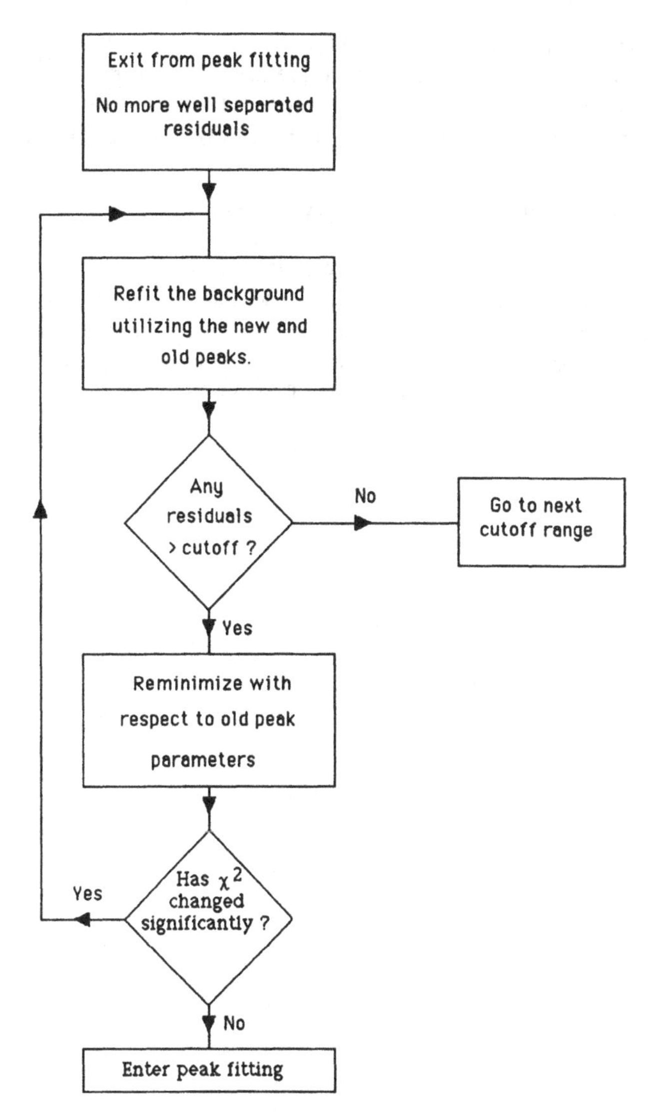

FIGURE 6.8
Flowchart of the peak adding sequence.

6.3 SPECTRA WITH MULTIPLE PEAK SHAPES

Section 6.2 outlined the mechanics of locating and fitting peaks; however, only a single standard was considered. The situation is complicated when the fit has to be performed with more than one standard. Now the residuals locator of Equation 6.3 must be run for each standard and the largest residual from all standards chosen as the peak position. To select the correct standard for this portion of the spectrum, ROBFIT performs a fit to this peak position for each standard. Only the parameters of the standard being fitted are allowed to vary. All peaks already present are fixed within explicitly defined ranges. This frequently allows otherwise unresolvable multiplets to be broken by giving ROBFIT extra information.

6.4 FITTING HEIGHT VARIATIONS ONLY

In some situations—for example, in analyzing repeated sections of a spectrum taken with differing statistics—it is useful to be able to hold the positions of the peaks and the shape of the background constant in order to study the variation in peak heights from spectrum to spectrum. The user can accomplish this by simply selecting the FIXK option from the ROBFIT menu (see Appendix A).

CHAPTER 7

Treatment of Errors

One of the most important features of ROBFIT is its ability to accurately determine the errors on fitted quantities. In this chapter we discuss the error calculations that are used and show how they are implemented in the code. We begin by giving a general introduction to error analysis, and then we extend this to the determination of the errors in the least-squares fitting constants. In the last three sections we describe the calculations involved in determining the errors in the parameters of a fit.

7.1 GENERAL ERROR ANALYSIS

Assume that there exists a function q of the data points that yields the exact result when applied to a set of data points containing no random error. When this function is used with N data points f_i, each containing differences Δf_i from their exact values, q_{ex}, then

$$q(f_1, f_2, \cdots, f_N) = q_{ex} + \sum_{i=1}^{N} \frac{\partial q}{\partial f_i} \Delta f_i \tag{7.1}$$

Although Δf_i is difficult to ascertain, in general, an ensemble of different but equivalent data points is expected to yield

$$\begin{aligned} \{\Delta f_i\} &= 0 \\ \{\Delta f_i \Delta f_j\} &= 0 \\ \{(\Delta f_i)^2\} &= \epsilon_i^2 \end{aligned} \tag{7.2}$$

where, in general, ϵ_i^2 is the expected variance in the ith data point. Thus,

$$\begin{aligned} \{(q(f_1, \cdots, f_N) - q_{ex})^2\} &= \sum_{i=1}^{N} \frac{\partial q}{\partial f_i} \sum_{j=1}^{N} \frac{\partial q}{\partial f_j} \{\Delta f_i \Delta f_j\} \\ &= \sum_{i=1}^{N} \left(\frac{\partial q}{\partial f_i}\right)^2 \epsilon_i^2 \end{aligned} \tag{7.3}$$

which reduces the error problem to that of finding the dependence of the calculated result on the input data.

In the following sections, Equation 7.3 will be examined in detail for its applications in ROBFIT, but it should be mentioned that errors can be calculated with Equation 7.3 as it stands. This is done by physically changing all data points and recalculating the q of interest. This yields

$$\frac{\partial q}{\partial f_i} = \frac{q(f_1, \cdots, f_i + \epsilon, f_N) - q_0}{\epsilon} \tag{7.4}$$

which, along with ϵ_i, is all that is needed for Equation 7.3.

7.2 COEFFICIENTS DETERMINED BY LEAST-SQUARES FITTING

The coefficients c_k are determined by minimizing

$$\begin{aligned} \chi^2 &= \sum_i \omega_i (f_i - f_{\text{th}}(\mathbf{c_n}, x_i))^2 \\ &= \sum_i \left(\frac{f_i - f_{\text{th}}(\mathbf{c_n}, x_i)}{\epsilon_i} \right)^2 \end{aligned} \tag{7.5}$$

with respect to c_j. The first derivatives of χ^2 with respect to c_k are

$$\frac{\partial \chi_0^2}{\partial c_k} = -2 \sum_i \omega_i (f_i - f_{\text{th}}(\mathbf{c_n}, x_i)) \frac{\partial f_{\text{th}}(\mathbf{c_n}, x_i)}{\partial c_k} \tag{7.6}$$

and the second derivatives are

$$\frac{\partial^2 \chi_0^2}{\partial c_l \partial c_k} = 2 \sum_i \omega_i \frac{\partial f_{\text{th}}(\mathbf{c_n}, x_i)}{\partial c_l} \frac{\partial f_{\text{th}}(\mathbf{c_n}, x_i)}{\partial c_k} \tag{7.7}$$

in which terms of order

$$\frac{(f_i - f_{\text{th}})}{\epsilon_i^2} \frac{\partial^2 f_{\text{th}}}{\partial c_l \partial c_k}$$

have been dropped owing to the relatively small size of

$$\frac{1}{N} \sum \frac{(f_i - f_{\text{th}}(x_i))}{\epsilon_i^2}$$

In general, the process of finding the coefficients involves expanding

$$\chi^2 = \chi_0^2 + \sum_k \frac{\partial \chi^2}{\partial c_{0k}} \delta_k + \frac{1}{2} \sum_{kl} \frac{\partial^2 \chi^2}{\partial c_{0l} \partial c_{0k}} \delta_k \delta_l$$

with $\delta_k = c_k - c_{0k}$, and solving for the δs that predict zero partials of χ^2 by solving

$$\frac{\partial \chi^2}{\partial \delta_l} = 0 = \frac{\partial \chi^2}{\partial c_{0l}} + \frac{1}{2} \sum_k \frac{\partial^2 \chi^2}{\partial c_{0l} \partial c_{0k}} \delta_l \tag{7.8}$$

$$\delta_l = - \sum_k \left(\frac{\partial^2 \chi^2}{\partial c_{0l} \partial c_{0k}} \right) lk^{-1} \frac{\partial \chi^2}{\partial c_{0l}} \tag{7.9}$$

so that the end of this process is a $\mathbf{c_0}$ for which $\partial \chi^2 / \partial c_{0l} = 0$ and an inverse matrix of second derivatives at $\mathbf{c_0}$. Having determined the coefficients that best describe the data, we now need to estimate the errors in these constants. Knowing these errors enables us to calculate uncertainties in other quantitites such as peak positioning, width, and area.

Following the general prescription of Section 7.1, move f_p to $f_p + \Delta_p$ to yield a new χ^2, given by

$$\chi^2 = \chi_0^2 + \frac{\partial \chi_0^2}{\partial f_p} \Delta_p \tag{7.10}$$

which can be expanded as

$$\chi^2 = \chi_0^2 + \sum_j \frac{\partial^2 \chi_0^2}{\partial f_p \partial c_j} \Delta_p \delta_j + \frac{1}{2} \sum_{ij} \frac{\partial^2 \chi_0^2}{\partial c_i \partial c_j} \delta_i \delta_j \tag{7.11}$$

where the fact that $\partial \chi_0^2 / \partial c_j = 0$ and $\partial^3 \chi_0^2 / \partial f_p \partial c_i \partial c_j = 0$ has been used. The quantity χ^2 calculated using the point $f_p + \Delta_p$ minimizes at values of δ that are different from those of zero already found for χ_0^2. These are found by setting $\partial \chi_0^2 / \partial \delta_l = 0$, which yields

$$\delta_l = - \sum_k \left(\frac{\partial^2 \chi^2}{\partial c_{0l} \partial c_{0k}} \right)^{-1} \frac{\partial^2 \chi^2}{\partial f_p \partial c_k} \Delta_p \tag{7.12}$$

from which we find

$$\frac{\partial c_l}{\partial f_p} = \frac{\delta_l}{\Delta_p} \tag{7.13}$$

Using Equation 7.3, the error in c_l is

$$\left\{ (c_l - c_{lx(l)})^2 \right\} =$$

$$\left[- \sum_k \left(\frac{\partial^2 \chi^2}{\partial c_{0l} \partial c_{0k}} \right)^{-1} * - \sum_m \left(\frac{\partial^2 \chi^2}{\partial c_{0l} \partial c_{0m}} \right)^{-1} \sum_p \left(\frac{\partial^2 \chi^2}{\partial f_p \partial c_m} \frac{\partial^2 \chi^2}{\partial f_p \partial c_k} \right) \epsilon_p^2 \right] \tag{7.14}$$

where the sum p over the data points has been interchanged with the matrix sums k and m. For this case of least-squares fitting, note that, from Equation 7.5,

$$\frac{\partial^2 \chi^2}{\partial f_p \partial c_j} = -2 \frac{\partial f_{\text{th}}(x_p)}{\partial c_j \epsilon_p^2} \tag{7.15}$$

so that the sum over p in Equation 7.14 is exactly twice $\partial^2\chi^2/\partial c_m \partial c_k$ as given in Equation 7.7, which allows the sum over m to yield δ_{lk} and a final result of

$$\left\{(c_l - c_{lx(l)})^2\right\} = \sigma_l^2 = 2\left(\frac{\partial^2 \chi^2}{\partial c_{0l} \partial c_{0l}}\right)^{-1} \tag{7.16}$$

To recap, we have described the fitted function in terms of a set of constants and calculated the uncertainties in each of these constants. We must now use this information to calculate the error in the fit.

7.3 ERROR IN THE FITTED FUNCTION

Equations 7.3 and 7.4, along with Equations 7.12 and 7.13, can now be used to find the errors in any function of the constants. One of the more interesting functions to consider is the fitted function itself. The standard deviations in the constants are frequently much larger than the error in the function. That is due to the fact that large positive fluctuations in some constants will always occur with large negative fluctuations in others.

To begin,

$$f(x) = f_{\text{th}}(\mathbf{c}, x) \tag{7.17}$$

from which we calculate

$$\frac{\partial f(x)}{\partial f_p} = \sum_k \frac{\partial f_{\text{th}}(\mathbf{c}, x)}{\partial c_k} \frac{\partial c_k}{\partial f_p} \tag{7.18}$$

Then, using Equation 7.3,

$$\left\{(f(x) - f_{lx}(x))^2\right\} = \sum_{kl} \frac{\partial f_{\text{th}}(\mathbf{c}, x)}{\partial c_k} \frac{\partial f_{\text{th}}(\mathbf{c}, x)}{\partial c_l} \sum_p \frac{\partial c_k}{\partial f_p} \frac{\partial c_l}{\partial f_p} \tag{7.19}$$

and using Equations 7.12 and 7.13,

$$= \sum_{kl} \frac{\partial f_{\text{th}}(\mathbf{c}, x)}{\partial c_k} \frac{\partial f_{\text{th}}(\mathbf{c}, x)}{\partial c_l} * \sum_m \left(\frac{\partial^2 \chi^2}{\partial c_{0k} \partial c_{0m}}\right)^{-1} * \sum_n \left(\frac{\partial^2 \chi^2}{\partial c_{0k} \partial c_{0n}}\right)^{-1} * \sum_p \frac{\partial^2 \chi^2}{\partial f_p \partial c_{0m}} \frac{\partial^2 \chi^2}{\partial f_p \partial c_{0n}} \epsilon_p^2 \tag{7.20}$$

Then, using Equation 7.15 and Equation 7.7, note that the sum over p is two times $\partial^2\chi^2/\partial c_{0m}\partial c_{0k}$. Then the sum over n becomes δ_{lm} and the sum over m can be performed to yield

$$\left\{(f(x) - f_{lx}(x))^2\right\} = 2\sum_{kl} \frac{\partial f_{\text{th}}(\mathbf{c}, x)}{\partial c_k} \frac{\partial f_{\text{th}}(\mathbf{c}, x)}{\partial c_l} \left(\frac{\partial^2\chi^2}{\partial c_k \partial c_l}\right)^{-1} \tag{7.21}$$

which may explain why the inverse matrix is sometimes called the error matrix. The off-diagonal terms do the correcting for the fact that large and small coefficients are correlated. Note that many elements of the error matrix will normally be negative and will subtract from, rather than add to, the total error.

7.4 THE ERROR IN THE AREA OF EACH PEAK

The area of each peak is given by height times width times a factor that is completely independent of statistics. Thus, only the error in $c^2w = c_1^2c_3$ needs to be determined here. Following the prescription of Equation 7.3, we find

$$\begin{aligned}\frac{\partial c^2 w}{\partial f_i} &= 2cw\frac{\partial c}{\partial f_i} + c^2\frac{\partial w}{\partial f_i} \\ &= 2c_1c_3\frac{\partial c_1}{\partial f_i} + c_1^2\frac{\partial c_3}{\partial f_i}\end{aligned} \tag{7.22}$$

We have assumed only a single peak, with the square root of its height being the first constant, its location being the second constant, and its width being the third constant. In practice, there may be up to 27 more constants representing other peaks and up to 4 more constants representing the background contained in the matrix. From this we can extract the $\partial c/\partial f_i$ and $\partial w/\partial f_i$ needed to find the error in the peak area.

$$\begin{aligned}\left\{(c^2w - (c^2w)_{\text{ex}})^2\right\} &= \sum_i \left(2c_1c_3\frac{\partial c_1}{\partial f_i} + c_1^2\frac{\partial c_3}{\partial f_i}\right)^2 \epsilon_i^2 \\ &= (2cw)^2\sum_{jk}\left(\frac{\partial^2\chi^2}{\partial c_1\partial c_j}\right)^{-1}\left(\frac{\partial^2\chi^2}{\partial c_1\partial c_k}\right)^{-1} 2\frac{\partial^2\chi^2}{\partial c_j\partial c_k} \\ &\quad + 2(2cw)c^2\sum_{jk}\left(\frac{\partial^2\chi^2}{\partial c_1\partial c_j}\right)^{-1}\left(\frac{\partial^2\chi^2}{\partial c_3\partial c_k}\right)^{-1} 2\frac{\partial^2\chi^2}{\partial c_j\partial c_k} \\ &\quad + c^4\sum_{jk}\left(\frac{\partial^2\chi^2}{\partial c_3\partial c_j}\right)^{-1}\left(\frac{\partial^2\chi^2}{\partial c_3\partial c_k}\right)^{-1} 2\frac{\partial^2\chi^2}{\partial c_j\partial c_k}\end{aligned} \tag{7.23}$$

so that

$$\left\{(c^2w-(c^2w)_{\text{ex}})^2\right\} = 2\left[(2cw)^2\left(\frac{\partial^2\chi^2}{\partial c_1\partial c_1}\right)^{-1} + 2(2cw)c^2\left(\frac{\partial^2\chi^2}{\partial c_1\partial c_3}\right)^{-1} + (c^2)^2\left(\frac{\partial^2\chi^2}{\partial c_3\partial c_3}\right)^{-1}\right] \tag{7.24}$$

Note that errors in the location of the peaks, in addition to the heights, locations, and widths of the other peaks, appear implicitly in the inverse matrix.

7.5 INPUTTING EXTRA WIDTH INFORMATION

The peaks of interest usually cannot be found to any great degree of accuracy by minimizing χ^2 in Equation 7.5. They are usually too close to other peaks or too close to background features, or simply too small. The same spectrum frequently contains other fitted peaks from which it is possible to find very accurate, but not perfect, estimates of the peak width. Since this peak width is, in general, due to detector response rather than the particular properties of the peak, it is quite reasonable to use this information in finding the position, height, and multiplicity of the peaks of interest. This is done by adding a penalty term to χ^2 so that it becomes

$$\chi^2 = \sum_i \left(\frac{f_i - f_{\text{th}}(\mathbf{c_n}, x_i)}{\epsilon_i}\right)^2 + \lambda(w_i - w_{ex})^2 \tag{7.25}$$

where w_{ex} is the expected width. Because w_{ex} has an error, we do not want to force w_i to become exactly w_{ex}. We want the predicted error in w_i to be set so that the error in the peak area predicted by Equation 7.24, and the error in the peak location predicted by Equation 7.16, will be approximately correct.

Start by setting

$$A = \left(\frac{\partial^2\chi^2(\lambda=0)}{\partial c_1\partial c_1}\right)$$

$$B = \left(\frac{\partial^2\chi^2(\lambda=0)}{\partial c_1\partial c_3}\right) \tag{7.26}$$

$$D = \left(\frac{\partial^2\chi^2(\lambda=0)}{\partial c_3\partial c_3}\right)$$

and note from Equation 7.25 that for λ not equal to zero

$$D + 2\lambda = \left(\frac{\partial^2\chi^2(\lambda = 0)}{\partial c_3 \partial c_3}\right) \tag{7.27}$$

For an isolated peak whose error comes from its height and width alone, it is useful to note that

$$\left(\frac{\partial^2\chi^2}{\partial c_i \partial c_j}\right) = \begin{pmatrix} A & B \\ B & D + 2\lambda \end{pmatrix} \tag{7.28}$$

and

$$\left(\frac{\partial^2\chi^2}{\partial c_i \partial c_j}\right)^{-1} = \frac{1}{A(D + 2\lambda) - B^2}\begin{pmatrix} D + 2\lambda & -B \\ -B & A \end{pmatrix} \tag{7.29}$$

This enables us to calculate λ by noting that

$$\delta^2 w = \frac{2A}{A(D + 2\lambda) - B^2} \tag{7.30}$$

and thus the error matrix is

$$\left(\frac{\partial^2\chi^2}{\partial c_i \partial c_j}\right)^{-1} = \begin{pmatrix} \frac{1}{A} + \delta^2 w \frac{B^2}{2A^2} & \frac{-B\delta^2 w}{2A} \\ \frac{-B\delta^2 w}{2A} & \frac{\delta^2 w}{2} \end{pmatrix} \tag{7.31}$$

This approximation used by ROBFIT is not exact. The penalty term in χ^2 is actually independent of f_i so that

$$\left(\frac{\partial^2\chi^2}{\partial f_i \partial c_j}\right) = \left(\frac{\partial^2\chi^2(\lambda = 0)}{\partial f_i \partial c_j}\right) \tag{7.32}$$

Thus, in Equation 7.14, the inverse matrices are $(\partial^2\chi^2(\lambda)/\partial c_l \partial c_m)$ while the sum over the data points produces $(\partial^2\chi^2(0)/\partial c_m \partial c_k)$, which prevents the final matrix reductions, so that our overall error matrix becomes

$$E_{ij} = \left(\frac{\partial^2\chi^2(\lambda)}{\partial c_i \partial c_l}\right)^{-1} \left(\frac{\partial^2\chi^2(\lambda)}{\partial c_k \partial c_l}\right)^{-1} \left(\frac{\partial^2\chi^2(0)}{\partial c_l \partial c_j}\right) \tag{7.33}$$

or

$$E = \frac{1}{(A(D + 2\lambda) - B^2)^2}\begin{pmatrix} A(D + 2\lambda)^2 - 2B^2(D + 2\lambda) + B^2 D & B^3 - BAD \\ B^3 - BAD & A(AD - B^2) \end{pmatrix} \tag{7.34}$$

rather than Equation 7.29. The λ in Equation 7.32 is quite different from the λ in Equation 7.29. Again, solving for λ by setting $\delta^2 w$ to $2E_{33}$, the error matrix in terms of $\delta^2 w$ becomes

$$E = \begin{pmatrix} \dfrac{1}{A} + \delta^2 w \dfrac{B^2}{2A^2} & \dfrac{-B\delta^2 w}{2A} \\ \dfrac{-B\delta^2 w}{2A} & \dfrac{\delta^2 w}{2} \end{pmatrix} \tag{7.35}$$

which is exactly the same as before. The only difference is the value of λ,

$$\lambda_a = \frac{B^2 - AD}{2A} + \frac{1}{\delta^2 w} \tag{7.36}$$

whereas for the true error matrix

$$\lambda_t = \frac{B^2 - AD}{2A} + \sqrt{\frac{2(AD - B^2)}{A\delta^2 w}}$$

which, because of the square root, is much smaller than the approximate value. Putting the approximate λ_a into Equation 7.34 for $\delta^2 w$ yields

$$\delta^2 w = \frac{(AD - B^2)}{A}\delta^4 w_a = \frac{\delta^4 w_a}{\delta_N^2}$$

which is the actual spread of width values that will be found when ROBFIT uses Equation 7.30 to adjust the penalty function.

In summary, width information is put into the error matrix by adding a penalty function. The resulting width error, as calculated from the inverse matrix of a 2×2 matrix involving the peak's height and width, then correctly predicts the 2×2 error matrix. It is assumed that this same property survives the much larger matrix inversions made in practice. The principle caveat is that the output widths from this procedure will not deviate from the input widths by the correct distribution, even though the error matrix will be correct.

CHAPTER 8

Application of ROBFIT

This chapter has been included to give the reader a feel for how ROBFIT is used to analyze a spectrum. The first section shows the accuracy that can be achieved in dealing with clearly separated peaks. It also outlines some of the practical considerations that must be addressed before fitting is started. Section 8.2 gives a detailed account of the reanalysis of the supernova data. It shows how ROBFIT can be tuned to a particular analysis method. The reanalysis illustrates how easy it is to fit a spectrum, even if peak shapes and widths are unknown prior to fitting. The final section gives a feel for how the code can be used on different kinds of spectra. The example we show is a study of solar seismology data. These spectra do not follow the usual Gaussian statistic fluctuations from channel to channel. We show that the code can effectively deal with this kind of data even though ROBFIT's minimization scheme has been designed around Gaussian statistics.

8.1 COMPUTER-GENERATED TEST DATA

The following computer-generated test cases have been performed to provide an insight into the operational accuracy of ROBFIT. They test the code's ability to correctly reproduce a single peak in both high and low background regions. The tests also study the effect of peak width on determination of peak parameters. The code's accuracy in detecting multiple peaks is also addressed. For the purpose of these tests, it is worth clarifying the definition of background. The background is defined as the underlying continuum on which the peaks sit. Any additional peaks introduced by fluctuations in this background continuum will be called spurious peaks even though they are in fact a background to the true peaks.

8.1.1 Peaks in Large Background Regions

A constant background level of 50 counts is used in the following test runs. Peaks are then added to this background by specifying their position, width, and total area. The number of counts in each channel has been randomly distributed using a Gaussian function. The standard deviation in a channel has been set equal to the square root of the number of counts in the channel.

TABLE 8.1
Results of five separate fits to large peaks

Fit	Position (channels)	Error (channels)	Width (channels)	Error (channels)	Strength (counts)	Error (counts)
1	500.55	0.39	58.38	0.82	9,649	135
2	500.24	0.41	59.69	0.86	10,118	145
3	500.28	0.38	60.91	0.80	10,216	134
4	499.80	0.37	59.33	0.78	9,871	129
5	499.56	0.40	58.44	0.85	9,932	145

First consider the large-peak case. A Gaussian-shaped peak has been generated in this test. It has a height approximately four times the background level. Each peak has been positioned at channel 500 in a 1,000-channel spectrum and has a width of 58.85 channels. The area under each peak, or strength, is 10,000 counts. Table 8.1 shows the results of fits to five separate data sets; the position and error in position, width and error in width, and strength and error in strength are given for each peak.

This simple test shows that the program can accurately reproduce the underlying peak function. The results of fit 1 are shown in Figure 8.1.

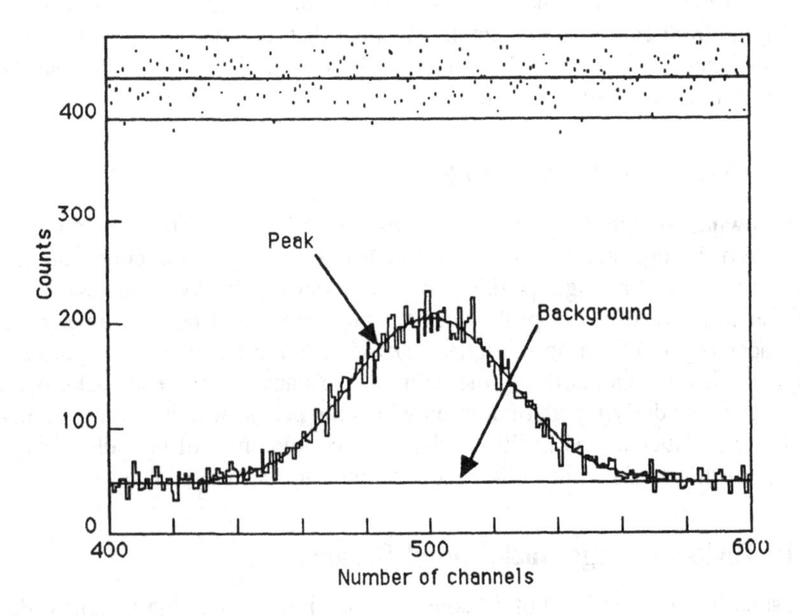

FIGURE 8.1
Result of a fit to a large peak on a large background. For peak parameters, see fit 1 in Table 8.1.

TABLE 8.2
Fit to peaks the same size as the background fluctuations

Fit	Position (channels)	Error (channels)	Width (channels)	Error (channels)	Strength (counts)	Error (counts)
1	504.54	3.31	51.83	5.12	612	74
2	494.59	3.95	45.00	6.96	667	122
3	503.16	3.23	62.64	5.16	853	83
4	497.46	3.40	55.63	5.00	653	75
5	497.75	3.30	52.65	5.62	704	84

Next, consider the small-peak case. To test the code's ability to detect small peaks on a large background, we show the results of five separate fits to the Gaussian-shaped peak. In these tests the total area underneath the peak has been reduced to 700 counts. This gives a peak height of 11 counts, approximately 1.6 times the background fluctuation level. Fitting peaks of this size requires a low cutoff; we have used a value of 2. At this level one anticipates finding a few spurious peaks. The number of acceptable spurious peaks is entirely dependent on the requirements of the data analysis. Here we have a maximum of one spurious peak in every 100 channels. The results of the five test runs are detailed in Table 8.2, and fit 1 is shown in Figure 8.2.

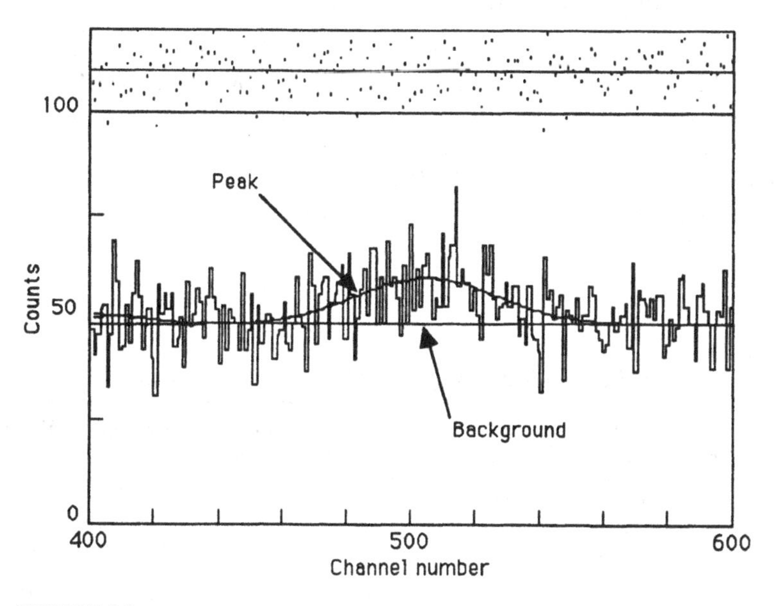

FIGURE 8.2
Result of a fit to small peak on a large background. For peak parameters, see fit 1 in Table 8.2.

8.1.2 Peaks in Low Background Regions

The following analysis examines the detection efficiency of ROBFIT when the background under the peak is small. In these tests a background level of 2 has been used. Again, peaks are added and the counts in each channel distributed according to a Gaussian distribution as described in the preceding discussion.

First consider the large-peak case. Here we have chosen a Lorentzian as the peak shape and generated three peaks at channel positions 167.6, 501, and 834.3. The peaks have half-widths of 11.11, 6.67, and 4.45 channels, and the areas under each peak are 6,824, 4,000, and 2,000, respectively. These parameters give each peak a height of approximately 200 counts. Table 8.3 shows the results of two fits to separately generated data sets, and Figure 8.3 shows the fit to the first set of three peaks.

Next consider the small-peak case. To test ROBFIT's ability to detect small peaks on a low background, we have generated three Lorentzian-shaped peaks in a manner similar to that just described. In these tests, each peak has the same area: 50 counts. This gives the peaks heights of 1.4, 2.4, and 3.6 counts. These peak heights are to be compared with a background level of 2 ± 1 count. The results of these low-peak, low-background tests are shown in Table 8.4 and in Figure 8.4.

8.1.3 Detecting Peaks with Narrow Widths

When dealing with histogrammed data, the user must always be aware of the effects of binning on the peak detection mechanism. In a typical nuclear physics experiment, data taking is usually organized so that spectral features have widths greater than 3 or 4 channels. Even at this binning the histogram does not contain all the information of the underlying deriving function.[23] However, it may be that a finer binning is not possible. This means the analysis technique must be capable of dealing with these narrow peaks.

TABLE 8.3
Fit to large peaks on a low background

Fit	Peak	Position (channels)	Error (channels)	Width (channels)	Error (channels)	Strength (counts)	Error (counts)
1	1	167.44	0.20	22.49	0.41	6,830	93
	2	501.18	0.16	13.10	0.32	3,791	70
	3	834.62	0.16	8.81	0.32	1,902	51
2	1	167.78	0.21	22.15	0.41	6,777	95
	2	500.88	0.15	12.52	0.30	3,982	73
	3	834.35	0.15	8.42	0.29	1,953	51

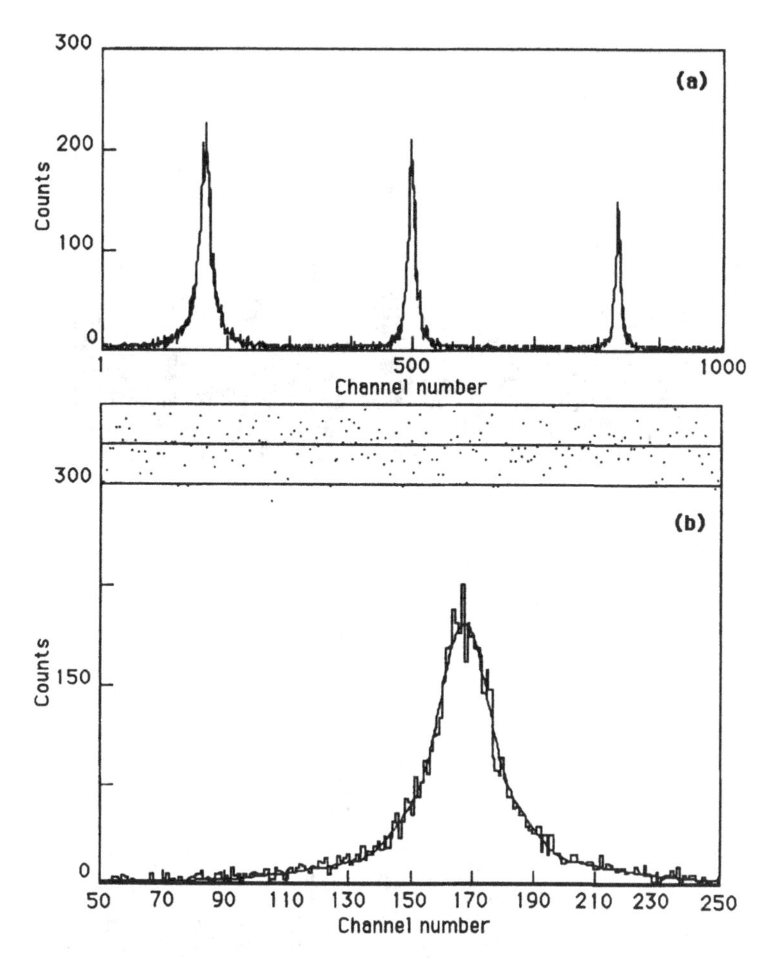

FIGURE 8.3
Results of fit 1 of Table 8.3. (*a*) Three large Lorentzian-shaped peaks on a low background. (*b*) The fit to the lowest channel region.

TABLE 8.4
Fits to small peaks on a low background

Fit	Peak	Position (channels)	Error (channels)	Width (channels)	Error (channels)	Strength (counts)	Error (counts)
1	1	165.79	4.23	26.40	5.50	70	18
	2	507.15	3.66	19.31	5.50	53	16
	3	836.70	1.42	7.88	3.50	42	13
2	1	171.50	5.00	23.20	5.59	50	17
	2	500.18	1.47	11.17	3.59	71	16
	3	834.53	0.94	6.58	2.27	54	13

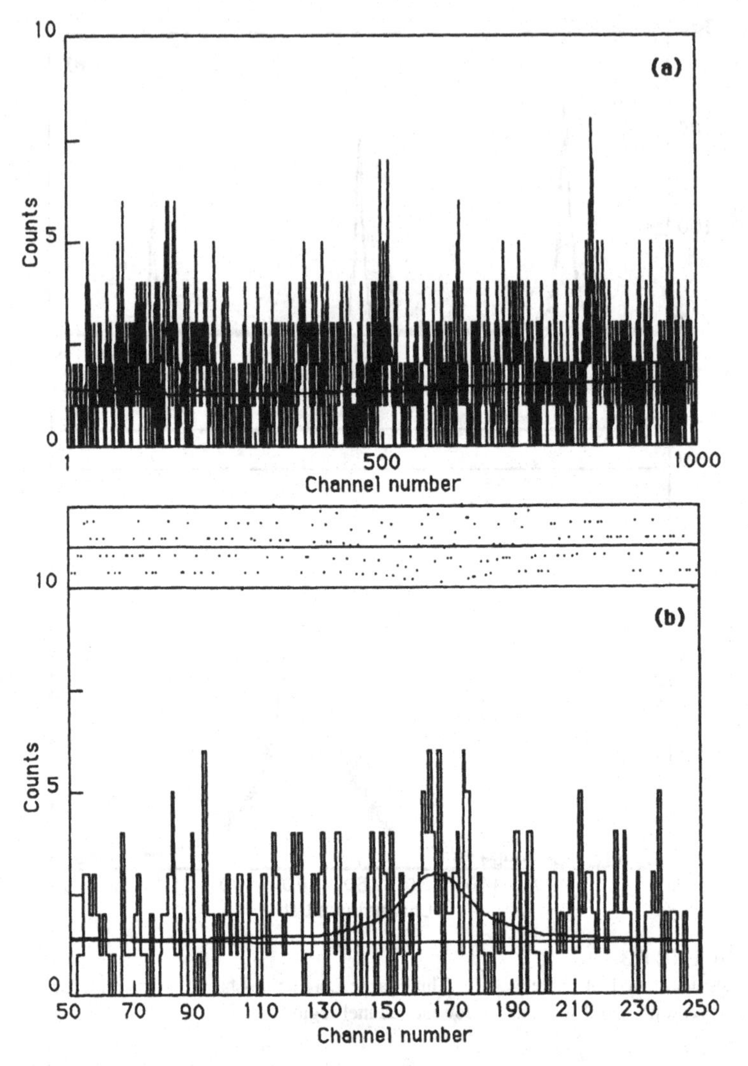

FIGURE 8.4
Results of fit 1 of Table 8.4. (*a*) Small Lorentzian-shaped peaks on a low background. (*b*) The fit to the lowest channel region.

TABLE 8.5
Fits to a peak centered on channel 500

Fit	Position (channels)	Error (channels)	Width (channels)	Error (channels)	Strength (counts)	Error (counts)
1	500.01	0.03	5.98	0.06	9,874	112
2	500.01	0.03	5.94	0.05	10,209	110
3	500.02	0.03	5.94	0.06	10,006	125

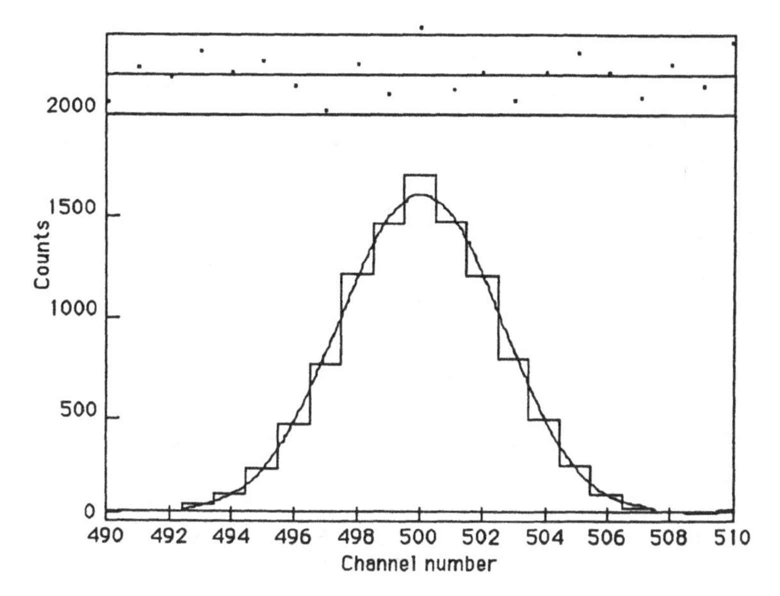

FIGURE 8.5
Fit to a peak centered at channel 500, using the data from fit 1 of Table 8.5.

This section shows how accurately ROBFIT can fit narrow peaks. A Gaussian-shaped peak with a strength of 10,000 counts and a width of 5.86 channels has been used as the test peak. The peak is placed upon a constant background level of 50 counts.

In the first test, the centroid of the peak is at channel 500. Table 8.5 shows the results of three separate fits to the data. The results of the first fit are shown in Figure 8.5.

To examine ROBFIT's sensitivity to shape changes due to the binning boundaries, the second test has the centroid of the peak shifted to position 500.3. Table 8.6 gives the results of three fits to these data. The results of fit 1 are shown in Figure 8.6. The peak detection performance of the code for these narrow peak widths is seen to be the same as for the greater widths of Sections 8.1.1 and 8.1.2.

TABLE 8.6
Fits to a peak centered on channel 500.3

Fit	Position (channels)	Error (channels)	Width (channels)	Error (channels)	Strength (counts)	Error (counts)
1	500.32	0.03	5.98	0.06	9,878	115
2	500.30	0.03	5.94	0.05	10,209	115
3	500.32	0.03	5.95	0.06	10,016	119

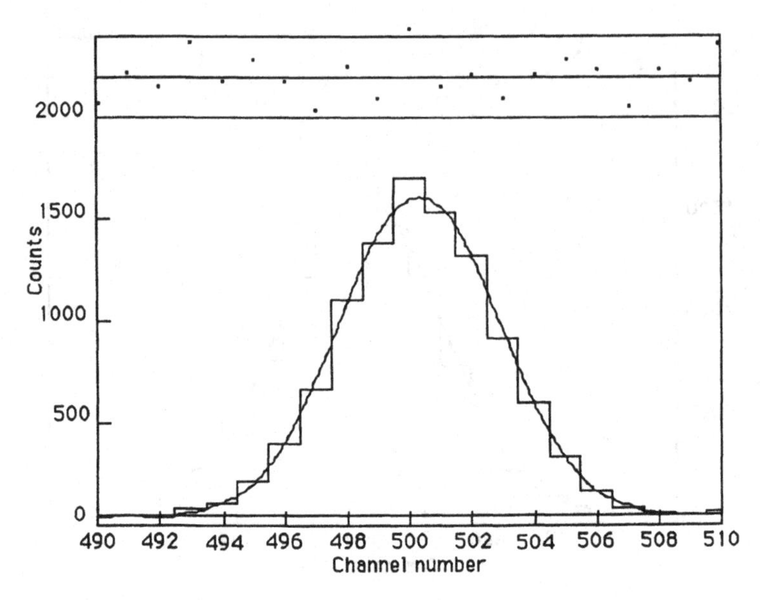

FIGURE 8.6
Fit to a peak centered at channel 500.3 using data taken from fit 1 of Table 8.6.

8.2 SUPERNOVA DATA

In this section we give a detailed account of the analysis of the supernova data. The results of this analysis are provided in Rester et al.[1] Here we give the step-by-step reduction of the raw data. This should provide the reader with an insight into how ROBFIT can be used in the analysis of real data. Figure 8.7 shows the

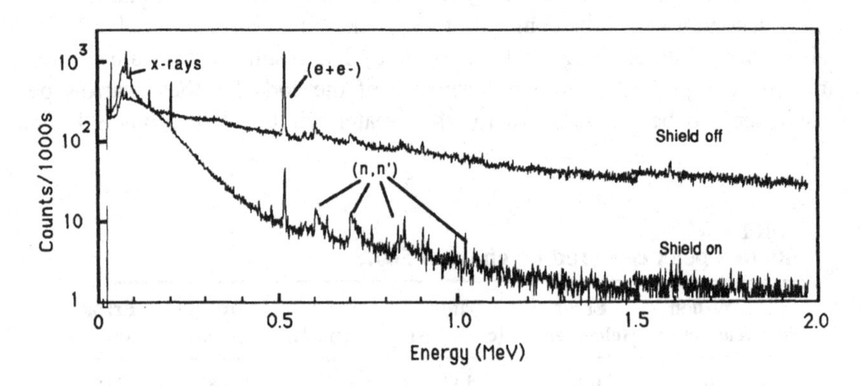

FIGURE 8.7
The supernova gamma-ray spectrum with the veto shield on and with the shield off. The analysis in this chapter has been performed on the "shield on" spectrum. Indicated in this figure are the four neutron-inelastic scattering peaks (n, n').

raw data. We have concentrated our analysis on the channel range 57 to 4,096; the first 57 channels are used for timing signals and threshold offsetting.

In this analysis we will need to set a number of startup parameters. These must be defined at the outset of each run. The first step of the data reduction will illustrate how these parameters are presented.

In the first step we are concerned only with getting a rough description of the peak shapes within the spectrum. To begin, we have generated a Gaussian standard containing five back-to-back cubic splines. This standard is then introduced to ROBFIT as the description of the peak shapes within the spectrum. From a brief inspection of the raw data we have estimated the peak widths, and for this first run we have set the width range to be from 5 ± 2 at channel 520 to 5 ± 2 at channel 1050. We expect to be able to locate all peaks to within 3σ of the background fluctuations by specifying a cutoff range from 10 to 3, taking 7 steps between these limits. At this point these values are just convenient starting levels. We do not expect to be able to identify all the peaks until we have representations of all the peak shapes in the data. There is one last parameter that must be specified, and it is the number of coefficients to be fitted to the background. Again, as a starting point, we will begin the background cycle with four coefficients and add to it a maximum of eight. The starting conditions are defined by:

4, 8	Background constant range
10, 3, 7	Cutoff range
5, 2, 520; 5, 2, 1050	The single standard width range
$w_L, E_L, p_L, w_H, E_H, p_H$	Additional width variations for other standards specified as low and high widths, errors, and positions

Figure 8.8 illustrates the types of peaks identified in this first trial run, and Table 8.7 lists all the fitted peak parameters. As can be seen, the single peaks at channels 1,022 and 1,050 have been well fitted. The broad regions between channels 860 and 890, and channels 1,080 and 1,090, where we know neutron-inelastic scattering peaks reside, have been fitted with multiple peaks. This multiple peak fitting is called braiding and generally signifies that we have an incorrect peak shape. Of course, if there really were only one peak shape present within the spectrum, this braiding would be justified, and the user would need to study these regions carefully to understand the overlapping peaks. For these data we know a priori that these regions actually require a fundamentally different peak shape. These regions are dominated by neutron-inelastic scattering peaks. The spectrum contains four such regions centered around channels 620, 720, 870, and 1,080. We must therefore generate a peak standard to take care of these shapes. This becomes important later in the analysis because one of the supernova peaks actually resides on top of the 870 neutron peak. How do we create the standard for these peaks? Because we have no analytic function for generating these peak shapes, we must create a shape from the data. For this purpose the neutron peak within the channel range 700 to 760 was fitted with five back-to-back

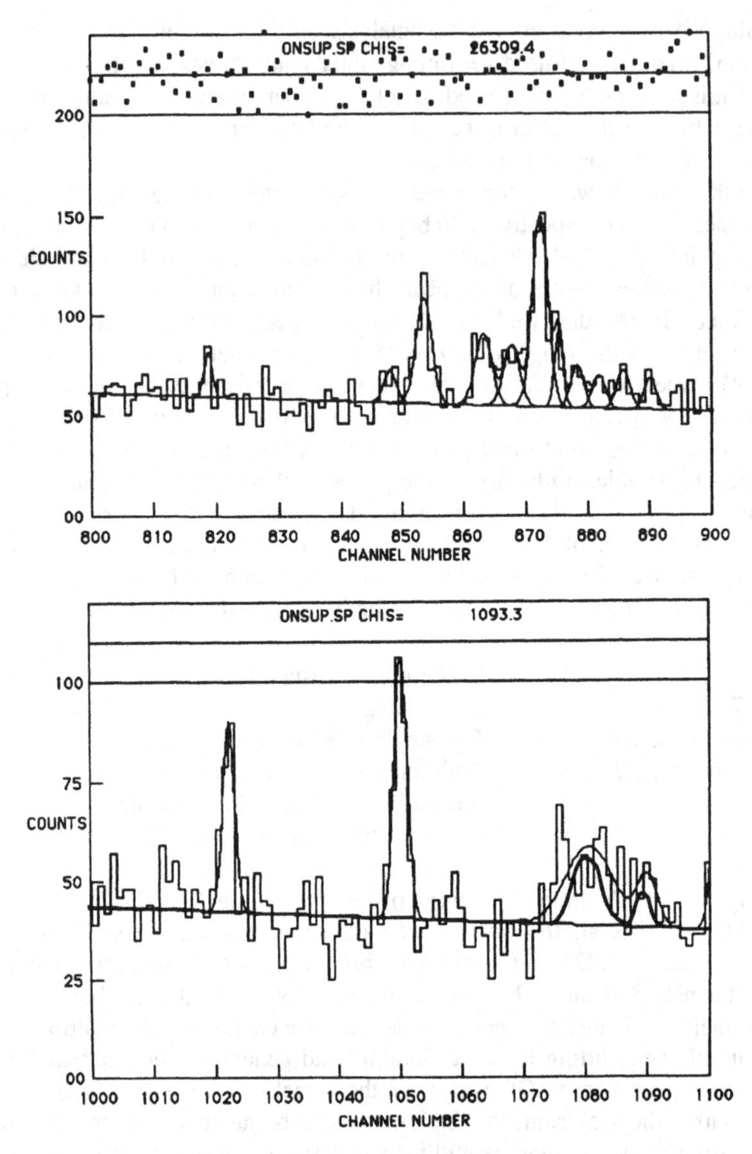

FIGURE 8.8
Fits to the 800–900 and 1,000–1,100 channel regions of the supernova data. The figures illustrate the two peak shapes that are present in the supernova data.

cubic splines. This region of the spectrum is shown in Figure 8.9. This second standard can now be entered into the fitting.

Before starting the second round of fitting, we must also think about our Gaussian-shaped standard. Is this shape ideal for the gamma-ray peaks? A well-defined peak shape produces a better fit. Before the supernova experiment was undertaken, a number of gamma-ray lines were measured with the detector that

TABLE 8.7
ROBFIT output from the first round of fitting

```
OUTPUT FROM ROBFIT
IP=    1
  27 PEAKS 1001 CH#N, CHIS=          1093. NITB=     8
CUTOFF=    3.00
 1.0672  1.0000
 526.356    .081      3.704    .170        1854.80      84.12  1
 586.144    .847      8.344   1.958         255.97      52.91  1
 618.965    .491      9.000   1.133         896.02      99.80  1
 631.662   1.194      9.000   3.873         339.37     122.93  1
 642.382   2.205      7.017   3.876         135.41      84.02  1
 648.958    .334      3.255    .778         221.32      54.28  1
 718.864    .413      7.862    .943         836.83      86.38  1
 726.379    .784      4.283   2.230         202.49     216.60  1
 732.597   1.821      9.000   5.748         440.47     281.39  1
 744.543   1.352      9.000   2.713         347.65     108.62  1
 777.033    .231      3.143    .536         168.98      25.42  1
 853.649    .245      2.926    .541         160.91      27.26  1
 864.857    .780      6.735   1.983         198.82      47.60  1
 872.591    .244      4.085    .574         352.49      47.99  1
 882.420   1.312      9.000   2.592         153.97      42.93  1
 912.615    .576      4.149   1.326          86.35      24.40  1
 927.202    .199      2.520    .445         171.22      27.87  1
 934.195   1.487      5.641   2.665          63.90      31.84  1
 942.405    .343      2.749    .774         102.38      25.83  1
1022.343    .200      2.197    .444         112.20      20.71  1
1050.151    .137      2.452    .299         173.82      19.94  1
1080.635    .932      9.000   2.363         191.62      41.12  1
1089.929    .911      3.767   2.120          51.03      28.29  1
1101.082    .289      2.311    .661          62.17      15.94  1
1157.174    .255      1.515    .489          39.22      12.49  1
1263.145    .399      2.205    .919          38.37      14.18  1
1418.439    .763      4.445   1.766          48.29      16.91  1
```

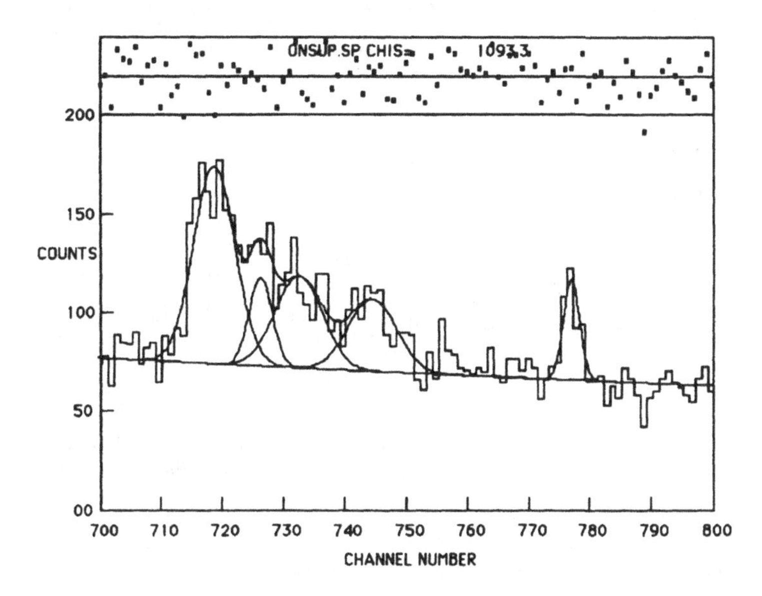

FIGURE 8.9
The region of the spectrum used to generate the neutron-inelastic scattering peak. The fit is from the first round of fitting.

was to be used in the experiment. In this way, a large number of counts could be accumulated under one gamma-ray peak. Then, using the standard fitting routine, a precise standard was made. This new standard is called the YT standard after the Yttrium spectrum, from which it was generated. The YT standard can now replace the Gaussian standard used in the first round of fitting.

The new standards are introduced in step 2 of the fitting. The starting parameters for this step are

4, 12	
10, 3, 7	
2.5, 0.5, 927, 2.5, 0.5, 1050	YT standard
11, 2, 620; 25, 2, 867	Neutron standard

We have allowed four more constants to be added to the background, and the cutoff range is held the same as in the first two runs. We have now included both standards in the fit. All the braiding peaks must be removed from the peak list before we begin the fit.

Figure 8.10 shows the fit to the neutron peak at channel 870. The braiding has now been corrected, but closer inspection shows that the width of the neutron-inelastic scattering peak is too large. Upon inspecting the other inelastic scattering peaks, we find that the peak in channel region 720 seems to be fitted reasonably well. Therefore, we take its width as the starting point for the next step.

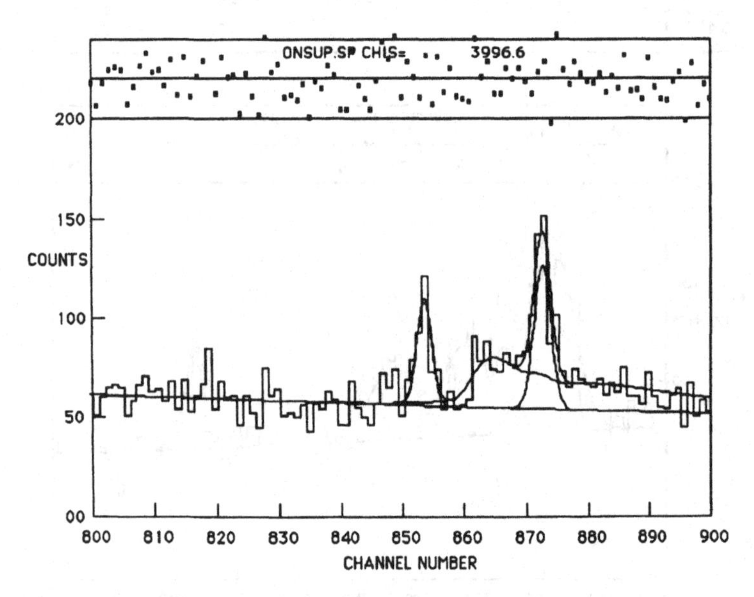

FIGURE 8.10
Third-round fit to the third inelastic scattering peak.

Determining the correct neutron peak widths is the objective of step 3. The starting parameters are

4, 12
10, 10, 1
2.61, 0.36, 927.2; 2.47, 0.28, 1050.2 YT standard
11, 1, 615.8; 11, 1, 864 Neutron standard

where we have redefined the neutron standard widths.

Because the previous run had a cutoff level of 3σ and thus found most of the peaks present in the spectrum, all we need to do in this run is refit the peaks that are already present. Consequently, we run the cutoff to a level of 10σ. The results of these refits are shown in Figure 8.11. It can be seen that the neutron peaks have been determined quite well. Most important for the supernova analysis is the fact that the 870 neutron peak has been well fitted. We are still having some problems with the 1,080 channel neutron peak; however, for the remainder of this analysis the precise fitting of this region is irrelevant.

In looking for a good fit, it is best to study the residuals plotted above each figure. For a good fit, the points must be equally scattered above and below the center line. The appearance of low points generally means the background level is too high and usually indicates undetected peaks within that region. As these peaks are identified, the background level will drop.

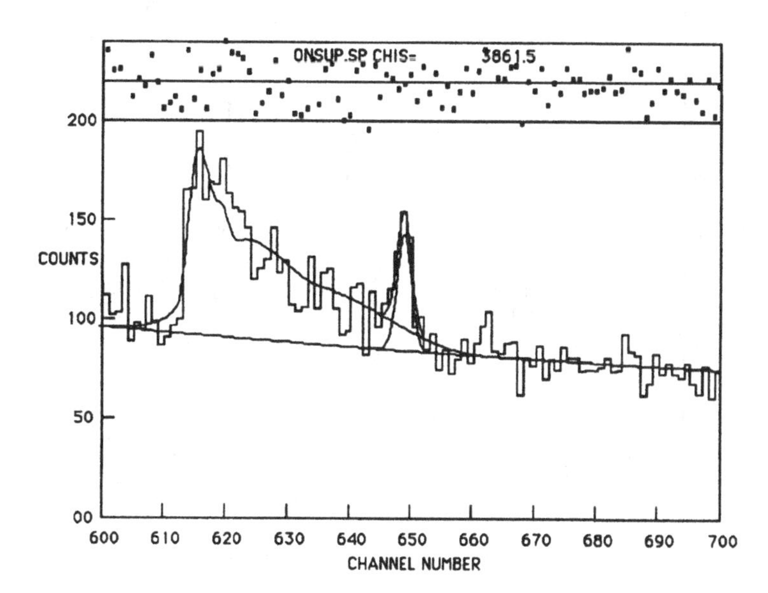

FIGURE 8.11a
The 620 channel neutron regions after the third round of fitting.

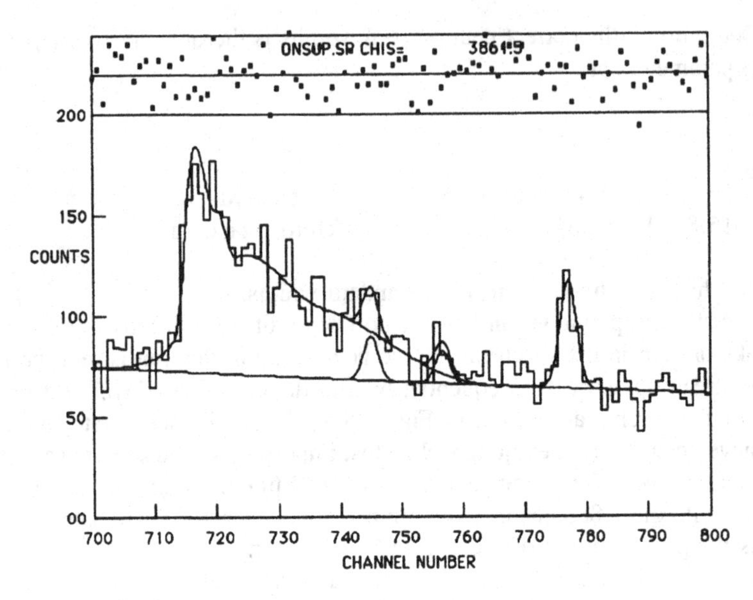

FIGURE 8.11b
The 720 channel neutron regions after the third round of fitting.

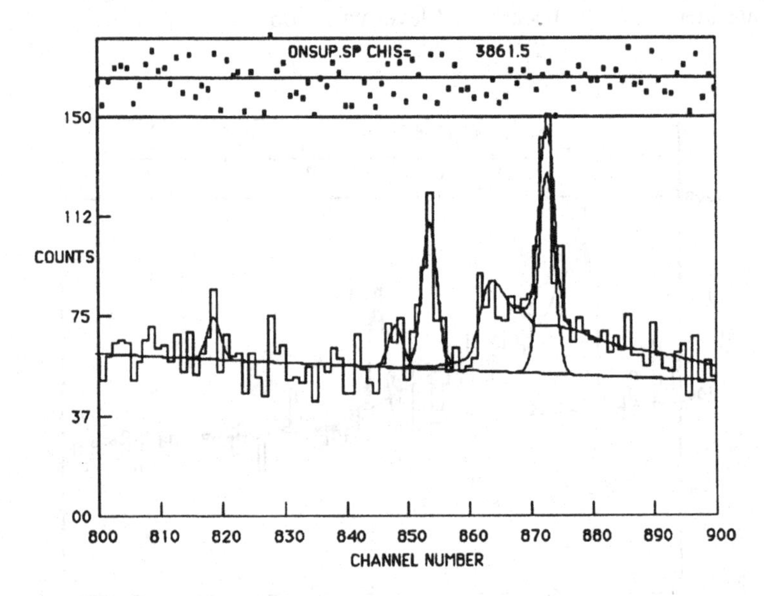

FIGURE 8.11c
The 870 channel neutron region after the third round of fitting.

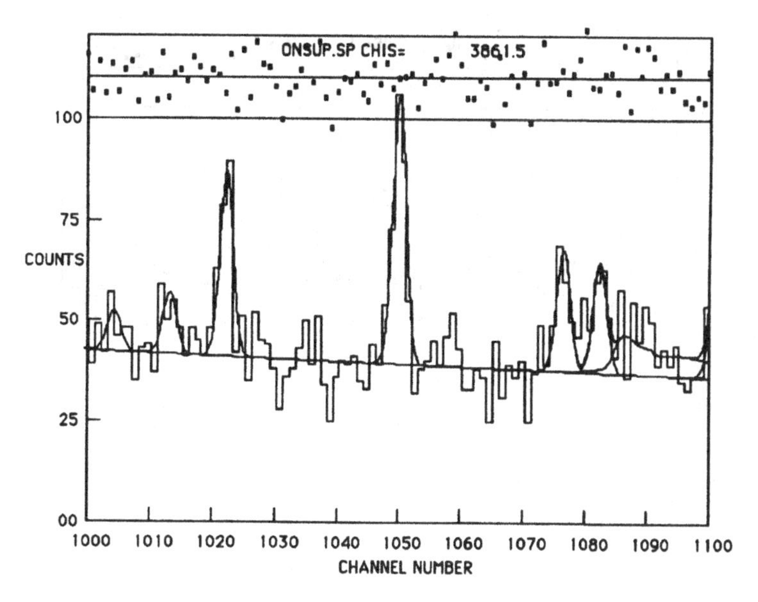

FIGURE 8.11d
The 1080 channel neutron regions after the third round of fitting.

Having fitted the neutron-inelastic scattering peaks and the gamma-ray peaks to reasonable accuracy, we can start the search for the supernova signals. Let us concentrate on the high channel region first. We expect to see a signal around the 1,260–1,290 channel region. The first thing we find is that we must reduce the cutoff level below 3σ to pick up any features in this region. Although running the cutoff below 3σ will allow spurious peaks to enter into the peak list, any fitted peaks will be supplied with an associated error. From this the user can decide if the peak is actually a real signal. Of course, the user must monitor the overall χ^2 value for goodness of fit, remembering that the χ^2 per number of degrees of freedom should be around 1. We do not want to introduce too many peaks, or spurious overfitting will occur.

Step 4 is performed to reduce the cutoff level; the starting values are

4, 12

10, 2, 8 for a first run; then 2, 1.9, 1 for the second

2.61, 0.36, 927.2; 2.47, 0.28, 1050.2

12.8, 0.6, 717; 12.8, 1.0, 863

We have used the neutron widths from step 3 to refine the width range. We have performed two succesive runs of this width setup; the first was to a 2σ cutoff, and the second ran the cutoff to an even lower level, 1.9σ. This was because, even at the 2σ level, the high supernova clump was not found. The 1.9σ fit is shown in Figure 8.12. The low supernova region at channel 870 is identical to the fit in Figure 8.11.

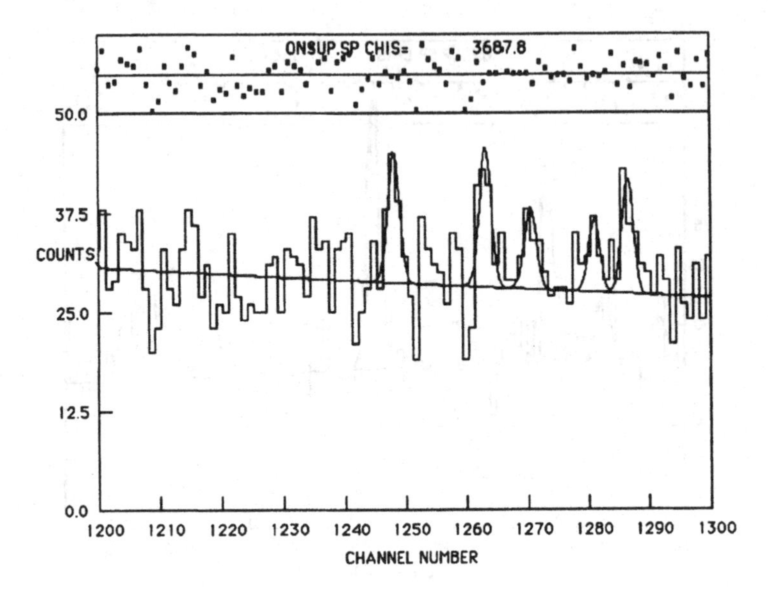

FIGURE 8.12
High supernova region after the fourth round of fitting.

At this point we could end the analysis, concluding that there is no supernova peak at the low region and that four gamma-ray lines have been identified at the high region. The rest of the analysis would then involve attempting to give meaning to these lines. However, from a second set of data taken when the experiment was not fully operational, but good enough to give some estimate of the shape of these supernova peaks, there was an indication that the supernova lines may be somewhat broader than the standard gamma-ray line. This is not too difficult to imagine, considering that the supernova is an explosion with fragments speeding in all directions. This motion would impart a spread of velocities on any gamma-ray line originating from this debri, which would broaden the line. The following analysis not only gives a deeper insight into the dynamics of the supernova, but also illustrates the flexibility of ROBFIT. Assuming that any peak due to the supernova is going to be broad in nature, we can generate a third standard, which we will call the wide-line standard, to take care of this shape. The standard is a Gaussian shape and is composed of three back-to-back cubic splines.

In step 5 we remove all the peaks that have been fitted to the high supernova region and introduce the wide-line standard. The starting parameters are:

4, 12
3, 2, 2
2.61, 0.36, 927.2; 2.47, 0.28, 1050.2
12.8, 0.6, 717; 12.8, 1.0, 863
6, 2, 1260; 6, 2, 1285 Wide-line standard

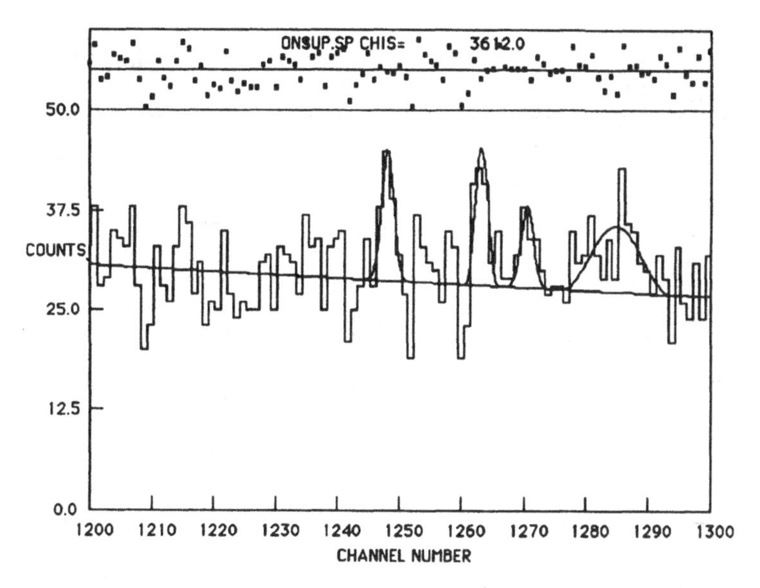

FIGURE 8.13
High supernova region after the fifth round of fitting. The code has started to find broad structures in this region.

where the width limits of the third standard have been estimated from the data structure around the 1,280 region. At this point they are just a convenient starting point.

Figure 8.13 shows the results of this fit. Now, even at the 2σ level, a broad peak has been fitted in preference to the narrow gamma-ray peak. What is not obvious from the fit is that the peak at channel 1,261 is also of the wide-line shape; however, ROBFIT prefers to split any structure in this region into two narrow peaks.

We could now cycle the code a few times to home in on the exact width of the peak, but we would like to throw another feature of the code into the analysis. If the broad peak at channel 1,285 is actually a supernova peak, then we must also see a signal around the 870 channel region. This now allows us to illustrate another feature of ROBFIT, the ability to add peaks to the peak list. From the last run, we find that the peaks within the 870 region have been fitted with the values given in Table 8.8.

It is necessary to introduce another peak, of the wide-line standard, into this region. The precise starting values for areas and widths are not that important at this stage, as the code will recalculate these when it refits the region. However, we must give the peak some starting area, or else the code will simply remove it. We must reduce the size of a nearby peak by the same amount because all the areas should sum to the same value as the previous fit. The new peak list is shown in Table 8.9. No errors need to be set at this point, as they will be recalculated during fitting.

TABLE 8.8
List of the peaks found in the 870 channel region in the fifth round of fitting

```
OUTPUT FROM ROBF1T
IP=    3
 175 PEAKS 4040 CHAN, CHIS=        3612. NITB=    12
CUTOFF=   2.00
 1.0682  1.0000  1.3065   .9993  1.0065  1.0000
```

"other peaks"

804.359	3.363	12.737	.862	69.68	46.75	2
818.784	.588	2.564	.409	41.44	17.67	1
848.065	.517	2.671	.397	49.88	17.04	1
853.661	.196	2.947	.399	172.36	22.98	1
863.647	.437	13.940	1.092	675.54	60.79	2
872.587	.179	3.456	.392	235.61	28.84	3
905.262	.681	2.754	.372	38.25	17.03	1

"other peaks"

Step 6 is to run the output files of step 5, plus the extra peak, back through the fitter. The starting values are:

4, 12

3, 3, 1

2.61, 0.36, 927.2; 2.47, 0.28, 1050.2

12.8, 0.6, 717; 12.8, 1.0, 863

8.7, 2, 1,260; 8.7, 2, 1286

We have redefined the width range of the fitted wide-line standard according to the width of the peak fitted in step 5. The results of this fit are illustrated in

TABLE 8.9
Modified list of peaks for input into round 6 of fitting

```
OUTPUT FROM ROBF1T
IP=    3
 175 PEAKS 4040 CHAN, CHIS=        3612. NITB=    12
CUTOFF=   2.00
 1.0682  1.0000  1.3065   .9993  1.0065  1.0000
```

"other peaks"

804.359	3.363	12.737	.862	69.68	46.75	2
818.784	.588	2.564	.409	41.44	17.67	1
848.065	.517	2.671	.397	49.88	17.04	1
853.661	.196	2.947	.399	172.36	22.98	1
863.647	.437	13.940	1.092	675.54	60.79	2
872.587	.179	3.456	.392	135.61	28.84	1
872.000	0	4	0	100	0	3
905.262	.681	2.754	.372	38.25	17.03	1

"other peaks"

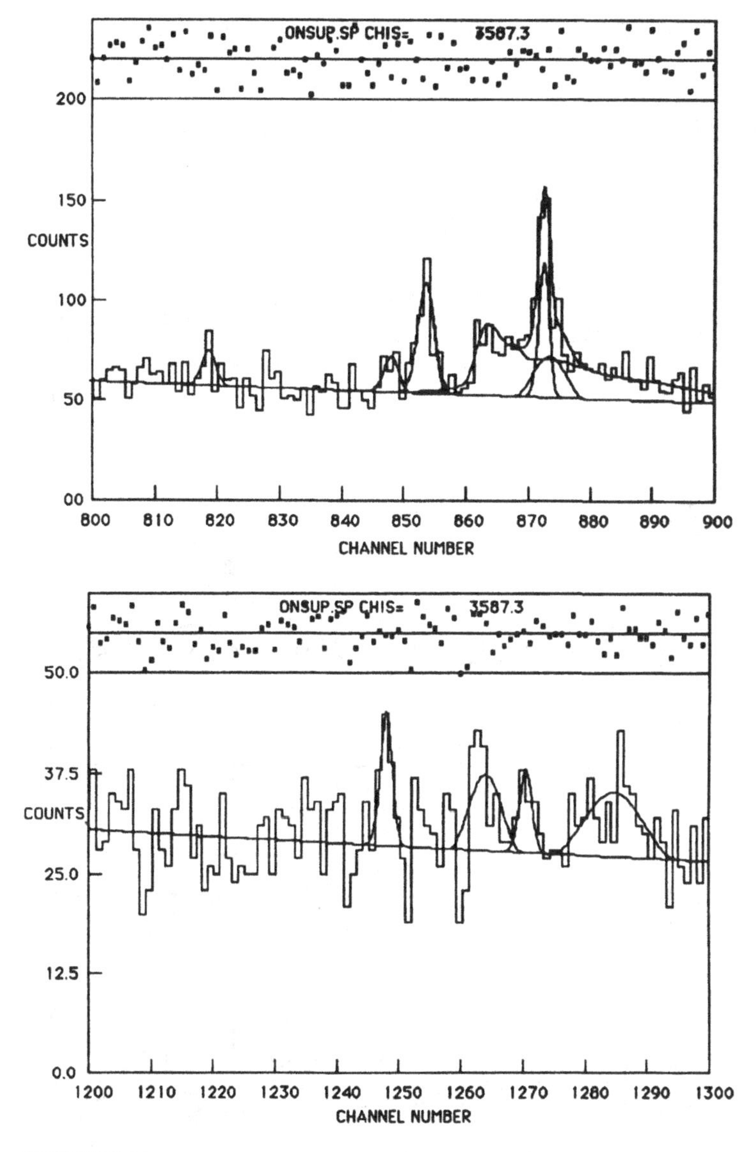

FIGURE 8.14
Low- and high-energy supernova regions after sixth round of fitting.

Figure 8.14, which shows the two supernova regions. From the figure we see that the code holds onto the wide-line peak at the low supernova region. It also moves the peak position to channel 873.4±1.1 with a width of 5.6±2.0 channels and an area 116±71. We also see some change at the high supernova region. The peak at channel 1,264 now looks dramatically wider. This indicates that the peak at channel 1,270 may be part of the same peak. This was also the case in our second data sample. We can perform a further cycle of the code with the 1,270 peak removed.

The final step is to perform this rerun. The starting values are:

4, 12
3, 3, 1
2.61, 0.36, 927.2; 2.47, 0.28, 1050.2
12.8, 0.6, 717; 12.8, 1.0, 863
5.6, 2.0, 873; 10.4, 2.0, 1284.

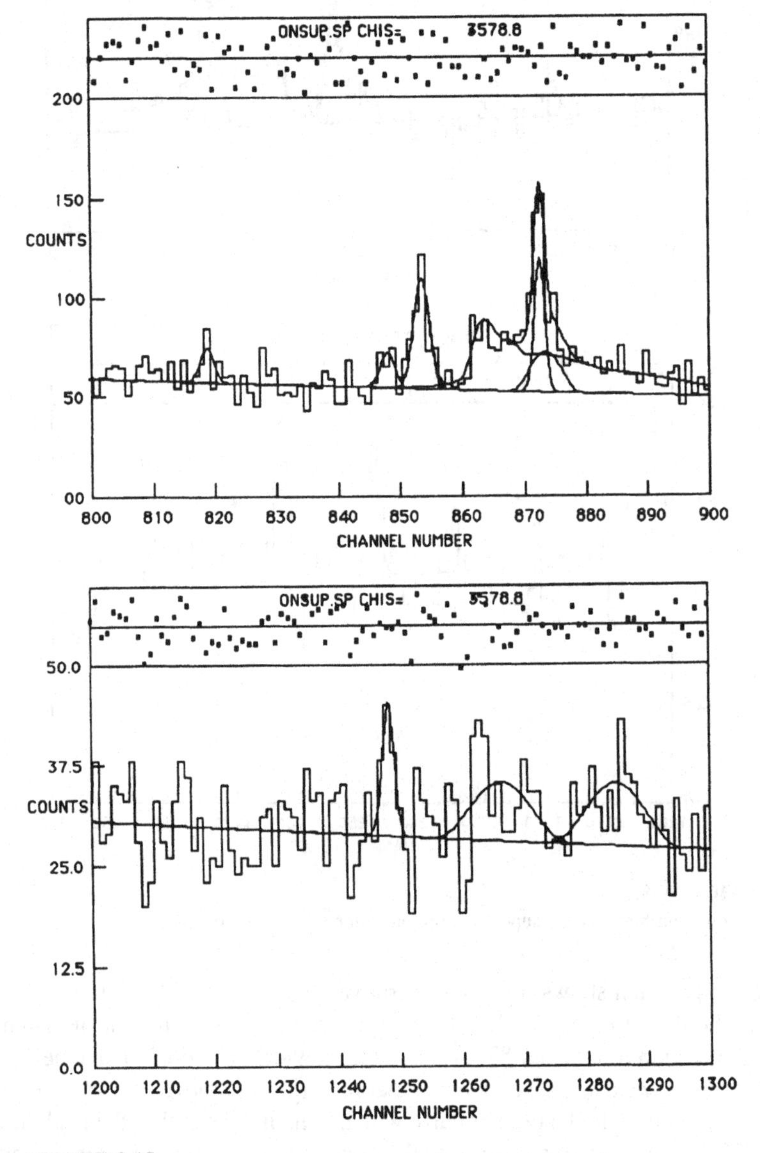

FIGURE 8.15
Results of the whole supernova analysis on the low and high energy regions.

where we have redefined the width range for the wide-line standard using the values output from the last fit. The final fits to the supernova regions are shown in Figure 8.15, and these are to be compared with the fits detailed in the supernova analysis.[1] The fitting presented in this section has been a complete reanalysis of the supernova data. It was somewhat reassuring for us to come up with the same answer in this second, completely separate analysis!

This concludes the analysis of the supernova data. We have pointed out the salient points of the analysis and highlighted some of the features of ROBFIT used to extract the signal. We will complete this section by showing a list of peaks found during this reanalysis. This list, shown in Table 8.10, contains the real peaks and a number of spurious peaks, which we have to live with. The identification of these spurious peaks is, as they say, left as an exercise for the reader.

TABLE 8.10
Final peak list for the supernova analysis

```
OUTPUT FROM ROBFIT
IP=    3
 184 PEAKS 4040 CHAN, CHIS=        3579. NITB=   12
CUTOFF=   3.00
 1.0682  1.0000  1.3065   .9993  1.0065  1.0000
  60.694  2.099      4.399   2.495         2897.44    2919.56  1
  62.952   .183      2.313    .638         3333.87    2749.00  1
  73.532   .217      3.513    .516         3655.14     463.66  1
  84.792   .211      2.299    .478         1104.03     204.94  1
 108.009  1.067      3.388    .729          363.94     195.20  1
 139.796   .066      2.266    .146         4844.12     284.06  1
 153.567  1.263      3.311    .551          194.14     125.82  1
 166.226   .802      3.309    .550          293.72     121.14  1
 177.388  1.195      3.185    .548          175.42     110.98  1
 189.385   .898      3.237    .547          225.76     106.54  1
 197.614  1.039      1.500   4.991          319.20     571.81  3
 200.631   .061      2.228    .150        10282.21     546.46  1
 243.772   .972      3.240    .539          152.18      78.17  1
 294.514  1.032      3.103    .532          101.48      57.52  1
 305.324   .825      3.308    .530          133.41      56.80  1
 330.201   .653      2.960    .527          127.57      48.31  1
 336.698   .800      3.052    .535          121.53      52.26  1
 340.727   .504      3.343    .559          219.19      56.20  1
 347.918   .989      3.107    .525           84.86      45.04  1
 352.694   .559      3.060    .530          143.17      45.61  1
 368.365   .826      3.058    .520           88.71      40.87  1
 376.466   .635      3.055    .519          111.68      40.35  1
 394.779   .840      3.130    .516           80.14      37.27  1
 404.166   .810      3.287    .514           88.19      37.13  1
 412.210   .717      2.908    .513           81.03      34.20  1
 439.814  1.041      3.051    .508           53.21      31.28  1
 451.496   .232      2.810    .506          224.86      37.05  1
 485.756   .210      2.365    .467          207.02      37.37  1
 514.597   .655      3.002    .544           76.73      29.52  1
 526.387   .057      3.636    .134         1807.44      63.07  1
 586.440   .387      3.908    .497          161.51      29.42  1
 600.108   .672      2.863    .483           61.61      23.66  1
 603.694   .736      2.743    .480           51.44      22.99  1
 615.727   .242     12.554    .315         1487.70      80.54  2
 620.838   .482      3.159    .476          109.76      33.94  1
 649.024   .217      2.926    .468          189.67      29.28  1
 662.622   .527      2.814    .459           62.68      20.83  1
```

(continued)

TABLE 8.10 (continued)
Final peak list for the supernova analysis

685.426	.876	2.754	.452	35.34	19.03	1
704.788	.653	2.936	.446	50.74	19.97	1
716.921	.196	13.115	.660	1918.03	89.06	2
721.061	.953	2.819	.443	45.50	30.43	1
744.972	.494	2.839	.437	72.23	24.24	1
756.777	.572	2.749	.433	51.53	21.73	1
763.702	.762	2.776	.429	37.82	18.20	1
768.704	.634	2.923	.427	48.95	18.67	1
777.099	.221	3.337	.423	194.00	25.25	1
818.771	.551	2.583	.445	48.54	18.66	1
848.060	.539	2.675	.417	50.22	17.88	1
853.661	.206	2.960	.418	172.89	24.13	1
863.650	.489	14.140	1.275	663.67	70.32	2
872.431	.184	1.898	.613	137.64	62.62	1
873.362	1.066	5.620	2.036	115.65	70.71	3
905.170	.721	2.758	.390	37.75	18.13	1
911.068	.634	2.502	.396	47.73	20.09	1
913.896	.509	2.528	.406	60.04	20.51	1
927.227	.167	2.601	.389	181.22	24.12	1
932.253	1.234	12.729	1.286	207.51	46.34	2
942.508	.454	2.534	.544	84.95	27.77	1
973.298	.499	2.445	.334	41.52	14.67	1
998.759	.764	2.520	.363	29.84	15.98	1
1004.391	.761	2.551	.358	31.67	16.05	1
1013.233	.540	2.589	.351	46.26	16.82	1
1022.349	.210	2.458	.345	126.59	20.75	1
1050.172	.139	2.571	.280	186.12	20.24	1
1058.730	.572	2.434	.273	32.18	13.38	1
1076.485	.264	2.536	.261	85.16	16.13	1
1082.278	.348	2.585	.263	75.44	17.50	1
1086.373	1.363	12.934	1.506	189.80	42.48	2
1101.149	.372	2.370	.235	52.40	15.19	1
1135.903	.799	2.372	.190	20.84	12.27	1
1140.699	.639	2.375	.183	26.73	12.53	1
1145.531	.764	2.346	.175	21.84	12.15	1
1151.266	.501	2.298	.165	33.22	12.55	1
1157.214	.339	2.338	.154	52.48	13.49	1
1171.208	.674	2.318	.121	23.51	11.76	1
1197.519	.849	2.290	.047	23.62	15.06	1
1248.133	.389	2.229	.196	39.89	12.52	1
1266.239	1.591	10.528	2.004	77.19	22.63	3
1284.750	1.511	10.648	2.004	83.13	23.17	3
1317.989	.521	2.126	.298	24.65	10.72	1
1345.036	.483	2.119	.335	26.03	10.68	1
1418.470	.387	2.273	.430	36.49	11.36	1
1429.899	.442	1.789	.444	22.21	9.67	1
1438.962	.616	2.040	.456	18.23	9.70	1
1464.780	.779	2.052	.489	14.34	9.34	1
1515.628	.658	1.773	.836	20.94	14.07	1
1531.591	.579	1.681	.578	13.66	8.43	1
1566.985	.545	1.885	.627	17.36	9.09	1
1668.445	.295	1.500	1.108	39.86	17.70	1
1683.262	.353	1.500	.813	21.49	9.90	1
1693.667	.292	2.479	.647	50.44	12.14	1
1699.163	.415	1.506	.767	16.38	8.53	1
1735.214	.596	1.500	.979	13.09	9.65	1
1765.917	.775	1.500	.800	8.59	7.25	1
1770.018	.624	1.927	.799	14.04	8.38	1

TABLE 8.10 (continued)

Final peak list for the supernova analysis

1845.229	.366	2.522	.844	34.59	10.18	1
1866.651	.374	1.500	.998	17.31	8.61	1
1875.819	.589	2.489	1.053	24.42	11.30	1
1884.998	.727	1.919	.871	10.92	7.52	1
1889.057	.692	1.839	.875	10.63	7.38	1
1925.358	.332	1.500	.970	17.94	8.19	1
1932.320	.473	1.500	1.082	13.59	8.48	1
1947.043	.815	1.674	.908	8.10	6.97	1
1955.803	.556	1.500	1.109	14.04	9.61	1
1988.858	.619	1.500	1.073	11.22	8.33	1
2046.298	.463	1.500	1.038	11.90	7.20	1
2075.263	.411	1.638	.964	14.81	7.67	1
2094.686	.435	1.500	1.051	13.33	7.49	1
2123.083	.714	2.268	1.010	13.34	7.89	1
2144.213	.596	1.733	1.284	13.31	9.34	1
2176.391	.658	2.648	1.037	18.89	8.78	1
2187.397	.590	1.500	1.192	8.84	7.06	1
2201.455	.526	1.693	1.169	12.71	8.05	1
2239.400	.364	1.500	1.129	13.90	7.18	1
2255.712	.582	1.761	1.080	11.46	7.18	1
2283.425	.347	1.500	1.135	14.26	7.19	1
2286.899	.529	1.517	.944	11.13	7.09	1
2290.207	.485	1.865	1.238	14.60	8.07	1
2293.437	.300	1.500	1.504	17.97	9.19	1
2296.064	.437	1.507	.751	14.31	7.53	1
2301.805	.587	1.500	1.183	11.18	8.08	1
2306.096	.791	3.134	1.285	22.10	10.55	1
2336.860	.818	1.500	1.249	7.04	6.71	1
2366.724	.493	2.149	1.084	17.52	8.09	1
2384.186	.643	2.397	1.180	15.73	8.17	1
2390.088	.436	1.500	.754	12.51	6.86	1
2398.710	.537	1.500	1.362	10.63	8.25	1
2405.666	.513	1.500	1.346	11.40	8.87	1
2410.413	3.684	18.163	2.097	61.55	21.56	3
2418.778	.793	1.500	1.283	7.06	7.42	1
2459.491	.405	1.500	1.180	10.99	6.39	1
2468.469	.430	1.500	1.190	10.16	6.22	1
2631.797	.790	1.500	1.591	8.64	8.45	1
2653.740	.918	3.068	1.317	13.80	7.82	1
2663.193	.535	1.500	1.062	10.51	7.25	1
2667.097	.513	1.500	.897	11.25	7.32	1
2685.464	.471	1.500	1.877	13.84	10.03	1
2692.832	.751	1.500	1.427	9.07	8.40	1
2767.213	.427	1.500	.877	11.32	6.31	1
2773.594	.557	1.500	1.605	8.86	6.87	1
2860.479	.418	2.821	.919	27.58	8.29	1
2872.928	.395	1.500	.681	11.60	5.73	1
2883.908	.482	1.500	.841	9.23	5.51	1
2931.375	.380	1.500	1.243	10.86	6.06	1
2934.854	.814	2.737	1.451	11.79	6.80	1
2942.055	.849	2.305	1.386	8.22	5.82	1
2979.661	.739	2.778	1.515	14.26	7.58	1
2992.118	.392	1.500	.703	10.66	5.40	1
3011.691	.522	1.500	1.310	7.78	5.31	1
3035.132	.428	1.654	.859	10.38	5.42	1
3054.996	.589	1.500	.997	10.88	8.12	1
3072.808	.647	2.465	1.434	11.75	6.32	1
3088.653	.624	1.500	1.733	7.51	6.32	1

(continued)

TABLE 8.10 (continued)
Final peak list for the supernova analysis

3125.981	.738	3.311	1.455	16.27	7.18	1
3138.504	.788	4.035	1.460	20.39	7.84	1
3169.787	.746	4.675	1.472	27.38	8.71	1
3216.957	.596	1.500	1.019	10.23	7.72	1
3269.666	.509	1.500	1.292	11.14	7.57	1
3334.729	.488	1.635	1.098	8.61	5.10	1
3362.401	.372	1.500	1.236	9.54	5.11	1
3376.493	.345	1.500	1.340	10.20	5.40	1
3384.263	.723	2.825	1.656	12.99	6.71	1
3388.465	1.025	2.351	1.633	6.62	5.34	1
3409.419	.554	1.847	1.324	8.19	5.10	1
3421.797	.697	1.502	1.406	5.22	4.51	1
3427.537	.770	3.583	1.575	16.22	6.98	1
3473.406	.465	1.500	1.619	7.11	4.80	1
3483.158	.578	1.500	1.084	5.96	4.41	1
3491.737	.403	1.500	.867	12.08	6.32	1
3528.078	.642	1.500	1.120	9.71	8.09	1
3552.519	.334	1.500	1.179	13.28	6.59	1
3565.152	.422	1.500	.788	8.42	4.64	1
3685.837	.689	1.500	1.298	5.04	4.29	1
3708.966	.930	1.500	1.554	3.57	4.05	1
3757.831	.413	1.500	.781	8.97	4.70	1
3823.580	.441	1.500	1.468	7.39	4.72	1
3832.792	1.017	3.841	1.701	12.34	6.30	1
3845.382	.747	2.852	1.640	11.16	5.89	1
3883.821	.604	1.500	1.157	7.97	6.10	1
3978.655	.838	2.270	1.747	6.51	4.81	1
4026.423	.490	1.500	1.786	6.02	4.37	1
4031.140	.409	1.500	.754	8.06	4.33	1
4064.213	.546	1.500	1.122	5.60	4.01	1
4073.579	.361	1.500	1.136	9.06	4.68	1

8.3 SOLAR SEISMOLOGY DATA

In this section we illustrate how ROBFIT can be used on data in which the underlying channel fluctuations do not follow Gaussian statistics. For gamma-ray spectra, the variation in an actual channel value is expected to have a standard deviation σ^2 equal to f_i, as shown in Figure 8.16.

ROBFIT uses this information in its minimization algorithm. The maximum-likelihood technique assumes that the underlying statistical fluctuations are Gaussian. This can be seen from the following discussion, using the terminology of Chapter 4. The probability of making an observation f_i, assuming a Gaussian distribution within the channel of standard deviation σ_i, is given by

$$\text{Prob} = \frac{1}{\sigma_i\sqrt{2\pi}}\exp\left(-\frac{1}{2}\left[\frac{(f_i - f_{\text{th}})^2}{\sigma_i^2}\right]\right)$$

Thus, the probability of making a set of measurements $i = 1, \ldots, N$ is given by

$$\text{Prob} = \prod_i \frac{1}{\sigma_i\sqrt{2\pi}}\exp\left(-\frac{1}{2}\left[\frac{(f_i - f_{\text{th}})^2}{\sigma_i^2}\right]\right)$$

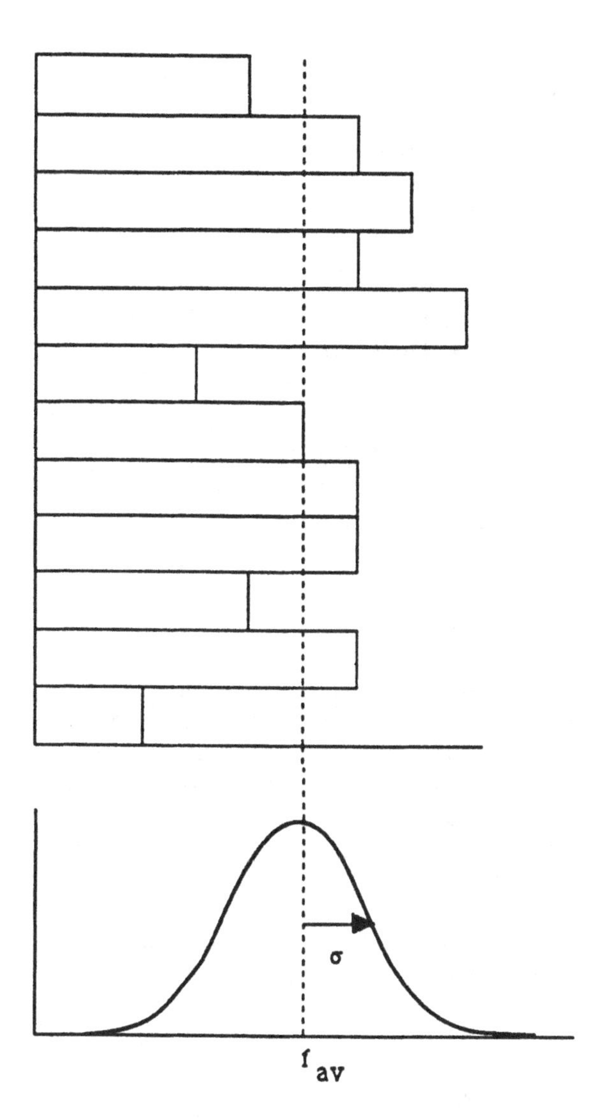

FIGURE 8.16
Illustration of the Gaussian fluctuation of the data points about the mean f_{av}.

for a given fit to the data f_{th}. The maximum likelihood method attempts to maximize this probability by varying the coefficients of the fit. Maximizing this probability is equivalent to minimizing the sum in the exponential. In other words, minimizing the χ^2 of the fit is identical to the set of procedures outlined in Chapter 4.

With certain data, however, the underlying probability distribution does not follow Gaussian statistics. Now the minimizing algorithm SMSQ is not operating on the correct probability function and is no longer obtaining the best fit to the data. In these situations, ROBFIT can still be used. Estimating the weights on each data point from fluctuations within the data can offset some of the effects of using an incorrect probability distribution. This section is intended to give the user confidence in using ROBFIT.

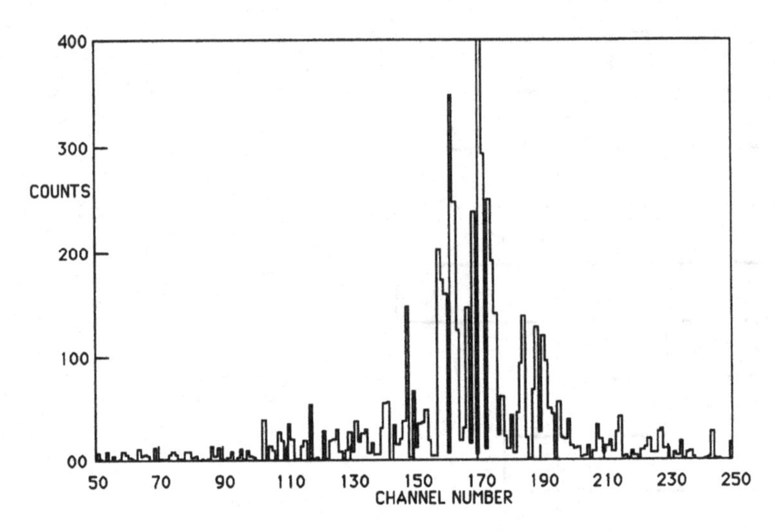

FIGURE 8.17
Illustration of the large channel-to-channel fluctuations in the solar spectrum. (See Figure 8.4*b* for comparison with Gaussian fluctuations.)

Solar seismology is the study of the vibrational modes of the sun. Observations of surface motions result in a frequency spectrum that contains peaks at resonant frequencies.[24] Analysis of these data requires the unambiguous determination of all these resonant modes. A more detailed view of some of these modes can be seen in Bamford et al.[25] These data contain channel-to-channel fluctuations that are distributed as χ^2 with 2 degrees of freedom (see Duvall, Harvey, and Pomerantz[26] and Anderson, Duvall, and Jefferies[27]) and are non-Gaussian. Figure 8.17 illustrates this situation.

The following tests use artificially generated data that contain channel-to-channel fluctuations calculated using the probability distribution of Equation 8.1. This distribution mimics the actual solar spectrum.

$$P(A) = \frac{1}{A_0} e - \frac{A}{A_0} \tag{8.1}$$

The tests have been organized as follows. Twenty-five individual spectra have been generated, and the data fluctuations have been randomized according to the probability distribution of Equation 8.1. Each of these spectra contain three Lorentzian-shaped peaks with positions, widths, and stengths given by:

Position (channels)	Width (channels)	Strength (area)
167.66	22.22	6,771
501.00	13.33	4,188
834.33	8.90	2,796

TABLE 8.11
Illustration of the effects of different weighting schemes on ROBFIT error analysis

	Standard deviation	Estimated standard deviation
	Weights = $\frac{1}{\text{Data value}}$	
Position and error:		
167.29± 0.28	1.40	0.82
501.09± 0.28	1.40	0.73
834.27± 0.21	1.05	0.57
Width and error:		
22.23± 0.86	4.10	1.69
15.58± 0.67	3.37	1.44
10.79± 0.92	4.60	1.35
Strength and error:		
7,207± 194	968	460
4,177± 123	613	285
2,672± 123	613	241
	Weights calculated from data fluctuations	
Position and error:		
167.37± 0.32	1.58	1.68
501.28± 0.23	1.17	1.29
834.34± 0.25	1.27	1.10
Width and error:		
21.76± 0.57	2.86	2.90
12.20± 0.59	2.96	1.84
8.290± 0.38	1.89	1.49
Strength and error:		
6,139± 136	681	756
3,980± 187	936	587
2,725± 151	756	458

Twenty-five separate fits have been performed on these data. The results are shown in Table 8.11. The table contains the results of two methods of analysis. One uses individual channel weightings corresponding to $\omega_i = 1/\text{Data value}$; the other estimates the weights from the fluctuation in the data. These weighting schemes were described in Section 4.1.3. Table 8.11 also shows how accurately the code calculates the errors on these fits. From the 25 fits, we can take each of the fitted values (position, width, and strength) and calculate a mean value. We can also deduce the standard deviation and error on the mean value. However, from each fit, we are given the estimated error on the values. This estimated error is effectively a measure of this standard deviation. We can average these estimates, and if our error analysis is correct, the average should be close to the standard deviation calculated from the spread in the values.

The table shows that the code correctly reproduces the positions, widths, and strengths in both weighting cases. However, when the weights are taken as the inverse of the data values, as in gamma-ray spectroscopy, the errors calculated from the fit are slightly underestimated. Selecting the option for calculating the weights from the fluctuations in the data is seen to improve the error estimates. Of course, this is still an approximation to the weights that should be used in this analysis.

On closer inspection, the non-Gaussian probability function that is to be minimized is[26,27]

$$S = N \ln(f_{\text{th}}) + \frac{1}{f_{\text{th}}} \sum_{i}^{N} f_i \qquad (8.2)$$

where the f_is are the data points, f_{th} is the theoretical model described in Chapter 4, and N is the total number of data points. But we can write an effective χ^2,

$$\chi^2 = \sum_i \frac{(f_i - f_{\text{th}})^2}{f_{\text{th}}^2}$$

which is in the same form as Equation 4.1. Here, however, the weights are given by $\omega_i = 1/f_{\text{th}}^2$. This function has a minimum when

$$\frac{\partial \chi^2}{\partial c} = -\sum_i \frac{2}{f_{\text{th}}^2}(f_i - f_{\text{th}})\frac{\partial f_{\text{th}}}{\partial c} = 0$$

which is in the same form as the minimum of Equation 8.2. Again, the minimum occurs when $f_i = f_{\text{th}}$. The minimization has been reduced to a standard least-squares with the weights altered as suggested. So a correct analysis of these data would involve altering the weighting scheme within the code so that it takes into consideration the $1/f_{\text{th}}^2$ dependence while minimizing. However, for the present, we wish only to illustrate the versatility of ROBFIT by showing that a good fit can be achieved with no modifications at all.

8.4 THE FINAL STAGE: CALIBRATION

The problem with finding lots of peaks is relating them all to something in the real world. This, as anyone who has tried it will confirm, is a nontrivial task. The technique is to start with a few large peaks and perform a preliminary calibration with these before moving to the smaller peaks. This process quickly leads to a situation in which many constants are being fitted to a calibration curve. ROBFIT contains a set of routines that enable just this kind of calibration; this code has been designed around the calibration of nuclear physics spectra and thus contains features specific to that kind of analysis. The heart of the calibration code is again based on cubic spline fitting.

For example, suppose we have a spectrum containing a number of peaks that we know correspond to gamma-ray lines of a certain energy. We wish to

TABLE 8.12
The input calibration data, supplied by the user and placed in the FILENAME.CA file

Calibration for the Sm decay chain of Europium
those in coincidence with 344KeV Gd (2-0) should not be here

CHANNEL	ERR	ENERGY	ERR	AREA	ERR	RI	ERR	COMMENT
1027.65	.09	244.699	.001	6049.	101.	359	6	Sm_2-1 E2
1243.28	.52	295.939	.008	669.	56.	21.1	.5	Sm_13-10 E1
1750.06	1.43	416.052	.006	212.	47.	5.3	.1	Sm_16-10
1864.03	.08	443.976	.005	6641.	103.	148	2	Sm_13-9
2051.58	.35	488.661	.039	958.	54.	19.5	.2	Sm_13-8 m1+e2 +5.6
2367.72	.42	564.021	.008	1297.	83.	22.4	.6	Sm_16-9 e1+(m2) +.07
2378.25	1.35	566.421	.008	371.	74.	6.2	.1	Sm_13-6 m1+e2 -.74
2756.01	.94	656.484	.012	405.	51.	6.9	.1	Sm_7-2
2833.13	.93	674.678	.003	466.	60.	8.0	.3	Sm_8-2 e1+(m2) -.02
2891.03	.21	688.678	.006	2306.	78.	40.0	.8	Sm_5-1
3020.17	.50	719.353	.006	910.	65.	15.6	.3	Sm_9-2 e2
3402.69	.40	810.459	.007	1137.	67.	15.2	.2	Sm_5-0 e2
3532.16	.94	841.586	.008	471.	56.	7.8	.1	Sm_6-1 e1
3641.20	.07	867.388	.008	10627.	131.	199	4	Sm_10-2 m1+e2 -6.5
3858.95	.39	919.401	.008	1157.	67.	20.9	.5	Sm_8-1 e1+(m2) +.02
3888.89	.61	926.324	.015	699.	62.	12.7	.4	Sm_11-2
4047.00	.03	964.131	.009	37271.	216.	693	9	Sm_9-1
4219.54	.19	1005.279	.017	2227.	69.	31	3	Sm_12-1
4558.29	.07	1085.914	.003	25250.	409.	475	7	Sm_9-0 e2
4668.34	.05	1112.116	.017	33833.	288.	649	9	Sm_10-1 m1+e2 -8.7
5091.65	.14	1212.950	.012	3323.	73.	67	.8	Sm_15-2 e1+(m2)
5425.99	1.11	1292.784	.019	233.	35.	4.9	.3	Sm_11-0
5910.68	.03	1408.011	.014	47819.	229.	1000	3	Sm_13-1 e1+(m2) +.043
6118.58	.33	1457.628	.015	1120.	54.	23.6	.5	Sm_15-1 e1+(m2) 0
6412.13	1.57	1528.115	.019	381.	103.	12.7	.3	Sm_16-5 e1+(m2) -.01
7424.29	3.10	1769.093	.047	24.	10.	.42	.03	Sm_19-0 e2

calibrate the spectrum to make sure that all the lines can be explained. Consider the calibration file shown in Table 8.12. The structure of the file is given in Appendix A. This calibration file is from a fit to the samarium decay chain of europium and will be used to calibrate a peak file.

The first step in performing the calibration is to create the files that contain the x-axis and efficiency calibration constants; these filenames have the extensions .XDA and .EFF, respectively. The results of the calibration are given in a file with the filename extension .FCA; for the present calibration, this file is shown in Table 8.13. The table shows the results of the calibrated energy and efficiency values for each of the fitted peaks.

The next step is to calibrate a peak file. The peak file to be calibrated is given in Table 8.14. This file is from a fit performed with the FIXK option for peak fitting; this is why the errors on the peak positions and widths are all zero. Using this option holds the peaks at the given input positions and widths. The list also gives the reader some idea of the number of peaks that can be fitted. A calibration can now be run using the .XDA file produced. The calibration produces a calibrated peak list, which can be found in the FILENAME.PCA file. For the preceding peak list, this is given in Table 8.15.

TABLE 8.13
Results of the calibration found in the FILENAME.FCA file

```
Calibration for the Sm decay chain of Europium
those in coincidence with 344KeV Gd (2-0) should not be here
 THE ENERGY CONSTANTS
   0.287469E+00   0.238169E+00  -0.164101E-07   0.576008E+04
       CHAN              ENERGY              ECAL              RES
 1027.65 +-  .09  244.67 +-  .03  244.70 +- 0.00      -.81        Sm_2-1 E2
 1243.28 +-  .52  296.06 +-  .13  295.94 +-  .01       .99        Sm_13-10 El
 1750.06 +- 1.43  416.83 +-  .34  416.05 +-  .01      2.29        Sm_16-10
 1864.03 +-  .08  443.99 +-  .02  443.98 +-  .01       .68        Sm_13-9 el+m2 +.025
 2051.58 +-  .35  488.68 +-  .08  488.66 +-  .04       .27        Sm_13-8 ml+e2 +5.6
 2367.72 +-  .42  564.02 +-  .10  564.02 +-  .01      -.05        Sm_16-9 el+(m2) +.07
 2378.25 +- 1.35  566.52 +-  .32  566.42 +-  .01       .32        Sm_13-6 ml+e2 -.74
 2756.01 +-  .94  656.53 +-  .22  656.48 +-  .01       .23        Sm_7-2 e2+ml+e0 e2/ml
 2833.13 +-  .93  674.91 +-  .22  674.68 +- 0.00      1.05        Sm_8-2 el+(m2) -.02
 2891.03 +-  .21  688.71 +-  .05  688.68 +-  .01       .54        Sm_5-1 e0+e2+ml e2/ml
 3020.17 +-  .50  719.47 +-  .12  719.35 +-  .01      1.02        Sm_9-2 e2
 3402.69 +-  .40  810.61 +-  .10  810.46 +-  .01      1.58        Sm_5-0 e2
 3532.16 +-  .94  841.46 +-  .22  841.59 +-  .01      -.58        Sm_6-1 el
 3641.20 +-  .07  867.43 +-  .02  867.39 +-  .01      2.34        Sm_10-2 ml+e2 -6.5
 3858.95 +-  .39  919.31 +-  .09  919.40 +-  .01      -.98        Sm_8-1 el+(m2) +.02
 3888.89 +-  .61  926.44 +-  .15  926.32 +-  .02       .81        Sm_11-2
 4047.00 +-  .03  964.11 +-  .01  964.13 +-  .01     -1.90        Sm_9-1 ml+e2+e0 e2/ml
 4219.54 +-  .19 1005.21 +-  .05 1005.28 +-  .02     -1.47        Sm_12-1 ml+e2+e0 e2/ml
 4558.29 +-  .07 1085.91 +-  .02 1085.91 +- 0.00      -.42        Sm_9-0 e2
 4668.34 +-  .05 1112.12 +-  .01 1112.12 +-  .02       .32        Sm_10-1 ml+e2 -8.7
 5091.65 +-  .14 1212.95 +-  .03 1212.95 +-  .01       .06        Sm_15-2 el+(m2)
 5425.99 +- 1.11 1292.59 +-  .26 1292.78 +-  .02      -.74        Sm_11-0
 5910.68 +-  .03 1408.03 +-  .02 1408.01 +-  .01       .82        Sm_13-1 el+(m2) +.043
 6118.58 +-  .33 1457.54 +-  .08 1457.63 +-  .02     -1.06        Sm_15-1 el+(m2) 0
 6412.13 +- 1.57 1527.46 +-  .37 1528.12 +-  .02     -1.76        Sm_16-5 el+(m2) -.01
 7424.29 +- 3.10 1768.52 +-  .74 1769.09 +-  .05      -.77        Sm_19-0 e2
 THE EFFICIENCY CONSTANTS
   0.425976E+01  -0.664569E-04  -0.944262E-05   0.123885E+04
  -0.230265E-06   0.305449E+04
    CHAN   ENERGY    CSTRENGTH  STRENGTH      RES
 1027.65  244.70   359.00   359.00 +- 12.48  0.00       Sm_2-1 E2
 1243.28  295.94    21.10    21.85 +-  2.22   .34       Sm_13-10 El
 1750.06  416.05     5.30     4.98 +-  1.11  -.29       Sm_16-10
 1864.03  443.98   148.00   147.15 +-  4.04  -.21       Sm_13-9 el+m2 +.025 + mix of
 2051.58  488.66    19.50    19.55 +-  1.15   .04       Sm_13-8 ml+e2 +5.6
 2367.72  564.02    22.40    23.90 +-  1.57   .96       Sm_16-9 el+(m2) +.07
 2378.25  566.42     6.20     6.82 +-  1.36   .45       Sm_13-6 ml+e2 -.74
 2756.01  656.48     6.90     7.01 +-   .89   .13       Sm_7-2 e2+ml+e0 e2/ml +4.3
 2833.13  674.68     8.00     8.04 +-  1.04   .03       Sm_8-2 el+(m2) -.02
 2891.03  688.68    40.00    39.72 +-  1.50  -.19       Sm_5-1 e0+e2+ml e2/ml +19
 3020.17  719.35    15.60    15.72 +-  1.15   .10       Sm_9-2 e2
 3402.69  810.46    15.20    20.14 +-  1.22  4.04       Sm_5-0 e2
 3532.16  841.59     7.80     8.41 +-  1.01   .61       Sm_6-1 el
 3641.20  867.39   199.00   191.21 +-  3.40 -2.29       Sm_10-2 ml+e2 -6.5
 3858.95  919.40    20.90    21.12 +-  1.25   .18       Sm_8-1 el+(m2) +.02
 3888.89  926.32    12.70    12.79 +-  1.14   .08       Sm_11-2
 4047.00  964.13   693.00   688.94 +-  8.23  -.49       Sm_9-1 ml+e2+e0 e2/ml -8.0 +6
 4219.54 1005.28    31.00    41.64 +-  1.35  7.88       Sm_12-1 ml+e2+e0 e2/ml -3.2 *
 4558.29 1085.91   475.00   482.87 +-  8.69   .91       Sm_9-0 e2
 4668.34 1112.12   649.00   651.76 +-  7.35   .38       Sm_10-1 ml+e2 -8.7
 5091.65 1212.95    67.00    65.84 +-  1.50  -.77       Sm_15-2 el+(m2)
 5425.99 1292.78     4.90     4.72 +-   .71  -.25       Sm_11-0
 5910.68 1408.01  1000.00  1000.47 +-  8.52   .05       Sm_13-1 el+(m2) +.043
 6118.58 1457.63    23.60    23.76 +-  1.16   .14       Sm_15-1 el+(m2) 0
 6412.13 1528.12    12.70     8.24 +-  2.23 -2.00       Sm_16-5 el+(m2) -.01
 7424.29 1769.09      .42      .56 +-   .23   .58       Sm_19-0 e2
```

TABLE 8.14

The peak list to be calibrated found in the FILENAME.PK file

```
OUTPUT FROM ROBFIT
IP=    2
 244 PEAKS 8001 CHAN, CHIS=   8293.       NITB=   28
CUTOFF=   3.60
 1.2132  1.0000  1.2132  1.0000
```

110.732	.000	16.664	.000	18.9333	6.197	1
160.904	.000	16.752	.000	31.1189	8.667	1
177.390	.000	16.771	.000	45.5337	9.825	1
205.937	.000	16.812	.000	35.7977	8.862	1
225.871	.000	16.842	.000	49.9562	10.80	1
247.490	.000	16.930	.000	70.1783	11.88	1
273.183	.000	16.996	.000	94.6036	13.07	1
302.396	.000	17.163	.000	151.120	15.27	1
333.108	.000	17.079	.000	101.579	14.29	1
360.501	.000	17.120	.000	113.810	17.09	1
378.133	.000	17.121	.000	102.741	16.97	1
406.071	.000	17.214	.000	132.301	17.21	1
433.138	.000	17.249	.000	173.290	19.61	1
460.006	.000	17.271	.000	136.675	18.65	1
498.896	.000	17.457	.000	280.373	28.63	1
525.407	.000	17.429	.000	604.315	40.68	1
543.795	.000	17.309	.000	1213.83	64.60	1
558.568	.000	17.328	.000	1801.77	76.46	1
574.767	.000	17.422	.000	1205.03	70.62	1
592.333	.000	17.568	.000	2074.78	76.72	1
610.498	.000	17.607	.000	1918.51	74.97	1
630.462	.000	17.638	.000	1530.91	72.58	1
649.791	.000	17.692	.000	1373.60	76.81	1
668.252	.000	17.809	.000	1522.89	74.54	1
691.149	.000	17.672	.000	1298.07	80.44	1
712.464	.000	17.645	.000	529.014	79.49	1
803.543	.000	17.832	.000	3072.83	136.8	1
833.031	.000	17.802	.000	1270.92	151.9	1
856.389	.000	17.856	.000	2118.18	172.7	1
880.790	.000	17.913	.000	1650.53	184.0	1
911.295	.000	17.950	.000	1801.68	205.3	1
937.986	.000	18.045	.000	4541.80	250.8	1
955.279	.000	17.950	.000	1630.46	265.0	1
975.646	.000	17.977	.000	1336.80	281.7	1
991.992	.000	18.016	.000	554.626	279.0	1
1060.878	.000	18.155	.000	2429.44	334.2	1
1078.805	.000	18.176	.000	2822.92	348.7	1
1112.177	.000	18.124	.000	2420.94	339.4	1
1161.119	.000	18.330	.000	2745.44	370.1	1
1190.376	.000	18.405	.000	2763.33	384.9	1
1234.315	.000	18.373	.000	2490.49	413.4	1
1257.549	.000	18.394	.000	4470.73	492.2	1
1272.319	.000	18.379	.000	2775.88	482.1	1
1322.938	.000	18.488	.000	3783.68	436.9	1
1353.085	.000	18.624	.000	5845.04	442.7	1
1391.853	.000	18.708	.000	6499.85	487.3	1
1411.679	.000	18.699	.000	6627.43	600.6	1
1425.202	.000	18.744	.000	10383.7	588.9	1
1465.712	.000	18.128	.000	216471.	790.9	1
1490.270	.000	18.647	.000	5190.03	490.8	1

(continued)

TABLE 8.14 (continued)
The peak list to be calibrated found in the FILENAME.PK file

1551.386	.000	18.828	.000	1689.47	491.5	1
1572.940	.000	18.746	.000	6570.14	498.1	1
1629.574	.000	19.097	.000	7849.47	487.5	1
1678.915	.000	19.072	.000	5629.07	513.5	1
1705.986	.000	19.259	.000	11286.4	532.3	1
1746.784	.000	21.727	.000	74582.5	634.1	1
1777.060	.000	17.787	.000	38013.8	560.3	1
1800.755	.000	19.147	.000	6678.34	539.0	1
1830.885	.000	19.487	.000	12240.9	560.6	1
1853.251	.000	19.266	.000	13240.0	575.0	1
1876.474	.000	19.111	.000	13016.1	555.1	1
1907.715	.000	18.983	.000	29424.5	565.4	1
1933.506	.000	18.928	.000	17832.7	542.0	1
1968.337	.000	20.146	.000	17000.1	554.3	1
2019.080	.000	19.538	.000	23927.7	600.1	1
2038.342	.000	19.437	.000	5790.58	622.9	1
2058.133	.000	19.311	.000	8118.22	597.9	1
2081.432	.000	19.177	.000	18033.5	610.5	1
2100.369	.000	19.386	.000	5973.41	579.7	1
2140.909	.000	19.814	.000	7140.50	560.2	1
2165.346	.000	20.238	.000	13097.5	571.2	1
2213.289	.000	19.796	.000	16747.1	590.6	1
2241.452	.000	19.795	.000	6191.43	629.2	1
2261.045	.000	19.770	.000	8136.08	708.0	1
2277.951	.000	19.795	.000	9879.05	721.0	1
2297.432	.000	19.947	.000	33452.0	686.7	1
2321.449	.000	19.884	.000	9722.71	681.2	1
2339.939	.000	19.956	.000	15958.6	753.7	1
2352.594	.000	19.854	.000	24324.9	952.8	1
2366.119	.000	19.880	.000	17526.2	856.0	1
2383.122	.000	19.944	.000	12059.0	963.4	1
2394.069	.000	20.115	.000	14367.0	946.4	1
2412.280	.000	20.373	.000	27265.4	740.2	1
2432.749	.000	25.046	.000	77035.3	824.6	1
2458.712	.000	20.030	.000	13990.6	905.4	1
2482.416	.000	21.017	.000	392640.	5591.	1
2486.882	.000	21.311	.000	577980.	5452.	1
2514.244	.000	19.965	.000	26362.3	710.5	1
2534.942	.000	20.094	.000	16550.0	1294.	1
2541.621	.000	20.314	.000	22892.8	1267.	1
2565.488	.000	20.049	.000	11112.5	567.4	1
2601.049	.000	20.197	.000	6953.17	527.2	1
2662.921	.000	20.337	.000	6739.28	515.5	1
2695.111	.000	19.708	.000	21032.2	539.7	1
2721.579	.000	20.198	.000	19408.6	602.9	1
2739.557	.000	20.345	.000	17852.4	589.6	1
2790.038	.000	19.169	.000	165503.	718.1	1
2822.642	.000	20.685	.000	6231.81	516.3	1
2863.975	.000	20.578	.000	8864.46	527.8	1
2896.163	.000	24.188	.000	58935.1	637.5	1
2927.611	.000	23.745	.000	48541.5	640.9	1
2951.176	.000	20.576	.000	27531.6	658.3	1
2968.469	.000	20.730	.000	7415.93	609.6	1
3000.919	.000	20.783	.000	68354.5	606.4	1
3027.653	.000	21.248	.000	30542.7	577.4	1
3055.213	.000	21.005	.000	8541.09	535.3	1
3092.829	.000	24.227	.000	44448.8	589.3	1
3146.468	.000	20.832	.000	4182.80	500.8	1

TABLE 8.14 (continued)

The peak list to be calibrated found in the FILENAME.PK file

3190.481	.000	21.142	.000	8608.81	508.0	1
3229.469	.000	21.225	.000	11906.2	564.6	1
3250.828	.000	22.830	.000	30355.3	607.4	1
3282.406	.000	21.204	.000	7957.26	554.3	1
3303.453	.000	21.215	.000	4576.53	542.2	1
3346.142	.000	20.211	.000	28342.6	529.4	1
3374.399	.000	20.959	.000	16344.8	547.0	1
3398.795	.000	21.384	.000	14390.7	540.1	1
3436.494	.000	21.890	.000	15215.8	521.7	1
3479.561	.000	21.738	.000	11803.5	538.8	1
3508.112	.000	21.498	.000	11854.2	690.0	1
3522.515	.000	21.486	.000	16097.6	731.7	1
3543.337	.000	21.475	.000	9839.47	580.0	1
3577.124	.000	21.699	.000	26319.4	554.8	1
3608.251	.000	21.467	.000	12412.8	540.4	1
3640.766	.000	21.116	.000	29310.2	584.6	1
3671.200	.000	22.372	.000	170965.	758.9	1
3705.994	.000	22.807	.000	28215.2	565.7	1
3750.800	.000	22.048	.000	15345.5	605.6	1
3775.110	.000	21.362	.000	28817.5	617.4	1
3821.971	.000	28.776	.000	55651.8	702.7	1
3860.724	.000	22.537	.000	15639.6	596.5	1
3893.961	.000	19.672	.000	38740.5	578.6	1
3944.591	.000	21.072	.000	22548.6	647.0	1
3962.902	.000	21.781	.000	7848.95	672.8	1
3988.182	.000	21.176	.000	25523.8	609.6	1
4021.550	.000	19.899	.000	26565.8	561.1	1
4060.820	.000	22.192	.000	14453.7	626.4	1
4086.407	.000	28.220	.000	52993.7	745.7	1
4120.861	.000	22.332	.000	7934.01	579.1	1
4179.888	.000	23.022	.000	15407.1	579.9	1
4224.045	.000	22.254	.000	2175.34	566.5	1
4255.805	.000	19.916	.000	46418.1	598.0	1
4284.047	.000	22.330	.000	3215.93	613.8	1
4305.914	.000	22.322	.000	8835.22	627.0	1
4332.559	.000	21.842	.000	24740.3	554.8	1
4380.346	.000	20.562	.000	73604.6	597.2	1
4414.647	.000	22.756	.000	13046.8	555.1	1
4447.721	.000	22.302	.000	20640.1	546.3	1
4526.525	.000	22.560	.000	4546.73	527.8	1
4558.196	.000	22.841	.000	12201.9	605.3	1
4579.538	.000	23.237	.000	24306.9	672.1	1
4601.171	.000	22.937	.000	15049.4	632.1	1
4625.563	.000	22.949	.000	11153.3	538.8	1
4657.643	.000	23.252	.000	10185.4	570.4	1
4685.378	.000	26.508	.000	34187.3	622.2	1
4726.152	.000	23.088	.000	4741.04	525.2	1
4763.545	.000	23.060	.000	8764.43	524.2	1
4824.153	.000	23.322	.000	8613.52	536.4	1
4857.687	.000	23.185	.000	14150.6	639.8	1
4879.085	.000	28.516	.000	48533.7	804.7	1
4902.801	.000	23.070	.000	5210.37	569.7	1
4928.527	.000	28.614	.000	69392.8	675.3	1
4969.584	.000	23.137	.000	13010.9	601.3	1
4988.273	.000	23.324	.000	20469.5	600.6	1
5054.073	.000	20.795	.000	106841.	623.6	1
5079.018	.000	25.212	.000	38526.2	612.1	1
5111.716	.000	23.519	.000	6886.06	523.9	1

(continued)

TABLE 8.14 (continued)
The peak list to be calibrated found in the FILENAME.PK file

5144.849	.000	23.910	.000	14414.2	538.7	1
5173.412	.000	23.350	.000	5882.75	510.8	1
5237.797	.000	23.075	.000	24454.7	510.3	1
5272.665	.000	23.496	.000	4307.57	488.0	1
5333.034	.000	23.638	.000	3799.50	479.3	1
5385.920	.000	23.745	.000	13669.2	518.4	1
5414.128	.000	23.730	.000	19950.9	710.2	1
5428.753	.000	23.787	.000	14277.4	764.1	1
5448.776	.000	23.868	.000	9440.54	571.6	1
5488.114	.000	23.289	.000	12598.5	487.9	1
5524.896	.000	23.759	.000	9805.74	633.1	1
5540.871	.000	23.756	.000	7402.89	768.6	1
5558.051	.000	23.803	.000	15684.0	704.9	1
5577.803	.000	23.844	.000	3700.50	570.5	1
5607.864	.000	23.895	.000	5243.60	494.0	1
5641.538	.000	24.049	.000	4502.56	475.4	1
5715.154	.000	24.004	.000	9850.46	500.3	1
5740.494	.000	23.923	.000	6389.25	540.1	1
5763.091	.000	24.095	.000	4385.78	468.5	1
5793.571	.000	24.128	.000	4318.02	482.8	1
5822.238	.000	24.110	.000	3255.42	471.1	1
5861.886	.000	24.444	.000	13686.9	475.6	1
5897.436	.000	24.233	.000	9717.55	506.8	1
5920.954	.000	24.276	.000	8755.03	551.8	1
5943.095	.000	24.568	.000	11800.7	521.8	1
5977.661	.000	24.477	.000	9743.21	463.6	1
6021.404	.000	24.294	.000	2778.49	461.2	1
6050.994	.000	24.370	.000	6502.60	505.3	1
6074.592	.000	24.409	.000	2662.12	502.3	1
6103.504	.000	24.437	.000	4004.05	477.1	1
6134.375	.000	24.431	.000	3650.82	537.7	1
6153.631	.000	24.479	.000	5010.83	568.6	1
6177.126	.000	24.565	.000	5635.28	499.1	1
6210.092	.000	24.626	.000	5774.47	452.7	1
6279.859	.000	24.586	.000	2136.06	495.8	1
6301.562	.000	24.141	.000	18270.7	538.1	1
6327.159	.000	24.591	.000	2914.61	471.6	1
6367.306	.000	24.890	.000	4834.08	454.4	1
6400.082	.000	24.725	.000	10699.4	453.6	1
6498.666	.000	25.040	.000	10198.9	443.0	1
6539.727	.000	25.017	.000	4787.09	441.4	1
6574.428	.000	25.025	.000	3025.24	436.0	1
6618.422	.000	25.107	.000	5049.78	455.0	1
6646.652	.000	25.065	.000	10143.9	504.4	1
6670.394	.000	25.039	.000	6067.38	508.5	1
6696.659	.000	25.060	.000	6422.64	539.0	1
6716.961	.000	25.090	.000	13549.3	539.4	1
6744.380	.000	25.073	.000	3377.52	474.2	1
6771.983	.000	24.980	.000	8380.07	449.3	1
6861.492	.000	25.262	.000	3298.54	473.3	1
6883.183	.000	25.284	.000	5927.07	423.5	1
6914.370	.000	25.260	.000	3620.02	461.2	1
6937.575	.000	25.345	.000	4752.55	456.6	1
6971.992	.000	25.339	.000	2317.85	425.0	1
7001.184	.000	25.328	.000	1263.51	412.3	1
7083.207	.000	25.450	.000	3801.31	428.8	1
7108.938	.000	25.506	.000	5801.64	460.1	1
7135.670	.000	25.686	.000	6816.22	445.9	1

TABLE 8.14 (continued)
The peak list to be calibrated found in the FILENAME.PK file

7167.345	.000	25.675	.000	8540.04	442.5	1
7191.902	.000	25.554	.000	3361.65	450.7	1
7224.664	.000	25.561	.000	4021.45	589.2	1
7238.664	.000	22.479	.000	23728.2	569.3	1
7269.750	.000	25.696	.000	5163.72	442.3	1
7295.128	.000	25.701	.000	3660.78	440.0	1
7326.390	.000	25.769	.000	4496.61	416.3	1
7360.528	.000	25.835	.000	2923.09	403.2	1
7401.741	.000	25.989	.000	10021.8	414.7	1
7435.621	.000	25.924	.000	3866.34	415.1	1
7468.753	.000	25.848	.000	3550.25	414.5	1
7500.568	.000	25.953	.000	3056.51	400.6	1
7584.857	.000	25.945	.000	1226.45	379.6	1
7681.581	.000	26.015	.000	1840.11	393.3	1
7711.848	.000	26.069	.000	1155.65	391.3	1

This completes the present calibration example; however, we note that the efficiency calibration data have not been used. A further step would be to use the efficiency calibration constants to recalculate the strength of each peak in Table 8.15. This would result in a placement of calibrated intensity value in the final two columns of the table.

TABLE 8.15
The calibrated peak list, found in the FILENAME.PCA file

```
 DATA FROM
PERPLPAR.PK
 CALIBRATION FROM
PERPLPAR.CA
NULL
NULL
 THE ENERGY CONSTANTS
   0.222787E+00   0.205719E+00  -0.119779E-05   0.463816E+04
   0.326843E-04   0.612943E+04  -0.318502E-04   0.615730E+04
  OUTPUT FROM ROBFIT
  IP=  2
   244 PEAKS 8001 CHAN, CHIS=   8293.      NITB=   28
  CUTOFF=   3.60
```

CHANNEL		ENERGY		FWHM		STRENGTH	
110.73 +-	.00	17.96 +-	.42	3.39 +-	.00	18.93 +-	6.20
160.90 +-	.00	28.41 +-	.40	3.40 +-	.00	31.12 +-	8.67
177.39 +-	.00	31.84 +-	.40	3.41 +-	.00	45.53 +-	9.82
205.94 +-	.00	37.79 +-	.39	3.42 +-	.00	35.80 +-	8.86
225.87 +-	.00	41.94 +-	.39	3.42 +-	.00	49.96 +-	10.80
247.49 +-	.00	46.44 +-	.38	3.44 +-	.00	70.18 +-	11.88
273.18 +-	.00	51.79 +-	.38	3.45 +-	.00	94.60 +-	13.07
302.40 +-	.00	57.87 +-	.37	3.49 +-	.00	151.12 +-	15.27
333.11 +-	.00	64.26 +-	.36	3.47 +-	.00	101.58 +-	14.29
360.50 +-	.00	69.96 +-	.36	3.48 +-	.00	113.81 +-	17.09
378.13 +-	.00	73.63 +-	.35	3.48 +-	.00	102.74 +-	16.97
406.07 +-	.00	79.44 +-	.35	3.50 +-	.00	132.30 +-	17.21
433.14 +-	.00	85.08 +-	.34	3.51 +-	.00	173.29 +-	19.61
460.01 +-	.00	90.67 +-	.34	3.51 +-	.00	136.68 +-	18.65
498.90 +-	.00	98.76 +-	.33	3.55 +-	.00	280.37 +-	28.63
525.41 +-	.00	104.27 +-	.32	3.55 +-	.00	604.32 +-	40.68
543.79 +-	.00	108.10 +-	.32	3.52 +-	.00	1213.83 +-	64.60
558.57 +-	.00	111.17 +-	.32	3.53 +-	.00	1801.77 +-	76.46
574.77 +-	.00	114.54 +-	.31	3.54 +-	.00	1205.03 +-	70.62
592.33 +-	.00	118.19 +-	.31	3.57 +-	.00	2074.78 +-	76.72
610.50 +-	.00	121.97 +-	.31	3.58 +-	.00	1918.51 +-	74.97
630.46 +-	.00	126.12 +-	.30	3.59 +-	.00	1530.91 +-	72.58
649.79 +-	.00	130.14 +-	.30	3.60 +-	.00	1373.60 +-	76.81
668.25 +-	.00	133.98 +-	.29	3.62 +-	.00	1522.89 +-	74.54
691.15 +-	.00	138.74 +-	.29	3.60 +-	.00	1298.07 +-	80.44
712.46 +-	.00	143.17 +-	.29	3.59 +-	.00	529.01 +-	79.49
803.54 +-	.00	162.09 +-	.27	3.63 +-	.00	3072.83 +-	136.80
833.03 +-	.00	168.22 +-	.26	3.63 +-	.00	1270.92 +-	151.90
856.39 +-	.00	173.08 +-	.26	3.64 +-	.00	2118.18 +-	172.70
880.79 +-	.00	178.14 +-	.26	3.65 +-	.00	1650.53 +-	184.00
911.29 +-	.00	184.48 +-	.25	3.66 +-	.00	1801.68 +-	205.30
937.99 +-	.00	190.02 +-	.25	3.68 +-	.00	4541.80 +-	250.80
955.28 +-	.00	193.62 +-	.24	3.66 +-	.00	1630.46 +-	265.00
975.65 +-	.00	197.85 +-	.24	3.66 +-	.00	1336.80 +-	281.70
991.99 +-	.00	201.24 +-	.24	3.67 +-	.00	554.63 +-	279.00
1060.88 +-	.00	215.54 +-	.23	3.70 +-	.00	2429.44 +-	334.20
1078.81 +-	.00	219.27 +-	.22	3.71 +-	.00	2822.92 +-	348.70
1112.18 +-	.00	226.19 +-	.22	3.69 +-	.00	2420.94 +-	339.40
1161.12 +-	.00	236.35 +-	.21	3.74 +-	.00	2745.44 +-	370.10
1190.38 +-	.00	242.42 +-	.21	3.75 +-	.00	2763.33 +-	384.90
1234.32 +-	.00	251.54 +-	.20	3.75 +-	.00	2490.49 +-	413.40
1257.55 +-	.00	256.36 +-	.20	3.75 +-	.00	4470.73 +-	492.20
1272.32 +-	.00	259.42 +-	.20	3.75 +-	.00	2775.88 +-	482.10
1322.94 +-	.00	269.92 +-	.19	3.77 +-	.00	3783.68 +-	436.90
1353.09 +-	.00	276.18 +-	.19	3.80 +-	.00	5845.04 +-	442.70
1391.85 +-	.00	284.22 +-	.18	3.82 +-	.00	6499.85 +-	487.30
1411.68 +-	.00	288.33 +-	.18	3.82 +-	.00	6627.43 +-	600.60

TABLE 8.15 (continued)
The calibrated peak list, found in the FILENAME.PCA file

1425.20 +- .00	291.13 +- .18	3.83 +- .00	10383.70 +- 588.90
1465.71 +- .00	299.53 +- .17	3.70 +- .00	216471.00 +- 790.90
1490.27 +- .00	304.62 +- .17	3.81 +- .00	5190.03 +- 490.80
1551.39 +- .00	317.29 +- .16	3.84 +- .00	1689.47 +- 491.50
1572.94 +- .00	321.76 +- .16	3.83 +- .00	6570.14 +- 498.10
1629.57 +- .00	333.49 +- .16	3.90 +- .00	7849.47 +- 487.50
1678.91 +- .00	343.71 +- .15	3.90 +- .00	5629.07 +- 513.50
1705.99 +- .00	349.32 +- .15	3.93 +- .00	11286.40 +- 532.30
1746.78 +- .00	357.77 +- .15	4.44 +- .00	74582.50 +- 634.10
1777.06 +- .00	364.04 +- .14	3.63 +- .00	38013.80 +- 560.30
1800.76 +- .00	368.95 +- .14	3.91 +- .00	6678.34 +- 539.00
1830.89 +- .00	375.19 +- .14	3.98 +- .00	12240.90 +- 560.60
1853.25 +- .00	379.82 +- .14	3.94 +- .00	13240.00 +- 575.00
1876.47 +- .00	384.63 +- .14	3.91 +- .00	13016.10 +- 555.10
1907.71 +- .00	391.09 +- .13	3.88 +- .00	29424.50 +- 565.40
1933.51 +- .00	396.43 +- .13	3.87 +- .00	17832.70 +- 542.00
1968.34 +- .00	403.64 +- .13	4.12 +- .00	17000.10 +- 554.30
2019.08 +- .00	414.14 +- .13	4.00 +- .00	23927.70 +- 600.10
2038.34 +- .00	418.13 +- .13	3.98 +- .00	5790.58 +- 622.90
2058.13 +- .00	422.22 +- .13	3.95 +- .00	8118.22 +- 597.90
2081.43 +- .00	427.04 +- .12	3.92 +- .00	18033.50 +- 610.50
2100.37 +- .00	430.96 +- .12	3.97 +- .00	5973.41 +- 579.70
2140.91 +- .00	439.34 +- .12	4.05 +- .00	7140.50 +- 560.20
2165.35 +- .00	444.40 +- .12	4.14 +- .00	13097.50 +- 571.20
2213.29 +- .00	454.31 +- .12	4.05 +- .00	16747.10 +- 590.60
2241.45 +- .00	460.13 +- .12	4.05 +- .00	6191.43 +- 629.20
2261.05 +- .00	464.18 +- .12	4.05 +- .00	8136.08 +- 708.00
2277.95 +- .00	467.68 +- .12	4.05 +- .00	9879.05 +- 721.00
2297.43 +- .00	471.71 +- .11	4.08 +- .00	33452.00 +- 686.70
2321.45 +- .00	476.67 +- .11	4.07 +- .00	9722.71 +- 681.20
2339.94 +- .00	480.49 +- .11	4.09 +- .00	15958.60 +- 753.70
2352.59 +- .00	483.11 +- .11	4.07 +- .00	24324.90 +- 952.80
2366.12 +- .00	485.90 +- .11	4.07 +- .00	17526.20 +- 856.00
2383.12 +- .00	489.42 +- .11	4.08 +- .00	12059.00 +- 963.40
2394.07 +- .00	491.68 +- .11	4.12 +- .00	14367.00 +- 946.40
2412.28 +- .00	495.44 +- .11	4.17 +- .00	27265.40 +- 740.20
2432.75 +- .00	499.67 +- .11	5.13 +- .00	77035.30 +- 824.60
2458.71 +- .00	505.04 +- .11	4.10 +- .00	13990.60 +- 905.40
2482.42 +- .00	509.93 +- .11	4.31 +- .00	392640.00 +- 5591.00
2486.88 +- .00	510.85 +- .11	4.37 +- .00	577980.00 +- 5452.00
2514.24 +- .00	516.51 +- .11	4.09 +- .00	26362.30 +- 710.50
2534.94 +- .00	520.78 +- .11	4.12 +- .00	16550.00 +- 1294.00
2541.62 +- .00	522.16 +- .11	4.16 +- .00	22892.80 +- 1267.00
2565.49 +- .00	527.09 +- .11	4.11 +- .00	11112.50 +- 567.40
2601.05 +- .00	534.43 +- .11	4.14 +- .00	6953.17 +- 527.20
2662.92 +- .00	547.21 +- .10	4.17 +- .00	6739.28 +- 515.50
2695.11 +- .00	553.85 +- .10	4.04 +- .00	21032.20 +- 539.70
2721.58 +- .00	559.32 +- .10	4.14 +- .00	19408.60 +- 602.90
2739.56 +- .00	563.03 +- .10	4.17 +- .00	17852.40 +- 589.60
2790.04 +- .00	573.44 +- .10	3.93 +- .00	165503.00 +- 718.10
2822.64 +- .00	580.17 +- .10	4.24 +- .00	6231.81 +- 516.30
2863.97 +- .00	588.70 +- .10	4.22 +- .00	8864.46 +- 527.80
2896.16 +- .00	595.34 +- .10	4.96 +- .00	58935.10 +- 637.50
2927.61 +- .00	601.83 +- .10	4.87 +- .00	48541.50 +- 640.90
2951.18 +- .00	606.69 +- .10	4.22 +- .00	27531.60 +- 658.30
2968.47 +- .00	610.25 +- .10	4.25 +- .00	7415.93 +- 609.60
3000.92 +- .00	616.94 +- .10	4.27 +- .00	68354.50 +- 606.40
3027.65 +- .00	622.46 +- .10	4.36 +- .00	30542.70 +- 577.40
3055.21 +- .00	628.14 +- .10	4.31 +- .00	8541.09 +- 535.30
3092.83 +- .00	635.89 +- .09	4.97 +- .00	44448.80 +- 589.30
3146.47 +- .00	646.95 +- .09	4.28 +- .00	4182.80 +- 500.80
3190.48 +- .00	656.02 +- .09	4.34 +- .00	8608.81 +- 508.00
3229.47 +- .00	664.05 +- .09	4.36 +- .00	11906.20 +- 564.60
3250.83 +- .00	668.45 +- .09	4.69 +- .00	30355.30 +- 607.40

(continued)

TABLE 8.15 (continued)

The calibrated peak list, found in the FILENAME.PCA file

3282.41 +-	.00	674.96 +-	.09	4.36 +-	.00	7957.26 +-	554.30
3303.45 +-	.00	679.29 +-	.09	4.36 +-	.00	4576.53 +-	542.20
3346.14 +-	.00	688.08 +-	.09	4.15 +-	.00	28342.60 +-	529.40
3374.40 +-	.00	693.90 +-	.09	4.31 +-	.00	16344.80 +-	547.00
3398.80 +-	.00	698.93 +-	.09	4.40 +-	.00	14390.70 +-	540.10
3436.49 +-	.00	706.69 +-	.09	4.50 +-	.00	15215.80 +-	521.70
3479.56 +-	.00	715.56 +-	.08	4.47 +-	.00	11803.50 +-	538.80
3508.11 +-	.00	721.43 +-	.08	4.42 +-	.00	11854.20 +-	690.00
3522.51 +-	.00	724.40 +-	.08	4.42 +-	.00	16097.60 +-	731.70
3543.34 +-	.00	728.68 +-	.08	4.42 +-	.00	9839.47 +-	580.00
3577.12 +-	.00	735.64 +-	.08	4.46 +-	.00	26319.40 +-	554.80
3608.25 +-	.00	742.04 +-	.08	4.42 +-	.00	12412.80 +-	540.40
3640.77 +-	.00	748.73 +-	.08	4.34 +-	.00	29310.20 +-	584.60
3671.20 +-	.00	754.99 +-	.08	4.60 +-	.00	170965.00 +-	758.90
3705.99 +-	.00	762.15 +-	.08	4.69 +-	.00	28215.20 +-	565.70
3750.80 +-	.00	771.36 +-	.08	4.54 +-	.00	15345.50 +-	605.60
3775.11 +-	.00	776.36 +-	.08	4.40 +-	.00	28817.50 +-	617.40
3821.97 +-	.00	786.00 +-	.08	5.92 +-	.00	55651.80 +-	702.70
3860.72 +-	.00	793.96 +-	.08	4.64 +-	.00	15639.60 +-	596.50
3893.96 +-	.00	800.80 +-	.07	4.05 +-	.00	38740.50 +-	578.60
3944.59 +-	.00	811.20 +-	.07	4.34 +-	.00	22548.60 +-	647.00
3962.90 +-	.00	814.97 +-	.07	4.49 +-	.00	7848.95 +-	672.80
3988.18 +-	.00	820.16 +-	.07	4.36 +-	.00	25523.80 +-	609.60
4021.55 +-	.00	827.02 +-	.07	4.10 +-	.00	26565.80 +-	561.10
4060.82 +-	.00	835.08 +-	.07	4.57 +-	.00	14453.70 +-	626.40
4086.41 +-	.00	840.34 +-	.07	5.81 +-	.00	52993.70 +-	745.70
4120.86 +-	.00	847.42 +-	.07	4.60 +-	.00	7934.01 +-	579.10
4179.89 +-	.00	859.54 +-	.07	4.74 +-	.00	15407.10 +-	579.90
4224.05 +-	.00	868.61 +-	.08	4.59 +-	.00	2175.34 +-	566.50
4255.81 +-	.00	875.13 +-	.08	4.11 +-	.00	46418.10 +-	598.00
4284.05 +-	.00	880.92 +-	.08	4.60 +-	.00	3215.93 +-	613.80
4305.91 +-	.00	885.41 +-	.08	4.60 +-	.00	8835.22 +-	627.00
4332.56 +-	.00	890.88 +-	.08	4.50 +-	.00	24740.30 +-	554.80
4380.35 +-	.00	900.69 +-	.08	4.24 +-	.00	73604.60 +-	597.20
4414.65 +-	.00	907.72 +-	.08	4.69 +-	.00	13046.80 +-	555.10
4447.72 +-	.00	914.51 +-	.09	4.60 +-	.00	20640.10 +-	546.30
4526.52 +-	.00	930.67 +-	.09	4.66 +-	.00	4546.73 +-	527.80
4558.20 +-	.00	937.17 +-	.09	4.71 +-	.00	12201.90 +-	605.30
4579.54 +-	.00	941.54 +-	.10	4.80 +-	.00	24306.90 +-	672.10
4601.17 +-	.00	945.98 +-	.10	4.73 +-	.00	15049.40 +-	632.10
4625.56 +-	.00	950.98 +-	.10	4.74 +-	.00	11153.30 +-	538.80
4657.64 +-	.00	957.56 +-	.10	4.80 +-	.00	10185.40 +-	570.40
4685.38 +-	.00	963.25 +-	.11	5.47 +-	.00	34187.30 +-	622.20
4726.15 +-	.00	971.61 +-	.11	4.76 +-	.00	4741.04 +-	525.20
4763.54 +-	.00	979.28 +-	.11	4.76 +-	.00	8764.43 +-	524.20
4824.15 +-	.00	991.72 +-	.12	4.81 +-	.00	8613.52 +-	536.40
4857.69 +-	.00	998.61 +-	.12	4.78 +-	.00	14150.60 +-	639.80
4879.08 +-	.00	1003.00 +-	.12	5.88 +-	.00	48533.70 +-	804.70
4902.80 +-	.00	1007.88 +-	.12	4.75 +-	.00	5210.37 +-	569.70
4928.53 +-	.00	1013.16 +-	.12	5.89 +-	.00	69392.80 +-	675.30
4969.58 +-	.00	1021.60 +-	.12	4.76 +-	.00	13010.90 +-	601.30
4988.27 +-	.00	1025.44 +-	.12	4.80 +-	.00	20469.50 +-	600.60
5054.07 +-	.00	1038.97 +-	.12	4.28 +-	.00	106841.00 +-	623.60
5079.02 +-	.00	1044.10 +-	.12	5.19 +-	.00	38526.20 +-	612.10
5111.72 +-	.00	1050.83 +-	.12	4.84 +-	.00	6886.06 +-	523.90
5144.85 +-	.00	1057.65 +-	.12	4.92 +-	.00	14414.20 +-	538.70
5173.41 +-	.00	1063.53 +-	.12	4.80 +-	.00	5882.75 +-	510.80
5237.80 +-	.00	1076.79 +-	.12	4.74 +-	.00	24454.70 +-	510.30
5272.66 +-	.00	1083.98 +-	.12	4.83 +-	.00	4307.57 +-	488.00
5333.03 +-	.00	1096.42 +-	.12	4.85 +-	.00	3799.50 +-	479.30
5385.92 +-	.00	1107.33 +-	.12	4.87 +-	.00	13669.20 +-	518.40
5414.13 +-	.00	1113.14 +-	.13	4.87 +-	.00	19950.90 +-	710.20
5428.75 +-	.00	1116.16 +-	.13	4.88 +-	.00	14277.40 +-	764.10
5448.78 +-	.00	1120.29 +-	.13	4.89 +-	.00	9440.54 +-	571.60

TABLE 8.15 (continued)

The calibrated peak list, found in the FILENAME.PCA file

5488.11	+-	.00	1128.41	+-	.13	4.77	+-	.00	12598.50	+-	487.90
5524.90	+-	.00	1136.01	+-	.13	4.87	+-	.00	9805.74	+-	633.10
5540.87	+-	.00	1139.31	+-	.13	4.87	+-	.00	7402.89	+-	768.60
5558.05	+-	.00	1142.85	+-	.13	4.88	+-	.00	15684.00	+-	704.90
5577.80	+-	.00	1146.93	+-	.13	4.88	+-	.00	3700.50	+-	570.50
5607.86	+-	.00	1153.14	+-	.14	4.89	+-	.00	5243.60	+-	494.00
5641.54	+-	.00	1160.10	+-	.14	4.92	+-	.00	4502.56	+-	475.40
5715.15	+-	.00	1175.32	+-	.15	4.91	+-	.00	9850.46	+-	500.30
5740.49	+-	.00	1180.56	+-	.16	4.89	+-	.00	6389.25	+-	540.10
5763.09	+-	.00	1185.24	+-	.16	4.93	+-	.00	4385.78	+-	468.50
5793.57	+-	.00	1191.54	+-	.17	4.93	+-	.00	4318.02	+-	482.80
5822.24	+-	.00	1197.48	+-	.17	4.93	+-	.00	3255.42	+-	471.10
5861.89	+-	.00	1205.68	+-	.18	5.00	+-	.00	13686.90	+-	475.60
5897.44	+-	.00	1213.05	+-	.19	4.95	+-	.00	9717.55	+-	506.80
5920.95	+-	.00	1217.92	+-	.20	4.96	+-	.00	8755.03	+-	551.80
5943.10	+-	.00	1222.50	+-	.21	5.02	+-	.00	11800.70	+-	521.80
5977.66	+-	.00	1229.67	+-	.22	5.00	+-	.00	9743.21	+-	463.60
6021.40	+-	.00	1238.73	+-	.24	4.96	+-	.00	2778.49	+-	461.20
6050.99	+-	.00	1244.87	+-	.25	4.97	+-	.00	6502.60	+-	505.30
6074.59	+-	.00	1249.76	+-	.26	4.98	+-	.00	2662.12	+-	502.30
6103.50	+-	.00	1255.76	+-	.27	4.98	+-	.00	4004.05	+-	477.10
6134.38	+-	.00	1262.16	+-	.28	4.99	+-	.00	3650.82	+-	537.70
6153.63	+-	.00	1266.14	+-	.28	5.03	+-	.00	5010.83	+-	568.60
6177.13	+-	.00	1270.98	+-	.27	5.05	+-	.00	5635.28	+-	499.10
6210.09	+-	.00	1277.76	+-	.26	5.07	+-	.00	5774.47	+-	452.70
6279.86	+-	.00	1292.11	+-	.24	5.06	+-	.00	2136.06	+-	495.80
6301.56	+-	.00	1296.57	+-	.24	4.97	+-	.00	18270.70	+-	538.10
6327.16	+-	.00	1301.84	+-	.23	5.06	+-	.00	2914.61	+-	471.60
6367.31	+-	.00	1310.10	+-	.22	5.12	+-	.00	4834.08	+-	454.40
6400.08	+-	.00	1316.84	+-	.21	5.09	+-	.00	10699.40	+-	453.60
6498.67	+-	.00	1337.12	+-	.19	5.15	+-	.00	10198.90	+-	443.00
6539.73	+-	.00	1345.57	+-	.18	5.15	+-	.00	4787.09	+-	441.40
6574.43	+-	.00	1352.71	+-	.17	5.15	+-	.00	3025.24	+-	436.00
6618.42	+-	.00	1361.76	+-	.16	5.16	+-	.00	5049.78	+-	455.00
6646.65	+-	.00	1367.57	+-	.16	5.16	+-	.00	10143.90	+-	504.40
6670.39	+-	.00	1372.45	+-	.15	5.15	+-	.00	6067.38	+-	508.50
6696.66	+-	.00	1377.85	+-	.15	5.16	+-	.00	6422.64	+-	539.00
6716.96	+-	.00	1382.03	+-	.14	5.16	+-	.00	13549.30	+-	539.40
6744.38	+-	.00	1387.67	+-	.14	5.16	+-	.00	3377.52	+-	474.20
6771.98	+-	.00	1393.35	+-	.14	5.14	+-	.00	8380.07	+-	449.30
6861.49	+-	.00	1411.76	+-	.13	5.20	+-	.00	3298.54	+-	473.30
6883.18	+-	.00	1416.22	+-	.13	5.20	+-	.00	5927.07	+-	423.50
6914.37	+-	.00	1422.64	+-	.13	5.20	+-	.00	3620.02	+-	461.20
6937.57	+-	.00	1427.41	+-	.13	5.21	+-	.00	4752.55	+-	456.60
6971.99	+-	.00	1434.49	+-	.14	5.21	+-	.00	2317.85	+-	425.00
7001.18	+-	.00	1440.50	+-	.14	5.21	+-	.00	1263.51	+-	412.30
7083.21	+-	.00	1457.37	+-	.15	5.24	+-	.00	3801.31	+-	428.80
7108.94	+-	.00	1462.67	+-	.15	5.25	+-	.00	5801.64	+-	460.10
7135.67	+-	.00	1468.17	+-	.16	5.28	+-	.00	6816.22	+-	445.90
7167.35	+-	.00	1474.68	+-	.16	5.28	+-	.00	8540.04	+-	442.50
7191.90	+-	.00	1479.73	+-	.17	5.26	+-	.00	3361.65	+-	450.70
7224.66	+-	.00	1486.47	+-	.17	5.26	+-	.00	4021.45	+-	589.20
7238.66	+-	.00	1489.35	+-	.18	4.62	+-	.00	23728.20	+-	569.30
7269.75	+-	.00	1495.75	+-	.18	5.29	+-	.00	5163.72	+-	442.30
7295.13	+-	.00	1500.97	+-	.19	5.29	+-	.00	3660.78	+-	440.00
7326.39	+-	.00	1507.40	+-	.20	5.30	+-	.00	4496.61	+-	416.30
7360.53	+-	.00	1514.42	+-	.21	5.31	+-	.00	2923.09	+-	403.20
7401.74	+-	.00	1522.90	+-	.22	5.35	+-	.00	10021.80	+-	414.70
7435.62	+-	.00	1529.87	+-	.23	5.33	+-	.00	3866.34	+-	415.10
7468.75	+-	.00	1536.69	+-	.24	5.32	+-	.00	3550.25	+-	414.50
7500.57	+-	.00	1543.23	+-	.25	5.34	+-	.00	3056.51	+-	400.60
7584.86	+-	.00	1560.57	+-	.27	5.34	+-	.00	1226.45	+-	379.60
7681.58	+-	.00	1580.47	+-	.30	5.35	+-	.00	1840.11	+-	393.30
7711.85	+-	.00	1586.70	+-	.31	5.36	+-	.00	1155.65	+-	391.30

CHAPTER 9

Conditions for Consideration of Other Standard Codes

In this chapter we outline some of the situations in which other spectral analysis codes may perform better than ROBFIT.

The first case is when the peak shapes and background for the spectrum are known analytically. The ROBFIT code builds up the peak shapes from back-to-back cubic splines and the background from cubic splines. These functions are approximations that introduce their own errors into the fits. Normally, the error introduced by the splines is very small compared to the statistical errors in the spectrum; however, they are always present. A code with the true shape and background built into it from the outset would more easily locate peaks within the spectrum. This assumes, of course, that the code can also correctly deal with complex backgrounds and overlapping peak regions.[4,5,28,29]

A second problem can arise when the spectrum contains very large peaks. The ROBFIT code assumes that each peak in the spectrum is exactly the same shape as the standard peak of its type. In gamma-ray spectra, the real peak shape has a high-energy leading edge and low-energy tailing. In most cases these inaccuracies, again caused by the peak shape approximation, do not concern the user, as they are masked by the errors introduced by the data fluctuations. However, in the case of a very large peak, where the statistics are better, there is a possibility that they will show up. The problem is identified by braiding of the large peak as the code tries to correct for the inaccuracy by adding more peaks. The user must be careful not to confuse this with braiding caused by incorrect peak width specification.

This problem can be reduced by using different standard peaks in different energy regions and by allowing more and more constants in the background. This is recommended if the object is to find small peaks in the vicinity of the large peaks. If, however, the object is to find properties involving the large peak, then there is no substitute for analytical information. Furthermore, in these situations the background is relatively unimportant, and the overlapping of peaks is probably negligible. In this case there is no reason to fit the entire spectrum, which is the principal advantage of ROBFIT.

Thirdly, for the code to pick out peaks that are in the noise, ROBFIT must have an accurate representation of the background across the peak region. ROBFIT uses the entire spectrum to define the background function and, for purposes of speed, compresses the spectrum by a factor of 16. This is normally not a problem since the background knots should be further apart than the full width at half-maximum of the peaks. However, if one tries to fit a spectrum of only 40 channels, no more than two constants can be defined in the background, and these rather poorly. Numerous other codes exist for treating short segments of spectra. The authors recommend using at least 32 channels per background constant; if this is not possible, ROBFIT is probably not the best code to use to analyze the data.[30,31,32,33]

Finally, a fair amount of user intervention is required to fit the data to the noise level. Peak shapes need to be determined from either the data or the same detector used on a calibration standard. The background constants frequently need to be either restricted or expanded in number to enable the code to avoid fitting peaks as background or vice versa. The widths of the standard peaks and especially their errors need to be input carefully in order to properly break up doublets. The standard type for each peak is not particularly well chosen by ROBFIT, and frequently the user will do much better to change it in the peak list. We consider this to be a small amount of user intervention compared to that involved in the truly interactive codes, but it is definitely not zero. The reward is a fit to noise level, and although we observe reasonable accuracy at higher levels, most of our testing has been to the noise level and we could have problems with poorly fitted spectra. Those potential users wanting truly black box data analysis are hereby advised to wait for release 10.5 or later of ROBFIT and to use other codes in the meantime.

One parting comment concerning background and foreground functions: We have assumed that the only relevant energy dependence of the peak shape is that the width depends on the square root of the energy. We have also assumed that the Compton profile is either a peak type or buried in the background which we have taken to be all parts of the spectrum which are slowly varying. The reader who requires information on the parts of the spectrum which we have buried in the noise is referred to a book by Debertin and Helmer[34] which gives a comprehensive review of the subject.

APPPENDIX A

ROBFIT users guide

1. Introduction

This appendix contains a users guide to the ROBFIT curve fitting code. ROBFIT is a software package which utilizes splines with optimized knot locations, model peak shapes, and background identification tools to provide robust fitting to complicated data such as encountered in spectral analysis. Here we concentrate on the IBM PC (and compatible) version of this code; however, ROBFIT operates in a similar manner on other operating systems. To date, we have versions that run on the IBM PC; the Macintosh, and also on IBM and VAX mainframe computers. On all these computer systems the menu driven input described in this appendix appears the same to the user. The major difference between these versions of ROBFIT is in the graphical routines, however, the code has been written in such a manner that all the system dependent parameters are contained within a few subroutines. This means that the code can be easily modified to suit any operating system.

The majority of the code debugging has been carried out on the North East Regional Data Center (NERDC) IBM 3090 600/VF machine. NERDC kindly supplied us with free computing time which aided the development of SMSQ, the minimization routine that enables the present code to run efficiently [see ref. 1,2]. Anyone running the code in such an environment should request the vectorized version of SMSQ which runs four times faster than the scalar version.

2. Overview of ROBFIT

The robust fitting code ROBFIT was initially conceived in 1979. Its first use was as a nuclear physics analysis package. Complex nuclear spectra with up to 500, frequently overlapping, peaks in 4000 channels, were analyzed on the NERDC IBM 3033 using overnight runs and batch processing. The code then migrated to various PC's for portability and further speedup. Running ROBFIT on a variety of machines has produced a compact and intrinsically efficient code. This early work helped develop a fitting technique that represents the peaks and background of a spectrum with the minimum number of cubic splines. An illustration of the use of the code has been the analysis of data taken on the Antarctic research program to search for supernova gamma rays [3,4]. In this experiment ROBFIT was run on a Macintosh II and provided a high performance nuclear analysis package.

Although the code was originally for use in the analysis of nuclear spectra, ROBFIT can be used for any general spectral analysis. The algorithms used by ROBFIT can be summarized as follows. It is assumed that the data consists of two distinguishable parts called background and foreground. The background consists of slowly varying features (e.g. bremmstrahlung and Compton scattering in nuclear physics spectra), while the foreground consists of rapidly varying features such as photopeaks, escape peaks, and backscatter peaks.

The background is fitted over the entire spectrum as a set of cubic splines with adjustable knots. Fitting over the entire spectral range allows these features to be fitted continuously with fewer constants and more accurate representations than is possible when they are fitted in short sections along with the foreground. The algorithms used are data compression, similar to that described in ref. 5, and second, a minimization algorithm SMSQ, an extension of the Marquardt procedure (also ref. 5, p. 523). This procedure allows efficient optimization of the knots in the exponentials of the splines as well as of all other more linear parameters.

The foreground typically consists of a relatively small number of peak types, that is, a few characteristic shapes which may be repeated many times. Each one of these shapes is fitted by a shape-finding routine as a collection so back-to-back cubic splines. These shapes are then, in turn, fitted to the spectrum by first finding the type of peak to try, then fitting the peak type to the data by varying the peak height, location, and width within user-specified limits. Clumps of overlapping peaks are fitted simultaneoulsly. The cpu time required, however, increases approximately as the fourth power of the number of peaks in each clump. The principle algorithms used are a fast residual locator for finding the peak positions, a routine for systematically controlling the allowed widths (ref. 2), and SMSQ (mentioned above).

2.1 General operation of ROBFIT

The routines collectively called ROBFIT have been developed to provide a general tool for spectral fitting. A typical spectrum may contain thousands of channels (individual histogram bins) with numerous peaks which vary in both strength (number of counts under the peak) and width. In addition these peaks can sit on top of a complex backgournd. ROBFIT has been developed to ease the analysis of such spectra. ROBFIT provides several modes of operation, three for curve fitting, three for display of the data and fits and one for calibration of the spectrum. For fitting, code is provided:

i. for generating a standard peak (STGEN),
ii. to fit the background alone (BKGFIT),
iii. to fit the complete spectrum (FSPFIT).

For display, code is provided:

i. to display the raw data (RAWDD),
ii. to display the standard peaks (STDIS),
iii. to display the full spectral fit (FSPDIS).

For calibration, code is provided:

i. to calibrate the x-axis (XCALIBER)

Figure 1 gives an overview of how ROBFIT is used in the analysis of a spectrum while section 3 provides a more detailed description of the above operation modes. Generally, the first thing a user wants to do at the outset of fitting is to have a quick look at the data. This is the job of the raw data display routine (RAWDD), it displays the spectrum without the fit. RAWDD allows the display to be zoomed to study certain regions of the spectrum for investigation of the peak shapes. After viewing the data, the next step is to generate a standard for each peak shape in the spectrum, this is performed by running the standard generating routine (STGEN). These standards are used in the full spectral fit to search for peaks of that particular shape. STGEN allows the user to either select one of the more common peak shapes (Gaussian, Lorentzian or Voigt) or create a standard from a peak within the raw data. Once a standard has been generated it can be displayed by running the standard display routine (STDIS). At this point the user may wish to add more constants to the peak representation to make it a closer copy of the actual peak shape. This would involve repeated runs of STGEN and STDIS until a satisfactory standard has been created. Generally, this is not a problem at the start of fitting, it is only once the full spectral fitting routines starts to find structure within a peak that the user needs to worry about the precise peak representation. Having looked at the data and created a few standards the user is now in a position to start fitting the spectrum. Before performing a full peak search it may be necessary to fit only the background features within the spectrum. This is done by running the background fitting routines (BKGFIT). This could, for example, be the case with a complex spectrum that has an underlying background that contains sharp steps in the background. With such a spectrum, a preliminary fit will help the full spectral fitter in its task of separating background from peaks. The full spectral fit routines (FSPFIT) does contain the background fitter and does not need a preliminary background fit for operation. FSPFIT cycles between background fitting and peak fitting to find successively smaller peaks within the spectrum. Once either BKGFIT or FSPFIT has been run the full spectral fit display routines (FSPDIS) can be run to view the fit, background, peaks and residuals. Undoubtedly, the user will need a display of intensity versus something other than channel number. In nuclear physics experiments this something else is usually energy. The XCALIBER routines convert a calibration file of peak channel numbers and energies into a file containing constants, which enables FSPDIS to display energy rather than channel number as its x-axis. XCALIBER also converts peak versus channel files into peak versus energy files and can also convert data files from intensity versus channel to intensity versus energy. This enables the user to add data with different calibrations.

During fitting, the user has control over the level to which the code will search for small peaks. The following sections illustrate how ROBFIT can be used to search for small overlapping peaks down to the noise level of the spectrum.

2.2 Interacting with an IBM PC (or compatible)

One feature that is required by all spectral analysis programs is some form of graphical representation of the spectrum. This fact has caused us considerable aggravation in our effort to design a machine independent spectral fitting routine. Although there are moves within the computer society as a whole to standardize computer graphics software, the situation at the present time is far from clear. To reduce the impact of having to redefine our graphics routines for different computers we attempted to place all the device dependent routines within a few subroutines. These can easily be modified for an operating system that has graphics capability available through a FORTRAN compiler.

On the personal computer version of ROBFIT we have chosen to use the Waterloo

Fortran Compiler WATFOR-77. This compiler allows us to link to graphics routines that are compatible with the Graphics Kernel System (GKS), an internationally accepted standard for computer graphics development. One of the stipulations for running ROBFIT on a personal computer is, therefore, that the user has access to the WATFOR-77 GKS support code. This support code is an executable program that must be run before GKS can be used. A copy of this code can be obtained from:

The WATCOM Group, Inc.
A15 Phillip Street
Waterloo
Ontatio
Canada N2L 3X2
Tel (519) 886-3700

A further requirement is that that personal computer has either EGA or VGA capability. Owing to its extremely valuable error checking WATFOR fortran is not the fastest way to run ROBFIT. About a factor of three speed up can be achieved by compiling FSPFIT and STGEN with either Microsoft or IBM Fortran.

2.3 Getting started

All that is required to get up and running with ROBFIT is to load onto your operating medium (e.g. hard disk) all files with .EXE and .MNU name extensions. These are the runnable codes. For those users wishing to really get their hands dirty, we do supply all source code. A set of test codes are provided with the code, we recommend working through these to familiarize oneself with the menu system. These test cases should also be copied onto the operating medium. The test cases are detailed later in Section 4 and Appendix I. The data formats that the code expects are given in section 2.4.2. After all the files have been copied across, the user is ready to operate ROBFIT. We recommend running the top-most menu page which gives a brief account of the available routines. Any one routine, such as RAWDD can be selected by typing in the line number on the menu page. The top-most menu page is accessed by typing ROBFIT <return>. This activates a batch file which will load and run any of the six ROBFIT modes. At this point, if the user wishes to graphically view the data or fit, then the WGKS.EXE support program from the WATCOM group must have been run. At a later date, once the user becomes more conversent with ROBFIT operation, the individual modes can be operated on a stand-alone-basis by simply typing the mode name, for example: STGEN <return> or FSPFIT <return>, etc. The mode menu pages are described in section 2.4.

2.4 Using the menu driver

Entry into ROBFIT is performed by running one of the executable codes discussed in section 2.3. Once a mode has been selected the user has access to that particular mode alone. At this point the mode menu pages can be operated, a number of pages become available each of which is described in full in the following Sections. From the top of the mode menu the user can either select a lower menu page, enabling ROBFIT parameters to be varied, or run/quit the code. The overall hierarchy of the menu is shown in Figure 2. Section 2.4.1 provides a fuller explanation of how to move around the mode menu pages and gives a description of each of the functions within the menus.

2.4.1 Mode menu page structures

In this section we describe the layout of each of the mode menu pages that can be operated under ROBFIT. In order to reach the mode menus the user must have followed the startup sequence of ROBFIT discussed in Section 2.3. The top-most session master menu page allows the user to select one of the modes of operation. The session master menu page has the following structure:

1 Display raw data (RAWDD)
2 Generate standard (STGEN)
3 Display standard (STDIS)
4 Fit background (BKGFIT)
5 Fit spectrum (FSPFIT)
6 Display spectral fit (FSPDIS)
7 Calibrate spectra (XCALIBER)

By entering the line number of the mode required, that particular mode can be selected for operation. On selecting a mode the following menu structure is presented. The master menu for each mode is the first page to be displayed. Here the user has a number of options to choose from. By entering 96 the user can call up a previous menu setup, this file need not have the extension .MNU. Entering 97 allows the user to save the current menu setup to a disk file. Entering 98 lets the user make a global change to the menu file. This is useful when changing the output file names for the background, peak and graphics data, these files usually all have the same filename, a global change can be made to rename all occurrences of the old filename to a new filename. Entering 99 will exit the current ROBFIT mode. If another mode is required ROBFIT is cycled by entering 99 at the mode master menu level which will return the user to the top-most page. Entering 100 will run the current mode. Entering 101 allows the user to make multiple passes of ROBFIT (ie batch like processing). In this mode the user must supply a file which contains a list of .MNU files, set up prior to running, ROBFIT will then cycle through each of these files in turn. The mode menus have the structure outlined in Figure 2.

From the mode master menu the user can select a sub-menu page related to that specific mode. The sub-menu pages contain the operating parameters for the mode, once in the sub-menu the user can alter any of the runtime parameters. Each sub-menu is accessed by entering the line number of that particular sub-menu. At the sub-menu level there are four ways to proceed.

1) The first is to enter 0,0 this will exit the sub-menu and return the menu driver to the mode master menu, from which the mode can be run or exited. <cr> is the default response which will start the run.

2) The second is to delete a line, this is done by entering LINE NUMBER,1. This option is used for deleting old standard information.

3) The third is to enter a new variable for one of the inputs, this is done by entering LINE NUMBER,10. The menu driver then asks for the new variable which must be entered on the next input line. After entry the driver will re-display the sub-menu with this change added.

4) The fourth and final option is to create a new line by entering LINE NUMBER (to create),100. This option is for use in adding in new standard information for a fit. Each standard must have its own specific set of initialization data as outlined later.

The following describes the mode menu and sub-menu pages. All data file formats are given in Section 2.4.2.

RAW DATA DISPLAY (RAWDD) MODE

The RAWDD mode enables the user to view the raw data to be input into ROBFIT. This may be for diagnostic purposes or simply to select out a region of interest for analysis. The menu pages are as follows:

Page RAWDD

1 General information for the display.
2 Difference plot information.
3 Corrected plot information.

By selecting one of the three options the user can change the display requirements. Line one shows details on file input and display plotting options. Line two shows difference plot options. These pages are used if a comparison between two spectra is required. The difference plot subtracts one spectrum from another and plots the residuals. Line three is used to correct bad data. If data corruption has occurred for some reason and the user wishes to modify a certain region of the spectrum then line three shows the options for performing the modification.

Page RAWDD1 (General information)

1	Files to display f1,f2	Fname1.UF(or SP),Fname2.UF(or SP)
2	Initial and final channels	1,default
3	Minimum and maximum peak heights	1,default
4	Log or linear y scale	log
5	Title for the graph	Title

On line one the names of the two files to compare are entered. If there is just a single file then a backslash (\) is entered for the second filename. Line two contains the limits in channel numbers for the display. The default option displays all the available channels. On line three the range of peak heights is input. The default option takes these values from the data. On line four the user specifies either a logarithmic or a linear Y scale. Line five contains the title that will be placed at the top of the display.

Page RAWDD 2 (Difference information)

1	Do you want a difference plot (y/n)	No
2	want to divide by the error (y/n)	No
3	the file name for sum	None
4	the file name for difference	None
5	min/max peak heights are	1,default
6	log or linear y scale	Log
7	normalization for the files	default,default

On line one the user selects the difference option by specifying YES or NO. Line two gives the option of dividing by the error, calculated from the difference between the files. This is for viewing purposes only; it helps reduce wide fluctuations that can occur when plotting the difference between two spectra. Line three is the name of the file in which the sum of the two data files is placed. This file is used in the full spectral fitting phase as the error on the difference data. The sum file is essential, ROBFIT goes wild in fitting difference data without it. Line four contains the name of the file for the difference data. Each file is specified as FILENAME.SP. Lines five and six specify the min-max peak heights and the log or linear choice for Y scale respectively. The default option for min-max is to the full range of the data.

Line 7 contains the scaling parameter for the two input files. Channel values for the second file are scaled by (first scale value)/(second scale value).

Page RAWDD 3 (Correction information)

1	Do you want to correct the data (y/n)	No
2	channel range to be corrected.	1,default/1,default
3	file name for corrected data.	None/none
4	beginning and ending heights.	1,default/1,default

On line one the user selects the correction option by specifying YES or NO. Line two contains the channel range over which the correction is to be performed. Corrections can be made to both input files if needed. Line three contains the files for the output data. Both files must be specified as FILENAME.SP. Line four contains the starting and ending channel heights for the selected region. The program will linearly interpolate over the troubled region between the selected heights on line four.

STANDARD GENERATION (STGEN) MODE

The standard generation mode allows the use to create a standard peak from the raw data or to use one of the common peak shapes held within the program.

Page STGEN

1 General data on the generation
2 Selection of the fit type details

Line one shows the options for the selected standard shape. Line two shows the detailed data for the Voigt and data fit options.

Page STGEN1 (Generation data)

1	Name of the output standards file	Filename.ST
2	Graphical output to be placed in file	Filename.GR
3	Fit type (Gaus,Lore,Voig or Fitd)	Gaus
4	Number of backgrnd coeffs and splines	1,1

Line one contains the name of the file into which the standard output is to be placed. The main spectral fitting routine will use this file in its peak searching phase. Line two contains the filename in which the display information is stored. This file is used in the standard display mode (STDIS) of ROBFIT. On line three the user can specify the standard required. If either a Gaussian (Gaus) or a Lorentzian (Lore) is specified, then no further input is required. If Voigt (Voig) or a section of data (Fitd) is used to generate a standard then further options have to be set on page STGEN2. On line four the user selects the number of background coefficients and the number of splines to be used in the standard.

Page STGEN2 (Type options)

1	Voigt fit ETA parameter	1
2	Data fit file name	Filename.UF (or SP)
3	beginning,ending channel numbers	1,default

Line one shows the Voigt ETA parameter that will be used in the generation. On line two the user specifies the filename in which the raw data resides and on line three the channel limits over which the standard is to be fitted. The default again being all the available range.

STANDARD DISPLAY (STDIS) MODE

The standard display mode allows the user to view the standard peak generated with STGEN.

Page STDIS

1 Information for the standard display
2 Viewing information for the display

Line one contains the general information for the display. Line two shows the options available during the display.

Page STDIS1 (General information)

1	Name of the input graphics file	Filename.GR
2	Beginning and ending channels	1,default
3	Desired min and max y scale	1,default
4	Log or linear y scale	Log

Line one contains the name of the file containing the input data for the plot. Line two shows the range of channels that will be displayed with the default being all channels. Line three shows the range over which the y scale will vary. The default option takes these values from the input data. Line four shows whether the plot will be linear or logarithmic in its Y scale.

Page STDIS2 (Viewing options)

1	View the curve fit (y/n)	Yes
2	View the splines (y/n)	Yes
3	View the background fit (y/n)	Yes

Answering yes to line one displays the curve fit. Similarly answering yes to line two shows the splines and line three shows the background under the standard.

BACKGROUND FIT (BKGFIT) MODE

The background fit allows the user to perform a fit to the background under the spectrum without the need to run the full spectral fitting code. This is especially useful for getting started in those cases where the actual background resembles a peak in some regions, frequently the case in the rollup low energy region of nuclear spectra.

Page BKGFIT

1 Input data information.
2 Output data information.

Selection of the appropriate line shows the details for either input or output from the background fitter.

Page BKGFIT1 (Input details)

1	Maximum number of background constants	100
2	Cutoff value set to	5
3	Input data file name	Filename.UF(or SP)
4	Beginning and ending channels for fit	1,default
5	Input background constants file	Filename.CN,Yes
6	Weights 0=sqrt(data),1=calc,2=file	0
7	File containing the weights	None

Line one shows the maximum number of constants that will be used in the fit. Line two shows the CUTOFF at which the routine will stop fitting. Line three contains the raw data filename and line four the range over which the fit is to be performed. Again the default is all the available channels. Line five contains the input background constants file. The qualifier for this line selects either a logarithmic (YES) or a linear (NO) background fit. Line six selects the weights that are placed on the data points. A 0 takes the error to be the square root of the data, 1 determines the error from the spread of the data points over the entire selected region, 2 reads the errors in from an external file whose name is entered on line seven. The format of the file must be either .UF or .SP formatted as outlined in Section 2.4.2.

Page BKGFIT2 (Output details)

1	Background constants output to	Filename.CN
2	Background fit	Filename.GR

Line one shows the file to which the constants for the background fit will be placed. Line two shows the file to which the background fit graphics information is sent.

FULL SPECTRAL FIT (FSPFIT) MODE

This part of ROBFIT provides the means to fit the complete spectrum of peaks and background.

Page FSPFIT

1 General data on the spectral fit.
2 Input data information.
3 Output data information.
4 Standards files.

Line one contains information which is most likely to be changed during a sequence of refits. Data such as the setting of the CUTOFF value and number of background constants can be found here. Line two holds the data which are to be input into the spectral fit. Similarly line three details where the output from the fit is to be placed. Line four shows a separate menu for standards files to be used with the fit, together with the width variations to be used for each standard.

Page FSPFIT1 (General fit information)

1	Maximum number of peaks	100
2	Starting,ending cutoffs, in N steps	2,1,2
3	VWID,FIXD and CONT,NONK,NOBF,FIXK	VWID,CONT
4	Maximum number of background constants	4, 10
5	Weights 0=sqrt(data),1=calc,2=file	0
6	File containing the weights	None

Line one shows the maximum number of peaks that will be included in the fit the largest value being 255. Line two specifies the CUTOFF range and the number of steps that will be taken between the two limits. For example the values listed above will have a starting CUTOFF value of 2 and an ending CUTOFF of 1 and will take 2 steps between these two limits. Line three shows how the program will handle the peak width variation and background fitting. For example VWID selects a variable width fit while FIXD selects the fixed width format. CONT,NONK,NOBF and FIXK select continuous knot addition, no knot addition, no background fit and fixed knot option respectively. Line four shows the minimum (starting) and maximum (ending) number of background constants that will be used in the fit. Line five selects the error that is to be placed on the data, 0 selects an error that is the square root of the data point at each channel, while 1 calculates the error from the spread of the data over the entire selected region, and 2 allows the user to read in the errors from an external file whose name is specified on line six. Frequently this is the sum file generated by RAWDD as mentioned in the discussion of the RAWDD menu.

Page FSPFIT2 (Input data)

1	Input data file name	Filename.UF(or SP)
2	Beginning and ending channels for fit	1,default
3	Input peak file	Filename.PK,YES
4	Input background constants file	Filename.CN,YES

Line one shows the file name in which the raw data reside. Line two defines the channel range for the fit, the default being the entire spectrum. The third line details the input peak file if there is one, NONE is specified if there is no peak file. If there is a peak file the qualifier determines if the peak widths are accurate or not. For example, for no input peak file, the user would specify NONE, NO whereas for a run with a peak file with accurate peak widths this would be specified as FILENAME.PK,YES while for input peaks at the correct location but wrong strengths this would be specified as FILENAME.PK,NO. Line four details the input parameters for the background. The input file name is defined first, again, NONE specifies no input background file. The qualifier for line four selects the type of background fit. A logarithmic fit to the background is selected by specifying YES while a linear fit is selected by NO. The CN file also contains a qualifier which overrides the YES or NO; to change from log to linear the user must restart with NONE,YES or NONE,NO.

Page FSPFIT3 (Output data)

1	Output file for background constants	Filename.CN
2	Output file for peak data	Filename.PK
3	Output file for graph data	Filename.GR

These three lines specify the files in which each of the fitted parameters are to be placed.

Page FSPFIT4 (Standards data)

1	First standard	Filename.ST
2	Type 1 width,error and channel no.	5,1,100
3	Type 1 width,error and channel no.	10,1,1000
4	Second standard	Filename.ST
5	Type 2 width,error and channel no.	20,5,10
6	Type 2 width,error and channel no.	20,5,2000
etc.		

This menu page details the standards to be used in the fit the files containing the standards are defined here. Following each standard file a pair of lines defines the limits over which that particular standard is to vary, for example, the first standard (type 1) has a peak width starting at 5+/-1 channels at around the 100 channel region and goes up to 10+/-1 channels at the 1000 channel region. Similarly the second standard (type 2) has its variation specified in lines five and six. Each new standard must have its width variation set on this menu page.

FULL SPECTRAL FIT DISPLAY (FSPDIS) MODE

These pages allow the full spectral fit to be displayed. The routine uses as its input the file specified on menu page FSPFIT3 line 3.

Page FSPDIS

1 Information on the spectral fit display
2 Viewing information for the display

The first line provides general information for the display while the second line shows the screen setup for the plotting routines.

Page FSPDIS1 (General information)

1	Name of the graphical file to be read	Filename.GR
2	Name of the calibration file (.XDA)	Filename.XDA
3	Beginning, ending channels for display	1.default
4	Vertical axis = COUNTS*SCALE FACTOR	1
5	Desired min and max y scale	1,default
6	Log or linear y scale	LOG

Line one contains the name of the file which has the graphical output. Line two contains the filename of the calibration file created by XCALIBER if a calibration has been performed, NONE is specified if no calibration has been performed. Line three specifies the channel range to be displayed. The default is specified as 1,DEFAULT and runs over the entire range of channels. Line four contains the scaling factor for the Y-axis of the graph. Line five sets the min and max y axis range the default being the min and max values found in the data. Line six shows if the display is to be logarithmic or linear in its y axis.

Page FSPDIS2 (Viewing options)

1	View the residuals (y/n)	Yes
2	View the curve fit (y/n)	Yes
3	View the peaks (y/n)	Yes
4	View the background fit (y/n)	Yes

These lines set up the screen that will be displayed the options can be switched off by selecting NO as the option.

CALIBRATION (XCALIBER) MODE

These menu pages allow calibrations of the x-axis of spectra, calibration of the peak files output from FSPFIT, and a determination of the detector relative efficiency versus channel number. It must be pointed out that due to the complex nature of any calibration, the majority of this section of ROBFIT is interactive in nature; only the most basic data are specified in these menu pages. Where necessary, the code asks relevant questions.

Page XCALIBER

1 Calibration constants files
2 Peak files to be calibrated

The first line provides the filenames for the various constants files that are used by other ROBFIT routines. The second line contains the peak (.PK) that are to be calibrated.

Page XCALIBER1 (Calibration files)

1	Calibration data file	Filename.CA
2	X-axis, cal constants (MAKE, Filename)	Make
3	Efficiency constants(MAKE, NONE, Filename)	None

Line one contains the name of the file for the user defined calibration. This file must have been created; no calibration can be performed without it. The file contains a list of peak positions as determined by FSPFIT, and user defined x-axis values corresponding to these positions. The format of this file is given in Section 2.4.2. Line two allows the user to create a file that can be read by FSPDIS to enable that routine to recalibrate the x-axis during display of the fit. The format of this filename is FILENAME.XDA. Specifying MAKE will create this file with the same filename as the .CA file. In addition, a file named "filename.FCA" will be created which allows the user to view the calibration. Line three contains the filename for the efficiency constants, NONE specifies no efficiency calibration. MAKE will create an efficiency constants file in a similar manner to the .XDA file. The format of this filename is FILENAME.EFF. If the user inputs pre-calibrated files on lines two or three then these will be used in preference to the calculated calibration using the .CA file.

Page XCALIBER2 (Peak files to calibrate)

1	First peak file to calibrate	Filename.PK
2	Second peak file	Filename.PK

In this sub-menu page the user can specify a number of peak files to calibrate using the calibration file defined on line one of menu page XCALIBER1. ROBFIT creates a calibrated peak file for each peak file specified, these calibrated files have the same filename but a different extension equal to .PCA. For example, specifying a peak file PFILE.PK would create a calibrated file PFILE.PCA.

2.4.2. Filenames and data input/output

The following describes the inputs and outputs that can be specified in each of the ROBFIT modes of operations.

2.4.2.1. Raw Data Display (RAWDD)

Input Data

RAWDD1

There are two input files to be specified in the raw data display mode. They are found on line one of the RAWDD1 sub-menu. Options for selecting either one or two files are discussed in Section 3.1 and will not be repeated here. Both files have the same format and must be specified as,

'Filename'.UF or 'Filename'.SP

where 'Filename' is chosen by the user. The UF qualifier to the filename indicates that the files are written unformatted. These files must have been created using the following format.

```
WRITE(8)NBDP,XOFF
WRITE(8)(DATA(J),J=1,NBDP)
```

Here the data are written to unit 8 which has been created as an unformatted file. NBDP is the number of data points and XOFF is an offset number of channels. As an example of the use of XOFF consider a case in which the first 10 channels were corrupted data. The XOFF would be set to 11 so the data fit now only considers the channels between 11 and NBDP. Using this file format means that the data is compacted and so takes up minimum disk space. However, it also means that the user cannot view the data with a text editor.

In order to give the user the option of viewing the raw data we have included a number of file formats that can be read by any text editor. We call these files collectively "speciality" files or SP files. Here we will describe only one of the formats in which the data is stored in an integer format. The other file formats are basically variations on this integer format and are used to compress the data. However, we believe the two file formats given here will be sufficient to cover most user needs. Anyone interested in the other formats is referred to the basic read subroutine BREAD which does all the raw data reading. The following shows the SP file format:

```
,...,...,...,...,...,...,...,...,...     (example of a header, anything can be written
,...,...,...,...,...,...,...,...,...           in place of the ,...)
,...,...,...,...,...,...,...,...,...
,...,...,...,...,...,...,...,...,...IDAT   ( NB. the IDAT must be in cols. 37 to 40)
     1     510                           ( NB. Beg. and ending channels [not usually used])
  2527   257   263   259   243   231   221   215   231   231
   215   255   195   277   301   285   259  3123 17829 22973
 26483  1249 55375 23217 20607 17337 14157 12167 11467 14991
 11289 11749 12389 12745 11887 11075  9715  7933  7117  6589
  6233  6115  6057  5489  5627  5523  5149  5175  4995  4653
  4563  4373  4347  4185  4231  4081  3949  3959  3749  3903
```

3829 3783 3825 3961 3951 3899 3547 3707 3691 3643
3565 3431 3551 3291 3341 3115 3109 2817 2947 2775
2859 2753 2675 2801 2635 2739 2693 2561 2477 2503
" ETC. TO END OF DATA "

The above SP File is read by the subroutine BREAD the top four lines can be used for file identification purposes. The only stipulation is that the last four characters on the fourth line contain IDAT, this selects the interger formatted read. This header information is written as four lines of 40 characters.

The next line contains the starting and ending channel values for the data. For example: the above file has 510 data points and the fit will begin at channel 1 and end at channel 510. The following lines in the file contain the data, these are written as 10I8 format.

Output Data

RAWDD 2

Here two files must be selected. One for the sum data, the other for the difference data. They are both specified as

'Filename'.SP

RAWDD 3

Here the files containing the corrected data must be selected. They are both specified as,

Filename'.SP

2.4.2.2 Standard Generation (STGEN)

Input Data

STGEN2

There is a single input to be defined in the standard generation code. If the user wishes to create a standard from the data then the file containing the raw data must be specified on line two of the STGEN2 sub-menu. The format of this file is identical to that of the RAWDD1 input files and the file is again specified in the sub-menu as

'Filename'.UF(or SP)

Output Data

STGEN1

Two files need to be specified here. The first is the file containing the standard information which is used in the spectral fitting mode. It is specified as

'Filename'.ST

in the sub-menu. We use the ST qualifier to label standard information. Standards files are structured as follows.

Spline number	Spline constants		
1	0.100E+01	0.496E-03	0.108E+01
2	0.663E-01	-0.852E+00	0.553E+00
3	0.665E-01	0.852E+00	0.549E+00

The second output of STGEN1 is the file containing the graphics information. This file is input into the STDIS mode to allow the generated standard to be viewed, the file is specified as,

'Filename'.GR

We specify graphics files by using a GR qualifier in the filetype.

2.4.2.3 Standard Display (STDIS)

Input Data

STDIS1

There is a single input into the standard display mode. It is specified in line one of the STDIS1 sub-menu. This file is the graphics output of STGEN1, and is specified as,

'Filename'.GR

2.4.2.4 Background Fit (BKGFIT)

Input Data

BKGFIT1

There are three input files to be specified on the BKGFIT1 sub-menu page. The first is the file containing the raw data which is input on line three of the sub-menu and is specified as,

'Filename'.UF(or SP).

It has a structure as outlined in the RAWDD inputs. The second input file contains background constants from a previous iteration of the code. In the first iteration no file can be input, however, subsequent cycles of the code use constants files that are specified as,

'Filename'.CN

where CN denotes a constants file. The structure of these files is shown in this section under BKGFIT2 outputs.

The third input file is mostly for subtracting data, it contains the error values for each of the channels within the spectrum. The file is specified as,

'Filename'.SP

Output Data

BKGFIT2

Here there are two files to be understood. Their structure is given in the following. The first file is specified as, 'Filename'.CN and contains the background constants for the fit. It has a structure as follows (where this file contains ten constants): The first line contains 0 for a linear or 1 for a logarithmic background fit.

1 (log [1] or linear [0] flag)
507.753
-0.961
0.608E-03
-0.127E-06
-0.122E-06
1575.717
0.611E-06
927.665
-0.602E-06
946.139

This file can be fed back into a further iteration of the background fitting if necessary, providing a starting point for the new fit. After fitting, the file can be renamed to any user specified filename. If, during fitting, the input and output constants files are the same, the code will read the old constants then overwrite the file with the new constants.

The second output file is the file that contains the graphics information for displaying the background fit. The file can be fed into the FSPDIS display pages as described in the main text. It is specified as,

'Filename'.GR.

2.4.2.5 Full Spectral Fit (FSPFIT)

Input Data

FSPFIT1

The file specified here contains the error values for each of the channels within the spectrum. The file is specifed as,

'Filename'.SP

FSPFIT2

Here there are three input files to be specified. The first is the raw data file which is input on line one of the FSPFIT2 sub-menu. This file is identical in structure to the RAWDD1 input files and is specified before as

'Filename'.UF (or SP)

The second file contains the peak data. These parameters are unknown on the first iteration of the code, so no file can be input, however, after the first cycle a peak file will be generated. Subsequent fitting can be speeded up by utilizing this information. Feeding this output file back into the fit provides FSPFIT with a starting point for the next iteration. Peak files are specified as,

'Filename'.PK

We use PK to indicate a peak file. The structure of these files is illustrated in the FSPFIT3 output files.

The third input file of FSPFIT2 contains the background constants. These data are identical in structure to that of the 'Filename'.CN files described in the BKGFIT input section. Again this information may be unknown on the first iteration of the code and so no file can be input. If, however, a BKGFIT or FSPFIT run has been performed, then a constants file will have been generated. This file can then be used as input into FSPFIT on the next fit. These files are again specified as

'Filename'.CN

FSPFIT4

Here the number of input files depends on the number of standards required to fit the data. Each file will have the same structure and is defined as

'Filename'.ST

These files are the standards that have been created with STGEN and their structure is outlined in the section on STGEN1 outputs.

Output Data

FSPFIT3

Here three output files need to be specified, namely,

Line 1 of sub-menu	'Filename'.CN	Background constants
Line 2 of sub-menu	'Filename'.PK	Peak parameters
Line 3 of sub-menu	'Filename'.GR	Graphics data

CN files have already been discussed in the background fitting section, the constants files output from FSPFIT are identical in structure to those in BKGFIT2 output.

The output peak files 'Filename'.PK contain information on the fitted peaks. Their structure is as follows,

```
      3 PEAKS 1000 CHAN, CHIS=923 . NITB=4
CUTOFF=  1.00
1.0560    1.0000
26.539    1.626    4.958    1.010    107.    73.    1
62.582    1.352    4.950    1.015    118.    67.    1
93 .161   1.448    4.569    1.012    90.     59.    1
```

etc. for every peak found.

The first three lines contain information on the fit. The following lines detail information on the individual peaks that were found. These numbers correspond to; peak position and error in position, width and error in width and strength and error in strength and a flag which identifies the standard the peak has been fitted with. Peak files can be edited by hand to change any of the peak parameters, for example, the position of a particular peak may be known before fitting. In this case the information on the position can be input into the fitting together with any information on width and strength estimates.

Finally, the file containing the graphics output must be specified. This file is for input into FSPDIS where viewing of the fit is performed.

2.4.2.6 Full Spectral Display (FSPDIS)

Input Data

FSPDIS1

There are two filenames that can be specified in the full spectral fitting display mode. The first is input on line one of the FSPDIS1 sub-menu and is the graphics output file of FSPFIT3. It is specified as

'Filename'.GR

The second is input on line four of the FSPDIS1 sub-menu and it is the calibration file output from XCALIBER. It is specified as,

'Filename'.XDA.

2.4.2.7 Calibration (XCALIBER)

Input data.

XCALIBER1

There are three files that can be specified on this sub-menu page. The first is the file which contains the user specified peak calibration data. ROBFIT uses this information to calibrate the peak files named on sub-menu page XCALIBER2 and to create the constants for the x-axis and efficiency calibrations. The file is specified as

'Filename'.CA

where CA designates a calibration file. These files have the following format,

```
Calibration for the Sm decay chain of Europium
those in coincidence with 344KeV Gd (2-0) should not be here
 CHANNEL  ERR  ENERGY   ERR   AREA     ERR   RI  ERR  COMMENT
 1027.65  .09 244.699  .001   6049.    101.  359   6  Sm_2-1 E2
 1243.28  .52 295.939  .008    669.     56.  21.1 .5  Sm_13-10 E1
 1750.06 1.43 416.052  .006    212.     47.   5.3 .1  Sm_16-10
 1864.03  .08 443.976  .005   6641.    103.  148   2  Sm_13-9
 2051.58  .35 488.661  .039    958.     54.  19.5 .2  Sm_13-8 m1+e2 +5.6
 2367.72  .42 564.021  .008   1297.     83.  22.4 .6  Sm_16-9 e1+(m2) +.07
 2378.25 1.35 566.421  .008    371.     74.   6.2 .1  Sm_13-6 m1+e2 -.74
 2756.01  .94 656.484  .012    405.     51.   6.9 .1  Sm_7-2 e2+m1+e0 e2/m1 +4.3
 2833.13  .93 674.678  .003    466.     60.   8.0 .3  Sm_8-2 e1+(m2) -.02
 2891.03  .21 688.678  .006   2306.     78.  40.0 .8  Sm_5-1 e0+e2+m1 e2/m1 +19
 3020.17  .50 719.353  .006    910.     65.  15.6 .3  Sm_9-2 e2
 3402.69  .40 810.459  .007   1137.     67.  15.2 .2  Sm_5-0 e2
 3532.16  .94 841.586  .008    471.     56.   7.8 .1  Sm_6-1 e1
 3641.20  .07 867.388  .008  10627.    131.  199   4  Sm_10-2 m1+e2 -6.5
 3858.95  .39 919.401  .008   1157.     67.  20.9 .5  Sm_8-1 e1+(m2) +.02
```

```
3888.89  .61  926.324  .015    699.     62.  12.7  .4  Sm_11-2
4047.00  .03  964.131  .009  37271.    216.  693    9  Sm_9-1
4219.54  .19 1005.279  .017   2227.     69.  31     3  Sm_12-1
4558.29  .07 1085.914  .003  25250.    409.  475    7  Sm_9-0 e2
4668.34  .05 1112.116  .017  33833.    288.  649    9  Sm_10-1 m1+e2 -8.7
5091.65  .14 1212.950  .012   3323.     73.  67    .8  Sm_15-2 e1+(m2)
5425.99 1.11 1292.784  .019    233.     35.  4.9   .3  Sm_11-0
5910.68  .03 1408.011  .014  47819.    229. 1000    3  Sm_13-1 e1+(m2) +.043
6118.58  .33 1457.628  .015   1120.     54.  23.6  .5  Sm_15-1 e1+(m2) 0
6412.13 1.57 1528.115  .019    381.    103.  12.7  .3  Sm_16-5 e1+(m2) -.01
7424.29 3.10 1769.093  .047     24.     10.  .42  .03  Sm_19-0 e2
```

Any text editor can be used to modify these files. The structure of the file is such that the first four lines are used for header information. The remaining lines contain the channel positions(plus error), user defined energies(plus error), areas(plus error) followed by the user estimates of the relative intensities(plus error) for the peaks in the calibration. Finally, a comment can be placed at the end of each line to help keep track of the peaks in the calibration. The user must construct each line in the above order, if efficiencies (relative intensities) are not being calibrated then zero's can be entered for these values, a space must separate each value on a line.

The second file in the XCALIBER1 sub-menu contains the constants for the x-axis calibration. This file is specified as

'Filename'.XDA

where XDA stands for x-axis data. This file can be read by FSPDIS in order to display a calibrated x-axis.

The third file contains the constants for the efficiency (strength) calibration and is specified as

'Filename'.EFF

A fuller description of both the x-axis and the efficiency calibration is given in Chapter 3.

XCALIBER2

Here the user must list all the peak files that are to be calibrated. Each file is specified as,

'Filename'.PK

Output data

There are two possible output files, the first is the x-axis calibration specified as

'Filename'.XDA

The second output file is the efficiency constants file specified as

'Filename'.EFF

The results of the above calibration are best viewed by looking at a summary file that is produced with each calibration; this file is named

'Filename'.FCA

and has the following structure

```
Calibration for the Sm decay chain of Europium
those in coincidence with 344KeV Gd (2-0) should not be here

 THE ENERGY CONSTANTS
  0.287469E+00  0.238169E+00  -0.164101E-07  0.576008E+04
     CHAN           ENERGY          ECAL         RES
 1027.65 +- .09  244.67 +- .03  244.70 +- 0.00   -.81   Sm_2-1 E2
 1243.28 +- .52  296.06 +- .13  295.94 +- .01    .99   Sm_13-10 E1
 1750.06 +- 1.43 416.83 +- .34  416.05 +- .01   2.29   Sm_16-10
 1864.03 +- .08  443.99 +- .02  443.98 +- .01    .68   Sm_13-9 e1+m2 +.025
 2051.58 +- .35  488.68 +- .08  488.66 +- .04    .27   Sm_13-8 m1+e2 +5.6
 2367.72 +- .42  564.02 +- .10  564.02 +- .01   -.05   Sm_16-9 e1+(m2) +.07
 2378.25 +- 1.35 566.52 +- .32  566.42 +- .01    .32   Sm_13-6 m1+e2 -.74
 2756.01 +- .94  656.53 +- .22  656.48 +- .01    .23   Sm_7-2 e2+m1+e0 e2/m1
 2833.13 +- .93  674.91 +- .22  674.68 +- 0.00  1.05   Sm_8-2 e1+(m2) -.02
 2891.03 +- .21  688.71 +- .05  688.68 +- .01    .54   Sm_5-1 e0+e2+m1 e2/m1
 3020.17 +- .50  719.47 +- .12  719.35 +- .01   1.02   Sm_9-2 e2
 3402.69 +- .40  810.61 +- .10  810.46 +- .01   1.58   Sm_5-0 e2
 3532.16 +- .94  841.46 +- .22  841.59 +- .01   -.58   Sm_6-1 e1
 3641.20 +- .07  867.43 +- .02  867.39 +- .01   2.34   Sm_10-2 m1+e2 -6.5
 3858.95 +- .39  919.31 +- .09  919.40 +- .01   -.98   Sm_8-1 e1+(m2) +.02
 3888.89 +- .61  926.44 +- .15  926.32 +- .02    .81   Sm_11-2
```

```
4047.00 +- .03 964.11 +- .01 964.13 +- .01 -1.90 Sm_9-1 m1+e2+e0 e2/m1
4219.54 +- .19 1005.21 +- .05 1005.28 +- .02 -1.47 Sm_12-1 m1+e2+e0 e2/m1
4558.29 +- .07 1085.91 +- .02 1085.91 +- 0.00 -.42 Sm_9-0 e2
4668.34 +- .05 1112.12 +- .01 1112.12 +- .02 .32 Sm_10-1 m1+e2 -8.7
5091.65 +- .14 1212.95 +- .03 1212.95 +- .01 .06 Sm_15-2 e1+(m2)
5425.99 +- 1.11 1292.59 +- .26 1292.78 +- .02 -.74 Sm_11-0
5910.68 +- .03 1408.03 +- .02 1408.01 +- .01 .82 Sm_13-1 e1+(m2) +.043
6118.58 +- .33 1457.54 +- .08 1457.63 +- .02 -1.06 Sm_15-1 e1+(m2) 0
6412.13 +- 1.57 1527.46 +- .37 1528.12 +- .02 -1.76 Sm_16-5 e1+(m2) -.01
7424.29 +- 3.10 1768.52 +- .74 1769.09 +- .05 -.77 Sm_19-0 e2
THE EFFICIENCY CONSTANTS
 0.425976E+01 -0.664569E-04 -0.944262E-05 0.123885E+04
-0.230265E-06 0.305449E+04
 CHAN ENERGY CSTRENGTH STRENGTH RES
1027.65 244.70 359.00 359.00 +- 12.48 0.00 Sm_2-1 E2
1243.28 295.94 21.10 21.85 +- 2.22 .34 Sm_13-10 E1
1750.06 416.05 5.30 4.98 +- 1.11 -.29 Sm_16-10
1864.03 443.98 148.00 147.15 +- 4.04 -.21 Sm_13-9 e1+m2 +.025 + mix of
2051.58 488.66 19.50 19.55 +- 1.15 .04 Sm_13-8 m1+e2 +5.6
2367.72 564.02 22.40 23.90 +- 1.57 .96 Sm_16-9 e1+(m2) +.07
2378.25 566.42 6.20 6.82 +- 1.36 .45 Sm_13-6 m1+e2 -.74
2756.01 656.48 6.90 7.01 +- .89 .13 Sm_7-2 e2+m1+e0 e2/m1 +4.3
2833.13 674.68 8.00 8.04 +- 1.04 .03 Sm_8-2 e1+(m2) -.02
2891.03 688.68 40.00 39.72 +- 1.50 -.19 Sm_5-1 e0+e2+m1 e2/m1 +19
3020.17 719.35 15.60 15.72 +- 1.15 .10 Sm_9-2 e2
3402.69 810.46 15.20 20.14 +- 1.22 4.04 Sm_5-0 e2
3532.16 841.59 7.80 8.41 +- 1.01 .61 Sm_6-1 e1
3641.20 867.39 199.00 191.21 +- 3.40 -2.29 Sm_10-2 m1+e2 -6.5
3858.95 919.40 20.90 21.12 +- 1.25 .18 Sm_8-1 e1+(m2) +.02
3888.89 926.32 12.70 12.79 +- 1.14 .08 Sm_11-2
4047.00 964.13 693.00 688.94 +- 8.23 -.49 Sm_9-1 m1+e2+e0 e2/m1 -8.0 +6
4219.54 1005.28 31.00 41.64 +- 1.35 7.88 Sm_12-1 m1+e2+e0 e2/m1 -3.2 *
4558.29 1085.91 475.00 482.87 +- 8.69 .91 Sm_9-0 e2
4668.34 1112.12 649.00 651.76 +- 7.35 .38 Sm_10-1 m1+e2 -8.7
5091.65 1212.95 67.00 65.84 +- 1.50 -.77 Sm_15-2 e1+(m2)
5425.99 1292.78 4.90 4.72 +- .71 -.25 Sm_11-0
5910.68 1408.01 1000.00 1000.47 +- 8.52 .05 Sm_13-1 e1+(m2) +.043
6118.58 1457.63 23.60 23.76 +- 1.16 .14 Sm_15-1 e1+(m2) 0
6412.13 1528.12 12.70 8.24 +- 2.23 -2.00 Sm_16-5 e1+(m2) -.01
7424.29 1769.09 .42 .56 +- .23 .58 Sm_19-0 e2
```

The first half of the file shows the results of the energy calibration. Each line contains information on a specific peak; first the channel position and error are given, then the calibrated energy and its error, followed by the input estimate of the energy and its error. The RES column contains the difference between the calibrated energy and the input estimate of this energy, divided by the standard deviation (which includes the input error and the curve fit error in its determination). Finally a comment can be specified at the end of the line. All this information is read in a free format which means that the user must leave a space between each of the quantities on any one line.

The second half of the file contains information on the efficiency (if performed). Again, each line contains specific peak data; first the channel position is given, this is followed by the user estimated energy and strength (intensity) values, then the calibrated strength and it's error are given. The RES quantity here is the difference between the calibrated efficiency and the input estimate of this efficiency divided by the standard deviation of this quantity (calculated in a similar manner to that mentioned above). Again the user can place a comment at the end of the line.

The RES quantity is used to indicate a bad fit to a peak in the list. If a peak is not fitting the calibration, then a large RES value will result. This may require the user to either remove the peak from the calibration or assign the peak a different energy.

XCALIBER2

For each peak file that is to be calibrated the code will produce a file with the same filename but an extension .PCA. These files contain the information on the calibration, the original peak file is not altered by the calibration. The format of the calibrated peak files is as follows,

```
 DATA FROM
PERPLPAR.PK
 CALIBRATION FROM
PERPLPAR.CA
NULL
NULL
 THE ENERGY CONSTANTS
  0.222787E+00  0.205719E+00 -0.119779E-05  0.463816E+04
  0.326843E-04  0.612943E+04 -0.318502E-04  0.615730E+04
 OUTPUT FROM ROBFIT
 IP= 2
 244 PEAKS 8001 CHAN, CHIS=  8293.    NITB=  28
 CUTOFF=  3.60
    CHANNEL          ENERGY          FWHM          STRENGTH
 110.73 +-  .00  17.96 +-  .42   3.39 +-  .00    18.93 +-   6.20
 160.90 +-  .00  28.41 +-  .40   3.40 +-  .00    31.12 +-   8.67
 177.39 +-  .00  31.84 +-  .40   3.41 +-  .00    45.53 +-   9.82
 205.94 +-  .00  37.79 +-  .39   3.42 +-  .00    35.80 +-   8.86
 225.87 +-  .00  41.94 +-  .39   3.42 +-  .00    49.96 +-  10.80
 247.49 +-  .00  46.44 +-  .38   3.44 +-  .00    70.18 +-  11.88
 273.18 +-  .00  51.79 +-  .38   3.45 +-  .00    94.60 +-  13.07
 302.40 +-  .00  57.87 +-  .37   3.49 +-  .00   151.12 +-  15.27
```

333.11 +- .00	64.26 +- .36	3.47 +- .00	101.58 +- 14.29
360.50 +- .00	69.96 +- .36	3.48 +- .00	113.81 +- 17.09
378.13 +- .00	73.63 +- .35	3.48 +- .00	102.74 +- 16.97
406.07 +- .00	79.44 +- .35	3.50 +- .00	132.30 +- 17.21
433.14 +- .00	85.08 +- .34	3.51 +- .00	173.29 +- 19.61
460.01 +- .00	90.67 +- .34	3.51 +- .00	136.68 +- 18.65
498.90 +- .00	98.76 +- .33	3.55 +- .00	280.37 +- 28.63
525.41 +- .00	104.27 +- .32	3.55 +- .00	604.32 +- 40.68
543.79 +- .00	108.10 +- .32	3.52 +- .00	1213.83 +- 64.60
558.57 +- .00	111.17 +- .32	3.53 +- .00	1801.77 +- 76.46
574.77 +- .00	114.54 +- .31	3.54 +- .00	1205.03 +- 70.62
592.33 +- .00	118.19 +- .31	3.57 +- .00	2074.78 +- 76.72
610.50 +- .00	121.97 +- .31	3.58 +- .00	1918.51 +- 74.97

etc to the end of the peak list.

These files show the channel number(plus error), the calibrated energy(plus error), the fullwidth at half maximum(plus error), and the calibrated strength(plus error) for each peak within the input peak list.

3. Procedures for running ROBFIT

Chapter 2 described the procedures required to load and initialize the ROBFIT routines. Here we show how to go about setting up the menus to perform a fit and cover the various operating modes of ROBFIT.

3.1 Raw data graphical display (RAWDD)

This section provides the procedure for viewing the raw input data, that is prior to any ROBFIT curve fitting or standard generation. The display is activated by running the RAWDD mode. In addition to being able to view the input data of a single spectrum, the user can opt to view two spectra simultaneously. The user can also opt to plot the difference between the two spectra. This option is useful for comparing data with a small signal in one of the spectra, an example being the detection of the supernova signal illustrated in reference 3. Another use of the raw data display is to select a region of the spectrum for use as a standard. If the spectrum contains peaks of unknown shape then a standard must be created from the data, for this purpose a relatively clean region of the spectrum containing a peak is required. The raw data display allows the user to scan the input data for such a region. Operation of the standard generation is outlined in Section 3.2.

There is a further option in the display which allows the user to reconstruct corrupted data. If for some reason the data has been affected by, for example, machine transfer problems the user can elect to smooth the data. Regions of the data can be removed by hand, the code interpolates across the selected region using user defined data.

The following gives an overview of the procedures that must be followed to set up the display menu for operation. All menu pages are described in full in Section 2.4.

Step 1. Setting up the general information (RAWDD1)

On entering the RAWDD mode menu, the user has the option of viewing three sub-menus. In the first instance the general information needs to be modified. On selecting the RAWDD1 sub-menu page by entering 1, the user is ready to proceed with the first step in setting up the display. Line one of the sub-menu requires file names to be entered, it is structured to accept two filenames, each of these files must contain data of the format outlined in Section 2.4.2.1. The second file is used only if the user wishes to compare two spectra, if there is only a single file to view then a "\" is entered for the second filename.

The next stage is to input the initial and final channels of the display region and the minimum and maximum peak heights, if known. If these values are not set then the display will select them from the data.

The final stage of setting up the display is to simply select the type of y-axis plot required, either logarithmic or linear, and set a title for the graph. Provided the user does not wish to correct data or plot a difference between two spectra the above is all that is required to initialize the display. The user would now go back up to the RAWDD mode master menu page and run the raw data display by entering 100.

Step 2. Setting up the difference information (RAWDD2)

This step must be followed if the user wishes to study the difference between two sets of input data. The difference plot is activated if YES is input into the first line of the RAWDD2 sub-menu page. If NO is entered then the following lines of the menu are ignored by the display driver.

Selecting the DIVIDE BY ERROR option will divide all individual channel residuals by the errors calculated from the sum of the two scaled data channels. This option is for viewing purposes only. It reduces the wide fluctuations of the data that can occur when taking the difference between two spectra.

Next the filename for the sum of the two raw data files is entered. The sum file is used in the full spectral fitting of the difference output, it contains the weights for each of the difference data channels. The file name for the difference data is then entered. This file contains the results of the subtraction of the two raw data files. The user then specifies the minimum and maximum peak heights, if known, and selects a logarithmic or linear fit.

Finally the user must specify the scaling parameter for each of the spectra. These values are used to normalize the two spectra. If there is no need to normalize then the default option selects 1 as the scaling for both files. Otherwise the code will scale the channel values of the second file by VALUE1/VALUE2.

Step 3. Setting up the correction information (RAWDD 3)

This third step is performed only if the data need correction. Again if NO is entered in line one of the RAWDD3 sub-menu then the following lines of the menu are ignored by the display driver.

The second line contains the range over which the correction is to take place. The third line contains the filename for the corrected data and the fourth line is the starting and ending points for the correction. ROBFIT will then linearly interpolate across the selected region. All corrected data must be stored in files which have a filename specified as 'Filename'.SP, these files can then be entered in the ROBFIT full spectral fitting routines.

3.2 Standard Generation (STGEN)

In its peak fitting phase, ROBFIT requires knowledge of the peak shapes it expects to find within the data. These prototype peaks, termed standards, can be of any shape. The standard generating mode allows the user to either create a standard from the raw data or to use one of the three most common shapes as provided with ROBFIT. Each standard is fitted to a polynomial plus a set of back-to-back cubic splines. The polynomial is assumed to represent the background under the standard while the splines represent the standard peak. Using this technique allows complex shaped standards to be generated.

Step 1. Setting up the general information (STGEN1)

On entering the STGEN mode menu pages the user must first select the general information page for initialization. The first stage is to select the filenames in which

i.) the standard coefficients are to be placed, and
ii.) the output in which the graphics data are to be placed.

The standards coefficients file will be read by the full spectral fit routines and used in the peak fitting stage while the graphics file is for input into the standard display mode STDIS.

The next stage is to select the standard type. Three special peak types are provided with ROBFIT, Gaussian, Lorentzian and Voigt (Ref. 6). This last shape is a Gaussian convoluted with a Lorentzian, where the ratio of the width of the Lorentzian to the Gaussian is supplied by the user, this option is explained in *Step 2*. If the data contain peaks of this shape, the user can generate the appropriate standard by selecting among GAUS, LORE or VOIG. The code will then proceed to generate the selected standard. If, however, the peak shapes within the real data are unknown, then a standard can be made from the raw data. FITD is specified to select a fit to the real data, this option is explained further in *Step 2*. All standards are normalized to have height equal to 1 and full width at half maximum equal to 1.

Finally the user must select the number of background coefficients (maximum 4) and splines (usually 3 to 5) to be used in the standard representation. Complex peaks and background will require more coefficients for a faithful reproduction of the shape. If the user is selecting a standard from the real data it may be that a number of iterations of fitting first a standard, then fitting the data with this standard will be required before a satisfactory standard can be found. Studying how the code fits a particular standard to the data shows how well STGEN has reproduced the true peak shape.

Step 2. Entering the fit type information (STGEN2)

This next step sets up some initialization parameters for the Voigt and data fit options of the standard generator. These only need to be set if the corresponding options have been selected in *step 1*, otherwise they are ignored by the standard generator. The first parameter to be set is the Voigt h (ETA), defined by $h = G_G/G_L$ where G_G is the width of a Gaussian of unit height that is convoluted with a Lorentzian also of unit height and width G_L.

The second stage is to set the raw data pointers. If the peak shapes within the raw data are unknown then a standard must be created from the data. In order to do this the user must select a relatively clean peak within the data. That is, a peak which is well separated from any other peaks or complicated background regions. This peak will then be fitted using a polynomial to represent the underlying background and a set of back-to-back cubic splines to represent the peak. Coefficients from this fit are then used in the full spectral fitting routine to determine where peaks of this shape reside. In order to fit to the real data the standard generator needs only to know the raw data filename and the channel range over which it has to fit the standard.

3.3 Displaying the standard (STDIS)

Once the standard has been created it can be viewed using the standard display option STDIS. With this display routine the user has the option of viewing the curve fit, the underlying background and the splines which make up the standard.

Step 1. Setting up the general information (STDIS1)

The first step is to enter the general information sub-menu STDIS1 and specify the name of the file containing the graphics output of the standard generator. The beginning and ending channel numbers and desired minimum-maximum y-scale range can then be entered. These parameters will default to the ranges within the data if they are not specified. Section 2.4. shows how to set up these default options. Finally the user must specify either a linear or logarithmic y-scale for the fit.

Step 2. Setting up the display options (STDIS2)

In this step the various plotting options are set. The user specifies either YES or

NO to select the viewing options.

3.4 Background fitting (BKGFIT)

Certain spectra are best analyzed by first determining an estimate of the underlying background function. By calculating the background shape the spectral fitting of the data will be enhanced, and in some cases a more accurate fit will result. This is especially so when complex backgrounds are involved where confusion between peaks and background can cause problems.

Step 1. Setting up the input parameters (BKGFIT1)

The first step in initializing the background fitter is to set the input parameters of sub-menu BKGFIT1. Here the user must first select the number of background-constants that are to be used in the fitting. This number will depend on the complexity of the background. The method recommended to determine the optimal number of constants is to perform a set of background-fits to the data, each run containing a few more constants. Once the optimal number has been reached the values of any additional constants will be small in comparison to the rest. By monitoring these numbers together with viewing the fitted background the user can determine the background shape to a level that will be satisfactory for input into the full spectral fitting code.

The next stage in setting up the input is to specify the CUTOFF parameter. This number is used to determine the degree of robustness in the fit (the height undetected peaks are expected to be above the background).

Now the filename of the raw data can be input together with the channel range over which the fit is to be performed. If there has already been some processing of the data and a file containing background-constants has been created, then these data can be used as a starting point for the new fit. All the user has to do is specify the filename containing the background-constants from the previous fit. If there is no background-constants file, as in the first iteration of fitting, NONE is input for the filename. In running the fit the data is usually best fitted on a logarithmic scale. This option being set by specifying YES on the line containing the background-constants filename or NONE. A NO will perform a linear fit.

The user next selects the weight to place on the data points. Selecting a 0 will take the error in a channel equal to $\sqrt{\text{data value}}$. If a 1 is selected then an averaging calculation is performed for the whole spectra and used to calculate the error at a given point. Selecting 2 allows the user to read the error values in from an external file. The filename containing the error values is then entered on the next line. This is only necessary if option 2 has been selected for the weights. The file format is given in Section 2.4.

Step 2. Checking the output parameters (BFGFIT2)

Here the output files in which the background-constants and fit information need to be defined. The first of the two files is the output file for the background constants and has a name FILENAME.CN. This file can be edited and/or renamed to any user specified name. If the file is to be used in further fitting of either the background or peaks, then this new filename must be input into these codes.

The second file named FILENAME.GR contains the graphical output required to display the standard when running FSPDIS.

3.5 Full Spectral Fitting (FSPFIT)

These routines make up the core of the fitting procedure, here a complete analysis of the data is performed. After cycling through this code the peak positions, widths and strengths are fully determined and the background shape will have been calculated. The code is able to start with or without information on the background and/or peak parameters. If, for example, the background has been determined as described in the previous section then this information can be entered into the full spectral fit at the beginning of the run. In doing so the code will not have to redetermine the background coefficients and the spectral fit will be speeded up. Alternatively the positions and/or widths of certain peaks may be known beforehand. In such a case, this information can also be linked into the program in a similar fashion, again resulting in operation speedup and possibly increasing the accuracy of the final peak parameters.

The code uses the standard peak shapes to search for additional peaks within the data. Each individual peak is determined by minimizing a weighted sum of differences between the data and the background with respect to the peak height, location and width along with up to nine neighboring peaks already present. After enough peaks have been found the weights and background are redetermined. Then the old peaks are refitted and new peaks are added and the cycle repeated until the largest 5 point sum is less than a user specified standard deviation. This iterative procedure is outlined in figure 3.

This section provides a brief outline of how to go about setting up a particular problem. Some of the steps in the following may be redundant if information on the data is known a priori. For example, peak widths may be known to a high accuracy or their variation with respect to channel number may be known. If this is the case then this information can be incorporated into the spectral fit at the outset.

Step 1. Setting up general fit information (FSPFIT1)

The first step is to define the maximum number of peaks that are to be found within the spectrum, on finding this number of peaks the code will stop.

The second stage is to define the range of CUTOFF values over which the fit will operate. Three values are specified on line two of the menu page. The first value being the starting CUTOFF, the second being the ending CUTOFF and the third being the number of steps to take between the two extremes. For example 2,1,2 will run the fit down to a CUTOFF of 2 in its first iteration, then 1.5 in its second and finally end at a CUTOFF of 1.

Next is the definition of how the routine is to handle the peak width variation and background fitting. Two parameters are defined on line three of the menu page; the first value specifies whether the fit is to proceed with variable width peaks, the variation being taken over the limits set on menu page FSPFIT4. This type of fit is initiated by specifying VWID. Alternatively a FIXD option can be specified where the widths are held constant at the values determined from menu page FSPFIT4. The second parameter sets up the background fitting operation. Four choices are available: CONT, NONK, NOBF and FIXK. CONT tells ROBFIT that on each background re-fit an additional knot is to be added to the background. NONK sets up a mode of operation where no new knots are to be added to the background. NOBF selects the no background fit mode of operation. The FIXK option allows the user to hold the knots stationary. This is useful for comparing data with the same underlying background structure. In this case, knot positions can be held constant while heights are allowed to vary, to account for differing normalizations. With this option comes the ability to fit certain backgrounds with fewer constants, thus increasing the speed of operation while maintaining the accuracy.

The next stage is to limit the number of background constants allowed for this particular fit. These values are entered on line four of the menu. For a complex background a large number of constants may be required, the code allows up to 100 constants to be used. The user selects the minimum and maximum number of constants. No peaks will be added until the background has been fitted with this minimum number of constants. Following this initial background fitting the code will cycle, succesively fitting peaks and adding background constants, until the maximum number has been reached.

The user then selects the weight to place on the data points. Selecting a 0 will take the error in a channel equal to $\sqrt{\text{data value}}$. If a 1 is selected then an averaging calculation is performed for the whole spectra and used to calculate the error at a given point. Selecting 2 reads the error values from an external file. The next line contains the filename of the file containing these error values, with the format of this file as shown in Section 2.4.

Step 2. Setting up the input data (FSPFIT2)

When starting off a particular fit the first thing to do is to set up the pointers to the information that will be input into the fit. On line one the user enters the name of the file containing the raw data. The format with which ROBFIT expects to find this data is given in Section 2.4.

Once the data file has been specified the user then has to select the channel range over which the fit will be performed. This range is entered on line two of the menu. The fit need not start at channel 1, any offset will automatically be taken care of.

The next stage is to specify whether there is a peak file that can be used in the fit. This would be the case after a number of cycles of the program where certain peaks have been accurately determined. The filename containing these values is entered on line three of the menu. Again the file must be specified as FILENAME.PK. Here PK specifies a peak input file. If there is no peak file then the user enters NONE for the filename. The format of the peak file is outlined in Section 2.4. Position, width and strength, and the corresponding errors of any peak can each be edited by hand before input into ROBFIT. During fitting ROBFIT will create a new peak file, this new file will contain the additional peaks found during the fit. It can be arranged so that the new peak file overwrites the input peak file, a procedure which will be outlined later in this section. A qualifier must be placed after the filename. If the strengths of the peaks are known accurately then YES is selected if not then NO. If the qualifier is NO then ROBFIT will first estimate the background using the .CN file, if any, and then calculate the strengths. If the qualifier is YES the epaks will be calculated as given before estimating the background.

The next stage in setting up the input files is to specify the file containing the background-constants. These have usually been calculated on a previous run. As above, if there is no input file then NONE is entered for the filename, otherwise the format is FILENAME.CN. If no file is specified then ROBFIT will recalculate the background. Depending on the complexity of the background this could add considerably to the execution time. In its operation ROBFIT will create a new background constants file. Again, it can be arranged so that the input constants file is overwritten as the program performs the fit. The qualifier that goes with this line specifies how the program is to fit the background. If a YES is selected then ROBFIT will perform a logarithmic fit to the background if NO then a linear fit will be performed.

Step 3. Setting up the output files (FSPFIT3)

The first stage in setting up the output files is to define the file in which the background-constants are to be placed. This file must have the name FILENAME.CN. It may be that the user wishes to overwrite the old input constants with constants from the new fit in which case the filename is set to the input filename. In doing this ROBFIT will initially read in the input constants and start the new fit with these values. Then on its output, after it has cycled through one complete fit, it will output the new constants to the old filename.

The next stage is to define the output peak file. After performing a fit the program writes out the peak positions, widths, strengths and their corresponding errors into the file that is specified on line two of the menu page. Again there is a file naming convention: the name must have the structure FILENAME.PK. As with the background constants, the program can be set up to iterate on a peak file and again the output file can be named as the input file.

Finally, the user must define the graphics output file. Again there is a fixed naming convention, the file must be named FILENAME.GR. The graphics file contains the information used in the display of the full spectral fit, it is this file which is specified in line one of menu page FSPDIS1.

Step 4. Setting up the standards (FSPFIT4)

With every fit there is a requirement that at least one standard be defined (without a standard ROBFIT can not function). Standards are generated using the standard generation mode of ROBFIT, see the STGEN menu pages. Again these standard files have a naming convention: all files are of the form FILENAME.ST. If the spectrum being fitted has within it a number of different peak shapes, each shape must have a standard defined. For example, in nuclear data, the photoelectric peaks are of one shape while the Compton background is of a different shape, hence each will require a standard to perform a correct fit.

On this menu page the user first defines the filename of the standard and then on the following two lines details the width variations allowed for this standard. The first line defines the starting values for this width variation the second line defines the ending values. See Section 2.4. for further details. Once this has been set up the standard definition is complete. If further standards are required the user can add in up to 4 additional standards .

3.6 Display of the curve fit/background fit (FSPDIS)

This mode of operation gives the user the ability to view a full spectral fit or a background fit. The structure of these menu pages is similar to that of the standard display option described above. Here the file containing the graphical output from FSPFIT or BKGFIT is input on line 1 of the FSPDIS1 sub-menu page. The user can also opt to calibrate the display in which case the XCALIBER file containing its calibration constants must be specified on line 2 of the FSPDIS1 sub-menu page. While line three contains the channel range for the display, line four contains the vertical scaling factor, line five contains the min and max y values and line six sets a logarithmic or linear y-axis.

FSPDIS also contains an option for helping the user determine the actual peak shapes in the spectrum. This part of the code is interactive in nature and proceeds as follows. When the fit is displayed, the user will frequently see a large peak above a well fitted background with other peaks too close to allow the large peak to be made a standard (see STGEN description). Do not give up yet! Entering W at the prompt will cause FSPDIS to ask for channel numbers of the peaks to remove (-1000 is entered if none are to be removed). Then FSPDIS will ask for a filename and the channels to output. The filename extension must be .WDA. The file will contain channels, intensities and weights of the data minus the remaining peaks and bakground that were displayed. STGEN will accept the .WDA file as data and will generate a standard fitted to this data.

3.7 Calibrate the Spectrum (XCALIBER)

By far the largest part of this mode of ROBFIT is interactive in nature. What is described here is the basic initialization of a calibration sequence. Calibration is performed at the end of peak fitting when the user wishes to convert a set of peak positions versus channel number to something that is more meaningful. In nuclear physics experiments this is usually a conversion to peak positions versus energy. In addition, the efficiency, number of counts under a peak, may follow a certain relationship, in which case, it to can be calibrated. All this helps the user to determine how the peaks in the fitted spectrum relate to the real world.

Setting up a calibration requires the user to create a calibration file FILENAME.CA, whose structure has been defined in Section 2.4, and then enter the filename on the first line of the XCALIBER1 sub-menu page. If no previous calibration has been run then MAKE is entered into line two of the XCALIBER1 sub-menu. The routines will then create x-axis calibration constants from the .CA file and place them into a file with the same filename but with an extension .XDA. If a previous calibration has taken place and the user wishes to use this set of constants then all that is required is to enter the name of the file containing these constants into line two; in this case the code will not create a new set of constants and will use the old constants in the calibration. In a similar manner an efficiency calibration is performed by placing the name for the efficiency constants on line three of the XCALIBER1 sub-menu. These files are specified as FILENAME.EFF. If no efficiency calibration is required then NONE is specified. Efficiencies are calibrated according to the user-defined

.CA file. The code also creates a summary file FILENAME.FCA where the user can view the results of the calibration.

Once the calibration constant filename and modes have been decided upon the next step is to simply list on the XCALIBER2 sub-menu all the peak (.PK) files that are to be calibrated ROBFIT creates a calibrated file for each peak file with the same filename as the peak file but with the extension .PCA.

4 Test cases, demonstration of the algorithm

This section describes a number of tests that have been performed on the spectral fitting routines. It is hoped that these tests will serve as a guide for the user in his or her analysis. The tests show the accuracy that can be achieved under certain circumstances and detail an analysis methodology that can be used on real data.

The four major areas that the testing was designed to probe are,

i). Reproducibility of a single generated peak.
ii). Peak detection level with respect to noise.
iii). Detection of peak separability.
iv). Complex background problems.

For each of these we have provided test cases which the user can run. Appendix I contains the details required to set up and run these test cases.

4.1 Single peak detection efficiency

In order to make these tests more realistic, generated peaks have been superimposed upon an exponentially decaying background. The first test was to generate a large peak in two separate regions of this background to show that under ideal circumstances the routines do accurately reproduce the underlying structure. The x-axis channel range runs from 1 up to 1000 channels and peaks are generated at two positions along the axis, one at channel 150 and the second at channel 850. The generated peaks have a width of 46.9 channels and a strength (total number of counts under the peak) of 10000 (which corresponds to a peak height of 200 counts). Both have Gaussian profiles. Results of ten separate runs for each position are shown in Table 1. As can be seen, at this level ROBFIT does indeed accurately reproduce the generated peak. A test case of the above single peak fitting is detailed in Appendix I.

The next stage of testing involved the reduction of the strength of the above peaks just discussed to monitor ROBFIT's ability to identify small peaks. An important factor in the ability to find a peak is the size of the peak with respect to the background fluctuations. The background has been generated as an exponential function with a corresponding error at a given channel equal to $\sqrt{\text{number of counts in channel}}$.

At the 150 position, the average level of the background is approx. 240 $\pm$ 15 counts while at the 850 position it is approx. 7 $\pm$ 3 counts. These represent two extreme regions of background and thus provide a realistic test for the routine. At each position a peak was generated with various strengths until the routine could no longer unambiguously detect a single peak. In general, during fitting, peaks are added until the ratio of their strength to the standard deviation in the counts under the peak is less than a CUTOFF value set by the user. In this test the CUTOFF was set at 4, which means that a chance positive fluctuation would be considered a peak only if it were larger than four standard deviations. This CUTOFF value is difficult to determine in real data. However, a technique of first determining large peaks within a spectrum with a large CUTOFF then working down to smaller peaks with successively smaller CUTOFF values works well in practice. Having set the CUTOFF level to 4, the peak strength variation was carried out with this value held constant.

Peak Position	Width (channels)	Strength (counts)
150.14 ± 0.62	46.40 ± 2.57	9357 ± 1173
149.76 ± 0.46	51.00 ± 1.13	11057 ± 226
149.06 ± 0.89	46.06 ± 3.01	9450 ± 1448
151.18 ± 0.61	43.04 ± 2.10	7439 ± 653
151.71 ± 0.60	44.76 ± 2.09	8258 ± 712
150.34 ± 0.70	47.07 ± 2.70	9500 ±1256
150.69 ± 0.59	47.34 ± 2.44	9904 ± 1136
150.61 ± 0.55	48.54 ± 1.79	9920 ± 632
152.81 ± 0.65	48.27 ± 1.83	8376 ± 442
150.61 ± 0.74	49.10 ± 2.62	10211 ± 1263
850.00 ± 0.22	46.67 ± 0.42	10018 ± 108
849.63 ± 0.22	46.96 ± 0.43	9997 ± 108
849.74 ± 0.22	47.15 ± 0.43	9947 ± 108
850.03 ± 0.22	46.66 ± 0.42	9926 ± 107
850.06 ± 0.22	47.30 ± 0.43	9825 ± 107
850.22 ± 0.22	47.48 ± 0.43	10067 ± 108
849.99 ± 0.22	47.65 ± 0.43	10062 ± 108
850.21 ± 0.22	46.50 ± 0.43	9986 ± 108
849.92 ± 0.22	46.83 ± 0.43	9974 ± 108
849.80 ± 0.22	47.05 ± 0.43	9903 ± 107

Table 1. Results of ten runs at two extreme regions of background. Peaks were generated at 150 (upper half of table) and 850 channels (lower half) with a width of 46.9 and a strength of 10000.

Peak Position (channels)	Width (channels)	Strength (counts)
149.11 ± 3.49	39.96 ± 9.67	905 ± 259
148.35 ± 3.48	47.66 ± 9.19	1168 ± 255
146.30 ± 2.44	38.58 ± 6.34	1199 ± 220
153.99 ± 3.22	30.00 ± 8.36	606 ± 186
No peak found	———	———
152.65 ± 2.92	42.50 ± 7.75	1155 ± 240

Table 2. Results of the six generated peaks with position = 150, width = 46.9 channels, and strength = 1150. In each run the peak is generated on a random, expoentially decaying background.

At the 150 channel position, peaks down to a strength of 1150 are detected, which corresponds to a peak height of 23 ± 5 counts. This is approx. 1.5 times the size of the background noise level. At this level roughly 80% of peaks could be found; the results of six separate runs are detailed in Table 2. As can be seen the code was able to reproduce the peak structure fairly well even at this level. At the 850 position, peaks could be distinguished unambiguously down to a strength of 300 or a peak height of 6 ± 2 counts again the order of 2 times the noise level. Test cases for the preceeding minimum detectable single peak fitting are detailed in Appendix I.

In order to study the effect of peak width on detectability a single peak of width 6 channels was generated at the 850 position. The fitting code was able to unambiguously select out a single peak when the strength was set to 70 which corresponds to a height of 11 ± 3 or approx. 4 times the noise level. The results of these runs are shown in Table 3. Depending on the peak width and the background level the routine can determine unambiguously peak positions and widths for peaks of height between 1.5 and 4 times the noise level. A test case is described in Appendix I.

Peak Position (channels)	Width (channels)	Strength (counts)
849.51 ± 0.58	7.17 ± 1.39	89 ± 18
849.72 ± 0.63	5.83 ± 1.55	61 ± 17
849.11 ± 0.64	5.09 ± 1.57	58 ± 19
851.16 ± 1.26	7.52 ± 3.17	40 ± 18
850.41 ± 0.59	5.47 ± 1.43	58 ± 16

Table 3. Results of five runs with a peak at position = 850, width = 6 channels, and strength = 70 counts.

The limits quoted above are conservative in that they require a very low level of spurious peak detection. If, however, we allow a certain number of spurious peaks to appear then we can run the CUTOFF value down to a lower level, thus enabling the fitting routine to pick up smaller peaks. With real data this is not uncommon in that the region of interest usually has a narrow channel range and the chance of spurious peak interfering is low. This, of course, depends entirely on the type of data being analyzed. For the present analysis the CUTOFF value could be taken down to a value of 1. At this level a spurious peak was found generally in every 120 channels, a frequency which we deemed acceptable. The strength of the peak was then reduced resulting in a new detection point for the peaks of width 6 channels. Now, peaks down to a strength of 25 or peak height 4 ± 2 channels, roughly 1.3 times the noise level, three times smaller than the previous limit, could be detected. These results can be seen in Table 4. Again the code faithfully reproduces the generated peak shape even at this low level. A test case of this narrow-width, single-peak fitting is detailed in Appendix I.

Peak Position (channels)	Width (channels)	Strength (counts)
848.29 ± 1.44	9.81 ± 3.42	49 ± 19
848.46 ± 2.30	7.14 ± 5.83	20 ± 18
848.27 ± 1.91	4.93 ± 1.17	16 ± 13
No peak found	----------	------
851.32 ± 1.18	3.64 ± 3.01	14 ± 12

Table 4. Results of five runs with a peak at position = 850, width = 6 channels, and strength = 25 counts.

Another important input to the spectral fit is the number of background constants one allows in the fit. The results above were taken with four background constants. As more constants are added the peak detectability will be degraded somewhat as the background tries to follow the data. Again this choice is highly dependent on the data being analyzed. If the background is complex then more constants will be required, this usually results in a compromise with peak detectability.

4.2 Multiple peak detection efficiency

The second topic addressed in our testing was the level at which the code could detect multiple peaks. Here we concentrated on generating peaks separated by a known channel spacing. From the preceeding results we generated two peaks close to the 850 channel position, each of strength 70. At this strength the program would have no problems in distinguishing the peaks independently at a CUTOFF of 4. The separation of the peaks was then varied until the program could just discern the two peaks. The peak widths were again set to 6 channels. The results of the limiting cases are shown in Table 5. Again the program correctly finds the peak shapes. A separation of 15 channels, roughly 2.5 times the peak full width at half maximum (FWHM), was required before the two peaks could be unambiguously determined. A test case of the above double peak fitting is detailed in Appendix I.

Peak Position (channels)	Width (channels)	Strength (counts)
849.64 ± 0.51	4.65 ± 1.23	43 ± 17
865.18 ± 0.41	6.25 ± 1.14	84 ± 17
849.60 ± 0.60	5.40 ± 1.10	58 ± 15
No peak found	-----	----
849.13 ± 0.65	5.11 ± 1.24	56 ± 17
864.86 ± 0.43	4.03 ± 1.06	54 ± 15
851.66 ± 0.89	5.49 ± 1.03	35 ± 12
863.93 ± 0.42	3.65 ± 0.98	45 ± 12
850.43 ± 0.57	5.27 ± 1.05	55 ± 14
865.37 ± 0.71	5.80 ± 1.04	49 ± 14

Table 5. Results of five separate runs to study the detectability of two peaks positioned at 850 and 865 (channel number) both with widths = 6 channels and strengths = 70 counts.

Peak Postition (channels)	Width (channels)	Strength (counts)
850.46 ± 0.61	7.00 ± 1.27	102 ± 19
857.98 ± 0.46	3.97 ± 1.03	57 ± 15
850.45 ± 0.73	6.02 ± 1.24	77 ± 18
857.17 ± 0.55	3.67 ± 1.26	44 ± 16
849.13 ± 0.74	5.32 ± 1.21	58 ± 17
856.38 ± 0.86	5.47 ± 1.18	52 ± 16
847.76 ± 2.22	5.10 ± 1.05	16 ± 12
855.53 ± 0.63	6.66 ± 1.28	90 ± 19
851.25 ± 0.81	5.00 ± 1.24	70 ± 19
857.79 ± 0.96	5.00 ± 1.24	46 ± 17

Table 6. Results of five separate runs to study the detectability of two peaks positioned at 850 and 857 (channel number) both with widths = 6 channels and strengths = 70 counts.

As with the previous section this level of separation is a conservative one and smaller separations can be picked up by running the CUTOFF level to a smaller value. With the CUTOFF set at 1.5 the routine could distinguish the above peaks when separated by 7 channels or approx. 1.2 times the FWHM. Here however, the user has to contend with the spurious peaks that are picked up at this low CUTOFF value. Table 6 shows the results of this analysis. A test case of the above minimum detectable separation for double peak fitting is detailed in Appendix I.

As in the previous section the strength of the peaks can be reduced to see how this affects the separability detection. At the CUTOFF of 1.5 level, peaks of strength 25 or peak height 4 can be detected provided they are separated by 2.5 times the width of the peaks. Thus, there is again a compromise between the separation detection and the strength of the peak with respect to the noise. A test case of this minimum detectable double-peak fitting is detailed in Appendix I. Additional testing has been carried out with three and four peaks separated by equal amounts, no deviations from the above results have been found in these cases.

4.3 Complex background fitting

The final test case considered is one which involves dealing with a complicated background. The function being fitted was an yttrium spectrum, which was chosen for its complexity. In order to fit this spectrum a large number of background constants will be needed. If the full spectral curve fitting code, FSPFIT, is run on data containing such a background, confusion between background and peaks will occur resulting with spurious peaks entering into the results. Taking out as much of the background as possible by first fitting it alone will therefore enhance the peak fitting. The level to which this compromise between background and peaks can be taken is dependent on the spectrum under analysis. The user must determine this level by deciding what information is needed from the curve fitting. Using a large number of constants to fit the background will degrade the detection of small peaks within the spectrum while using only a few background constants may leave some background contribution in the peak fitting. In this example we show how a large number of background coefficients can be fitted to the yttrium spectrum.

First the background fitting mode, BKGFIT, is entered and all relevant input parameters are set as outlined in Chapter 3. The fit is started with the number of background constants set to 10, once the background fitter has added all 10 constants it will exit the fitting. The fit will be performed with a CUTOFF value of 20, a low value for this data, this will reduce the peak contributions to the background calculation. The code adds in the constants two at a time, after new constants have been added and the fit re-adjusted to take into account these new constants the code cycles and add a further set of constants. This process is repeated until all ten constants have been added, whereupon the fitter will exit and write the constants to the FILENAME.CN file. The magnitudes of all 10 constants are then noted by editing the FILENAME.CN file. The next stage is to re-run the background fitter and add in a further 10 constants by raising the number of fitted constants to 20. Now the old background constants can be read in at the outset of fitting. This speeds up the fitting process as the code does not have to re-calculate the first ten constants. Once the 20 constants have been added and fitting stops, the constants file is again examined and the constants noted. For general purposes the process of adding more constants is then repeated until the following criteria are reached.

At the outset the constants will be relatively large. However, as more and more are added, the magnitude of the higher constants will slowly drop off. A sign that the optimum number of constants has been reached is a rapid drop in size of the added constant. A second indication is oscillation of the size of the constants from one to the next. At this point adding more constants will not improve the fitting. Here we detail only the background fitting up to the point at which 98 constants have been added.

A test case of the above complex background fitting is detailed in Appendix I. As can be seen by viewing the final fit, more constants would be needed to determine the shape fully. Usually such a spectrum would be fitted with a combination of peak fitting and background fitting. As the object of this test case was to illustrate the power of the background fitting under extreme circumstances, peak fitting was omitted.

5. Conclusions

ROBFIT is a powerful spectral analysis package which can be operated from any personal computer system. All code control and graphical viewing of results can be performed in a simple manner in the PC environment. Because of its development on PC systems the ROBFIT code is intrinsically efficient in operation. The code turns a personal computer (PC XT, AT, PS/2 or compatible) into a powerful spectral analysis workstation with the minimum of effort.

References

1. R. L. Coldwell, "An iterative peak finding code," in Radiative properties of Hot Dense Matter, 315-349, Sarasota, FL (World Scientific, 1983) Davis, Hooper et al editors.

2. R. L. Coldwell, "Iterative codes for fitting complete spectra," Nucl. Inst. and Methods in Physics Research A242 (1986) 455-461

3. A. C. Rester, R. L. Coldwell, F. E. Dunnam, G. Eichhorn, J. I. Trombka, R. Starr and G. P. Lasche, "Gamma-Ray Observations on Supernova 1987A from Antarctica," Astrophysical Journal Letters (submitted), Table 1.

4. G.J. Bamford R.L. Coldwell, and A.C. Rester, "ROBFIT: A General Purpose Spectral Analysis Package", A.I.P. Conference Proceedings, Astrophysics in Antarctica 198 (1989).

5. W. H. Press, B. P. Flannery, S. A. Teukolsy, W. T. Vetterling, Numerical Recipes, the Art of Scientific Computing, (Cambridge Univ. Press, 1986) pp 539-46

6. for example, see Dimitri Mihalas Stellar Atmospheres (Freeman, 1978) p. 279

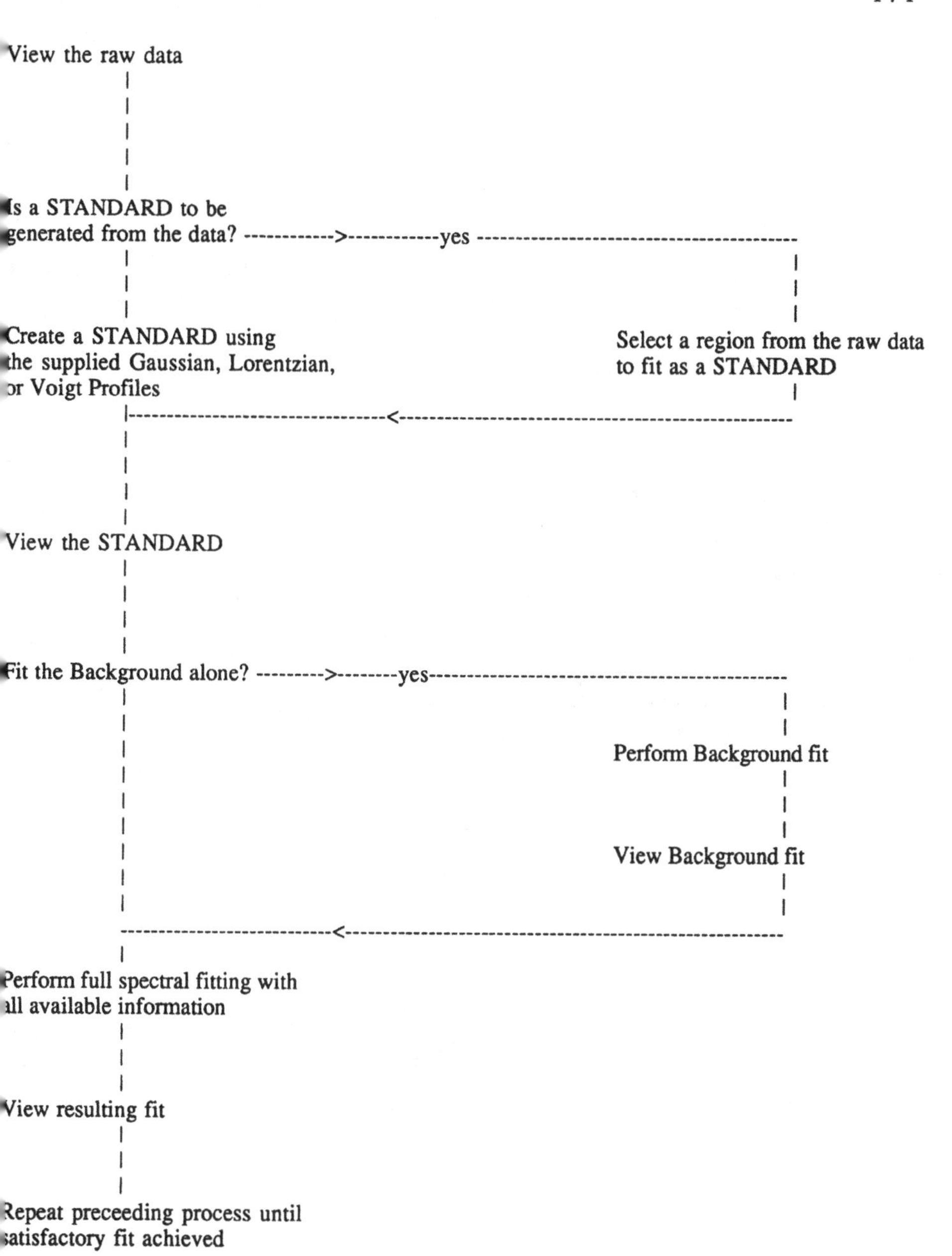
View the raw data
Is a STANDARD to be generated from the data?
yes
Create a STANDARD using the supplied Gaussian, Lorentzian, or Voigt Profiles
Select a region from the raw data to fit as a STANDARD
View the STANDARD
Fit the Background alone?
yes
Perform Background fit
View Background fit
Perform full spectral fitting with all available information
View resulting fit
Repeat preceeding process until satisfactory fit achieved

Figure 1. Overview of ROBFIT

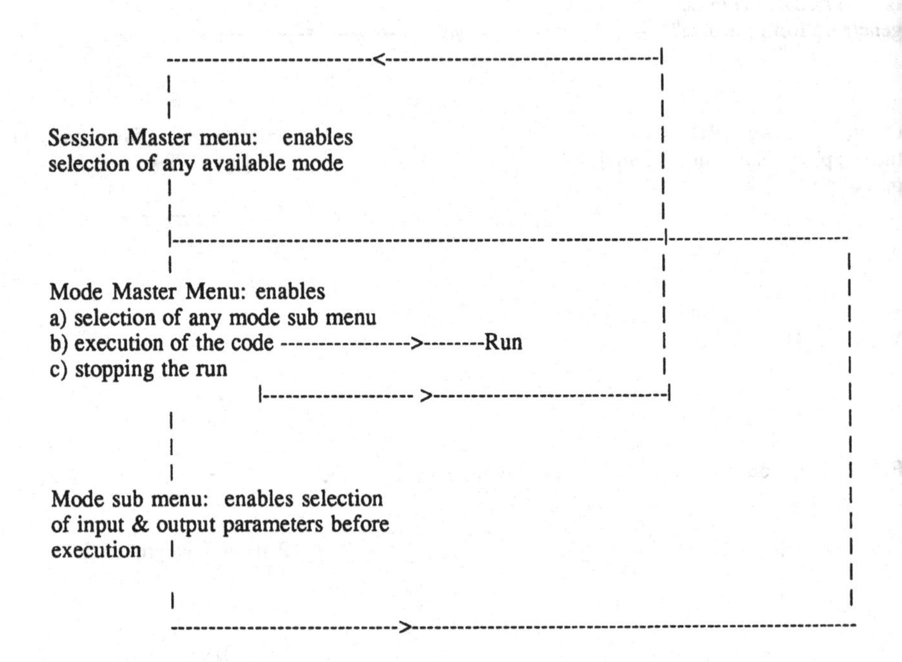

Figure 2. Overview of the ROBFIT menu driver

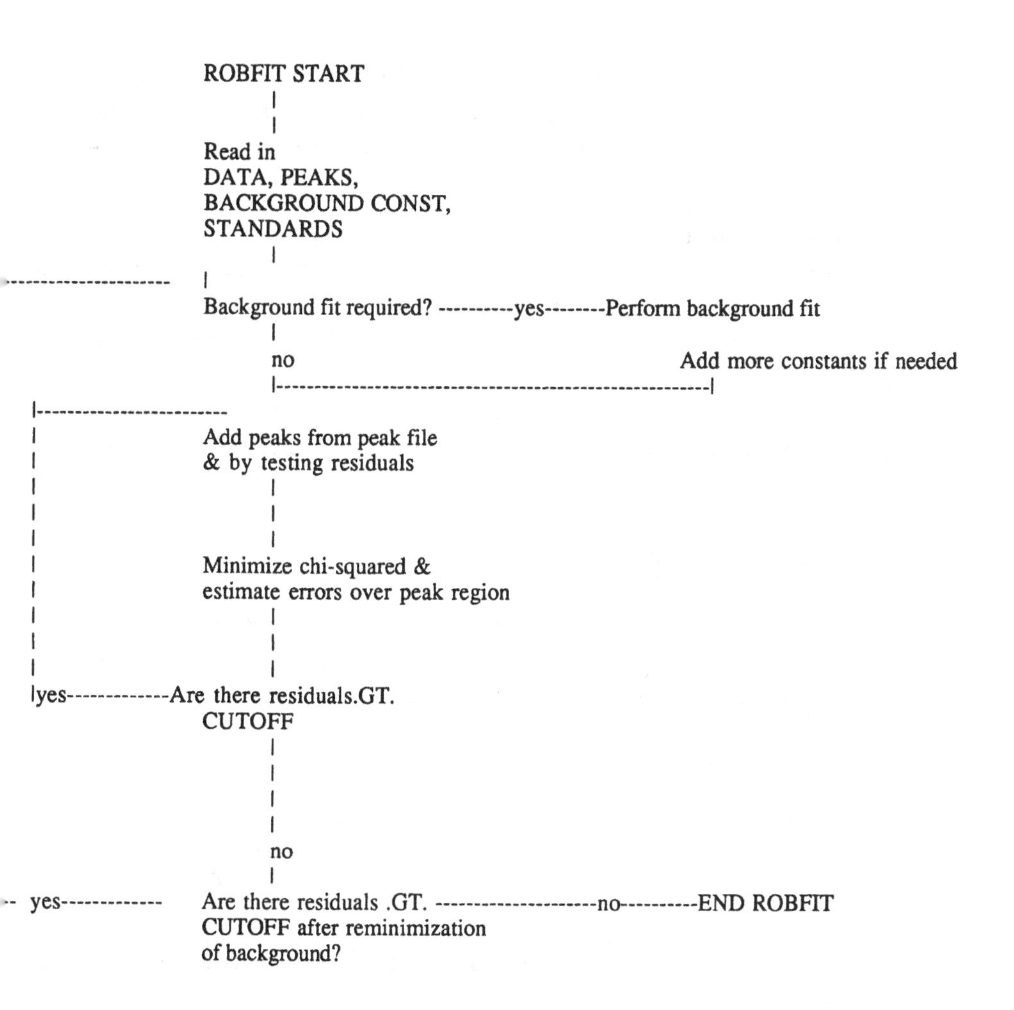

Figure 3. Overview of ROBFIT minimization procedure

APPENDIX I. User test cases

The test cases which follow have been selected to familiarize the user with the operation of ROBFIT. Each case has been selected from demonstration runs used to monitor the accuracy of the ROBFIT algorithms. All cases have been discussed in section 4 of this guide.

D1 Single peak fitting

Here a single gaussian peak has been generated and placed on an exponentially decaying background, as described in section 4.1.

The raw data reside in the ZTCASE1.SP file and can be viewed using the RAWDD mode.

Before running the fitting mode a standard must be generated using STGEN. We have supplied the gaussian standard used in the section 4 tests, ZTCASE1.ST, however, the user may wish to run the standard generator using the following setup;

Page STGEN1 (Generation data)

1	Name of the output standards file	TEST1.ST
2	Graphical output to be placed in file	TEST1.GR
3	Fit type (Gaus,Lore,Voig or Fitd)	Gaus
4	Number of background coeffs and splines	0,3

Page STGEN2 (Type options)

1	Voigt fit ETA parameter	1
2	Data fit file name	None
3	and beginning/ending channel nos	1,default

Once the standard has been created FSPFIT can be run, the FSPFIT mode menu pages must be set as follows;

Page FSPFIT1 (General fit information)

1	Maximum number of peaks.	100
2	Starting,ending cutoffs, in N steps	5,4,1
3	VWID,FIXD and CONT,NONK,NOBF,FIXK	VWID,CONT
4	Maximum number of background constants	4,4
5	Weights 0=SQRT(data),1=calc,2=file	0

6	File containing the weights	NONE

Page FSPFIT2 (Input data)

1	Input data file name.	ZTCASE1.SP
2	Beginning and ending channels for fit.	1,1000
3	Input peak file.	NONE, NO
4	Input background constants file.	NONE, YES

Page FSPFIT3 (Output data)

1	Output file for background constants	TCASE1.CN
2	Output file for Peak data	TCASE1.PK
3	Output file for graph data	TCASE1.GR

Page FSPFIT4 (Standards data)

1	First standard	ZTCASE1.ST
2	Type 1 width, error and channel no.	50,10,10
3	Type 1 width, error and channel no.	50,10,900

On output the TCASE1.PK, TCASE1.CN and TCASE1.GR files will contain all the fit data. TCASE1.PK should be identical to the ZTCASE1.PK file similarly TCASE1.CN will be identical to ZTCASE1.CN where the ZTCASE files are the outputs of our original runs.

The results of the curve fitting can be viewed by running FSPDIS with the following menu page setup;

Page FSPDIS1 (General information)

1	Name of the graphical file to be read	TCASE1.GR
2	Name of the calibration file (.XDA)	NONE
3	Beginning,ending channels for display	1,1000
4	Vertical axis = COUNTS*SCALE FACTOR	1
5	Desired min and max y scale	1,default
6	Log or linear y scale	LOG

Page FSPDIS2 (Viewing options)

1	View the residuals (y/n)	Yes
2	View the curve fit (y/n)	Yes
3	View the peaks (y/n)	Yes
4	View the background fit (y/n)	Yes

D2 Minimum detectable peak

i) Unambiguous peak determination in high noise region.

Here the ability of the fitting to find unambiguously, a small peak within the exponential background has been tested. The peak has been generated in the high noise region of the data as detailed in section 4.1.

Page FSPFIT1 (General fit information)

1	Maximum number of peaks.	100
2	Starting,ending cutoffs, in N steps	5,4,1
3	VWID,FIXD and CONT,NONK,NOBF,FIXK	VWID,CONT
4	Maximum number of background constants	4,4
5	Weights 0=SQRT(data),1=calc,2=file	0
6	File containing the weights	NONE

Page FSPFIT2 (Input data)

1	Input data file name.	ZTCASE2.SP
2	Beginning and ending channels for fit.	1,1000
3	Input peak file.	NONE, NO
4	Input background constants file.	NONE, YES

Page FSPFIT3 (Output data)

1	Output file for background constants	TCASE2.CN
2	Output file for Peak data	TCASE2.PK
3	Output file for graph data	TCASE2.GR

Page FSPFIT4 (Standards data)

1	First standard	ZTCASE1.ST
2	Type 1 width,error and channel no.	50,10,10
3	Type 1 width,error and channel no.	50,10,900

On output the TCASE2.PK, TCASE2.CN and TCASE2.GR files will contain all the fit data. TCASE2.PK should be identical to the ZTCASE2.PK file similarly TCASE2.CN will be identical to ZTCASE2.CN where the ZTCASE files are the outputs of our original runs.

The results of the curve fitting can be viewed by running FSPDIS with the following menu page setup;

Page FSPDIS1 (General information)

1	Name of the graphical file to be read	TCASE2.GR
2	Name of the calibration file (.XDA)	NONE
3	Beginning,ending channels for display	1,1000
4	Vertical axis = COUNTS*SCALE FACTOR	1
5	Desired min and max y scale	1,default
6	Log or linear y scale	LOG

Page FSPDIS2 (Viewing options)

1	View the residuals (y/n)	Yes
2	View the curve fit (y/n)	Yes
3	View the peaks (y/n)	Yes
4	View the background fit (y/n)	Yes

ii) Unambiguous peak determination in low noise region.

A similar test to i). above has been performed in the low noise region of the data, again the details are outlined in section 4.1. Fitting and display setups are;

Page FSPFIT1 (General fit information)

1	Maximum number of peaks.	100
2	Starting,ending cutoffs, in N steps	5,4,1
3	VWID, FIXD and CONT, NONK, NOBF, FIXK	VWID, CONT
4	Maximum number of background constants	4,4
5	Weights 0=SQRT(data),1=calc,2=file	0
6	File containing the weights	NONE

Page FSPFIT2 (Input data)

1	Input data file name.	ZTCASE2B.SP
2	Beginning and ending channels for fit.	1,1000
3	Input peak file.	NONE, NO
4	Input background constants file.	NONE, YES

Page FSPFIT3 (Output data)

1	Output file for background constants.	TCASE2B.CN
2	Output file for Peak data.	TCASE2B.PK
3	Output file for graph data	TCASE2B.GR

Page FSPFIT4 (Standards data)

1	First standard	ZTCASE1.ST
2	Type 1 width,error and channel no.	50,10,10
3	Type 1 width,error and channel no.	50,10,900

On output the TCASE2B.PK, TCASE2B.CN AND TCASE2B.GR files will contain all the fit data. TCASE2B.PK should be identical to the ZTCASE2B.PK file. Similarly TCASE2B.CN will be identical to ZTCASE2B.CN where the ZTCASE files are the outputs of our original runs.

The results of the curve fitting can be viewed by running FSPDIS with the following menu page setup.

Page FSPDIS1 (General information)

1	Name of the graphical file to be read	TCASE2B.GR
2	Name of the calibration file (.XDA)	NONE
3	Beginning,ending channels for display	1,1000
4	Vertical axis = COUNTS*SCALE FACTOR	1
5	Desired min and max y scale	1,default
6	Log or linear y scale	LOG

Page FSPDIS2 (Viewing options)

1	View the residuals (y/n)	Yes
2	View the curve fit (y/n)	Yes
3	View the peaks (y/n)	Yes
4	View the background fit (y/n)	Yes

iii) Effect of peak width on detectability.

Here the width of the generated peak was made narrow to study its effect on detectability. Fitting and display setups are,

Page FSPFIT1 (General fit information)

1	Maximum number of peaks.	100
2	Starting,ending cutoffs, in N steps	5,4,1
3	VWID, FIXD and CONT, NONK, NOBF, FIXK	VWID, CONT
4	Maximum number of background constants	4,4
5	Weights 0=SQRT(data),1=calc,2=file	0
6	File containing the weights	NONE

Page FSPFIT2 (Input data)

1	Input data file name.	ZTCASE3.SP
2	Beginning and ending channels for fit.	1,1000
3	Input peak file.	NONE, NO
4	Input background constants file.	NONE, YES

Page FSPFIT3 (Output data)

1	Output file for background constants	TCASE3.CN
2	Output file for Peak data	TCASE3.PK
3	Output file for graph data	TCASE3.GR

Page FSPFIT4 (Standards data)

1	First standard	ZTCASE1.ST
2	Type 1 width, error and channel no	5,3,10
3	Type 1 width, error and channel no.	5,3,900

On output the TCASE3.PK, TCASE3.CN and TCASE3.GR files will contain all the fit data. TCASE3.PK should be identical to the ZTCASE3.PK file. Similarly TCASE3.CN will be identical to ZTCASE3.CN where the ZTCASE files are the outputs of our original runs.

The results of the curve fitting can be viewed by running FSPDIS with the following menu page setup;

Page FSPDIS1 (General information)

1	Name of the graphical file to be read	TCASE3.GR
2	Name of the calibration file (.XDA)	NONE
3	Beginning,ending channels for display	1,1000
4	Vertical axis = COUNTS*SCALE FACTOR	1
5	Desired min and max y scale	1,default
6	Log or linear y scale	LOG

Page FSPDIS2 (Viewing options)

1	View the residuals (y/n)	Yes
2	View the curve fit (y/n)	Yes

3	View the peaks (y/n)	Yes
4	View the background fit (y/n)	Yes

iv) Absolute limit of peak detection.

Here the cutoff value has been reduced to a level where spurious peaks are being picked up. At this level, however, ROBFIT can determine peaks which are burried well within the noise. Fitting and display setups are;

Page FSPFIT1 (General fit information)

1	Maximum number of peaks	100
2	Starting,ending cutoffs, in N steps	2,1,1
3	VWID,FIXD and CONT,NONK,NOBF,FIXK	VWID,CONT
4	Maximum number of background constants	4,4
5	Weights 0=SQRT(data),1=calc,2=file	0
6	File containing the weights	NONE

Page FSPFIT2 (Input data)

1	Input data file name.	ZTCASE3B.SP
2	Beginning and ending channels for fit.	1,1000
3	Input peak file.	NONE, NO
4	Input background constants file.	NONE,YES

Page FSPFI3 (Output data)

1	Output file for background constants	TCASE3B.CN
2	Output file for Peak data	TCASE3B.PK
3	Output file for graph data	TCASE3B.GR

Page FSPFIT4 (Standards data)

1	First standard	ZTCASE1.ST
2	Type 1 width,error and channel no.	5,3,10
3	Type 1 width,error and channel no.	5,3,900

On output the TCASE3B.PK, TCASE3B.CN and TCASE3B.GR files will contain all the fit data. TCASE3B.PK should be identical to the ZTCASE3B.PK file. Similarly TCASE3B.CN will be identical to ZTCASE3B.CN where the ZTCASE files are the outputs of our original runs.

The results of the curve fitting can be viewed by running FSPDIS with the following menu page setup;

Page FSPDIS1 (General information)

1	Name of the graphical file to be read	TCASE3B.GR
2	Name of the calibration file (.XDA)	NONE
3	Beginning,ending channels for display	1,1000
4	Vertical axis = COUNTS*SCALE FACTOR	1
5	Desired min and max y scale	1,default
6	Log or linear y scale	LOG

Page FSPDIS2 (Viewing options)

1	View the residuals (y/n)	Yes
2	View the curve fit (y/n)	Yes
3	View the peaks (y/n)	Yes
4	View the background fit (y/n)	Yes

D3 Multiple peak detection

i) Unambiguous peak and separation determination.

Here the ability to determine unambiguously a peak at a given separation has been studied. These tests are outlined in section 4.2. Fitting and display setups are,

Page FSPFIT1 (General fit information)

1	Maximum number of peaks.	100
2	Starting,ending cutoffs, in N steps	5,4,2
3	VWID,FIXD and CONT,NONK,NOBF,FIXK	VWID,CONT
4	Maximum number of background constants	4,4
5	Weights 0=SQRT(data),1=calc,2=file	0
6	File containing the weights	NONE

Page FSPFIT2 (Input data)

1	Input data file name.	ZTCASE4.SP
2	Beginning and ending channels for fit.	1,1000
3	Input peak file.	NONE, NO
4	Input background constants file.	NONE, YES

Page FSPFIT3 (Output data)

1	Output file for background constants	TCASE4.CN
2	Output file for Peak data	TCASE4.PK
3	Output file for graph data	TCASE4.GR

Page FSPFIT4 (Standards data)

1	First standard	ZTCASE1.ST

2	Type 1 width, error and channel no.	5,1,10
3	Type 1 width, error and channel no.	5,1,900

On output the TCASE4.PK, TCASE4.CN and TCASE4.GR files will contain all the fit data. TCASE4.PK should be identical to the ZTCASE4.PK file similarly TCASE4.CN will be identical to ZTCASE4.CN where the ZTCASE files are the outputs of our original runs.

The results of the curve fitting can be viewed by running FSPDIS with the following menu page setup;

Page FSPDIS1 (General information)

1	Name of the graphical file to be read	TCASE4.GR
2	Name of the calibration file (.XDA)	NONE
3	Beginning,ending channels for display	1,1000
4	Vertical axis = COUNTS*SCALE FACTOR	1
5	Desired min and max y scale	1,default
6	Log or linear y scale	LOG

Page FSPDIS2 (Viewing options)

1	View the residuals (y/n)	Yes
2	View the curve fit (y/n)	Yes
3	View the peaks (y/n)	Yes
4	View the background fit (y/n)	Yes

ii) Effect of cutoff value on separation determination.

Here we lower the cutoff value and detail its effects on detecting separated peaks. The fitting and display setups are;

Page FSPFIT1 (General fit information)

1	Maximum number of peaks.	100
2	Starting,ending cutoffs, in N steps	2,1.5,1
3	VWID,FIXD and CONT,NONK,NOBF,FIXK	VWID,CONT
4	Maximum number of background constants	4,4
5	Weights 0=SQRT(data),1=calc,2=file	0
6	File containing the weights	NONE

Page FSPFIT2 (Input data)

1	Input data file name.	ZTCASE5.SP
2	Beginning and ending channels for fit.	1,1000
3	Input peak file.	NONE, NO
4	Input background constants file.	NONE, YES

Page FSPFIT3 (Output data)

1	Output file for background constants	TCASE5.CN
2	Output file for Peak data	TCASE5.PK
3	Output file for graph data	TCASE5.GR

Page FSPFIT4 (Standards data)

1	First standard	ZTCASE1.ST
2	Type 1 width,error and channel no.	5,1,10
3	Type 1 width,error and channel no.	5,1,900

On output the TCASE5.PK, TCASE5.CN and TCASE5.GR files will contain all the fit data. TCASE5.PK should be identical to the ZTCASE5.PK file. Similarly TCASE5.CN will be identical to ZTCASE5.CN where the ZTCASE files are the outputs of our original runs.

The results of the curve fitting can be viewed by running FSPDIS with the following menu page setup.

Page FSPDIS1 (General information)

1	Name of the graphical file to be read	TCASE5.GR
2	Name of the calibration file (.XDA)	NONE
3	Beginning,ending channels for display	1,1000
4	Vertical axis = COUNTS*SCALE FACTOR	1
5	Desired min and max y scale	1,default
6	Log or linear y scale	LOG

Page FSPDIS2 (Viewing options)

1	View the residuals (y/n)	Yes
2	View the curve fit (y/n)	Yes
3	View the peaks (y/n)	Yes
4	View the background fit (y/n)	Yes

iii) Absolute minimum peak separation determination.

This test shows the limit to which peak separation determination can be driven. Fitting and display setups are as follows,

Page FSPFIT1 (General fit information)

1	Maximum number of peaks.	100
2	Starting,ending cutoffs, in N steps	2,1,2
3	VWID,FIXD and CONT,NONK,NOBF,FIXK	VWID,CONT
4	Maximum number of background constants	4,4
5	Weights 0=SQRT(data),1=calc,2=file	0
6	File containing the weights	NONE

Page FSPFIT2 (Input data)

1	Input data file name.	ZTCASE6.SP
2	Beginning and ending channels for fit.	1,1000
3	Input peak file	NONE, NO
4	Input background constants file.	NONE, YES

Page FSPFIT3 (Output data)

1	Output file for background constants	TCASE6.CN
2	Output file for Peak data	TCASE6.PK
3	Output file for graph data	TCASE6.GR

Page FSPFIT4 (Standards data)

1	First standard	ZTCASE1.ST
2	Type 1 width,error and channel no.	5,1,10
3	Type 1 width,error and channel no.	5,1,900

On output the TCASE6.PK, TCASE6.CN and TCASE6.GR files will contain all the fit data. TCASE6.PK should be identical to the ZTCASE6.PK file. Similarly TCASE6.CN will be identical to ZTCASE6.CN where the ZTCASE files are the outputs of our original runs.

The results of the curve fitting can be viewed by running FSPDIS with the following menu page setup.

Page FSPDIS1 (General information)

1	Name of the graphical file to be read	TCASE6.GR
2	Name of the calibration file (.XDA)	NONE
3	Beginning,ending channels for display	1,1000
4	Vertical axis = COUNTS*SCALE FACTOR	1
5	Desired min and max y scale	1,default
6	Log or linear y scale	LOG

Page FSPDIS2 (Viewing options)

1	View the residuals (y/n)	Yes
2	View the curve fit (y/n)	Yes
3	View the peaks (y/n)	Yes
4	View the background fit (y/n)	Yes

D4 Complex background fitting

Here we show how the background fitter can perform on extremely complicated functions. We take as a data sample the Yttrium spectrum and fit it according to the procedure outlined in section 4.3. The background fitter is initially set up as follows;

Page BKGFIT1 (Input details)

1	Maximum number of background constants	10
2	Cutoff value set to	20
3	Input data file name	ZTCASE7.SP
4	Beginning and ending channels for fit	1,default
5	Input background constant file	NONE, YES
6	Weights 0=SQRT(data),1=calc,2=file	0
7	File containing the weights	NONE

Page BKGFIT2 (Output details)

1	Background constants output to	BKGCONS.CN
2	Background fit	BKGCONS.GR

The output constants for the first iteration are to be compared with those in ZY10.CN. Further cycles of the fitter can be performed by increasing the number of constants on line 1 of BKGFIT1 and feeding the old constants BKGCONS.CN back into the fit at line 5 of BKGFIT1. Each of the constants files up to a total of 98 are provided with the code and can be used for comparison with the test case output. These constants files have the naming convention ZY'no of constants' CN. The test case output can be viewed by feeding the BKGCONS.GR file into the FSPDIS display mode as described in section 3.6.

D5 Fitting a complete spectrum

The following is a fit to data that was used to calibrate the GRAD experiment used on the Antarctic supernova expedition ref 3. The spectrum is characterized by large numbers of peaks of varying shape all sitting on a complex background. The user must be warned that fitting this spectrum will take a considerable amout of CPU time. To increase the fitting speed the constants file ZTCASE8.CN can be read in on line 4 of FSPFIT2 on the menu pages.

Page FSPFIT1 (General fit information)

1	Maximum number of peaks.	255
2	Starting,ending cutoffs, in N steps	40,10,4
3	VWID,FIXD and CONT,NONK,NOBF,FIXK	VWID,CONT
4	Maximum number of background constants	4,32
5	Weights 0=SQRT(data),1=calc,2=file	0
6	File containing the weights	NONE

Page FSPFIT2 (Input data)

1	Input data file name.	ZTCASE8.SP
2	Beginning and ending channels for fit.	50,4096
3	Input peak file.	NONE, NO
4	Input background constants file.	NONE, YES

Page FSPFIT3 (Output data)

1	Output file for background constants	TCASE8.CN
2	Output file for Peak data	TCASE8.PK
3	Output file for graph data	TCASE8.GR

Page FSPFIT4 (Standards data)

1	First standard	YT.ST
2	Yttrium start width,error and channel no.	2.07,0.05,800
3	Yttrium end width,error and channel no.	2.68,0.1,1900
4	Compton standard	COMP.ST
5	Compton width,error and channel no.	80,20,50
6	Compton width,error and channel no.	80,20,4000

On output the TCASE8.PK, TCASE8.CN and TCASE8.GR files will contain all the fit data. TCASE8.PK should be identical to the ZTCASE8.PK file. Similarly TCASE8.CN will be identical to ZTCASE8.CN where the ZTCASE files are the outputs of our original runs.

The results of the curve fitting can be viewed by running FSPDIS with the following menu page setup;

Page FSPDIS1 (General information)

1	Name of the graphical file to be read	TCASE8.GR
2	Name of the calibration file (.XDA)	NONE
3	Beginning,ending channels for display	1,1000
4	Vertical axis = COUNTS*SCALE FACTOR	1
5	Desired min and max y scale	1,default
6	Log or linear y scale	LOG

Page FSPDIS2 (Viewing options)

1	View the residuals (y/n)	Yes
2	View the curve fit (y/n)	Yes
3	View the peaks (y/n)	Yes
4	View the background fit (y/n)	Yes

APPPENDIX B

ROBFIT (ver. 1.0) code listing

This Appendix is organized so that each major ROBFIT program (mode: eg RAWDD) is listed separately along with the routines that are specific to that particular program. Routines that are used by more than one of these programs are given at the end of the program listings, with the exception of MENURD which for clarity is listed with VRMAIN even though it is called by all modes. In addition we have included the *.BAT and *.MNU files that are updated whenever ROBFIT is operated.

The process starts when the user enters the command ROBFIT which
starts the file ROBFIT.BAT

```
ECHO OFF
VRMAIN.EXE
ROB2.BAT
```

which calls into execution the following code, VRMAIN, which writes the file ROB2.BAT e.g. if standard generation is chosen, ROB2.BAT is

```
ECHO OFF
STGEN.EXE
```

which is executed next.

```
        PROGRAM VRMAIN
C*******************************************************************
C      The session master menu driver
C*******************************************************************
C *** VRMAIN runs the master menu where the user chooses which ROBFIT
C *** mode to run. It accomplishes this by writing the appropriate
C *** executable filename into a second batch file called ROB2.BAT.
C *** The last operation in the ROBFIT.BAT is to run the ROB2.BAT
C *** file.
      DIMENSION FFCOM(600),FFVAR(600)
      DIMENSION FFFCOM(600),FFFVAR(600)
      CHARACTER*64 VALU,NA
      CHARACTER*1 CSTR
      COMMON/FCALL/NSTR,NCMENU
      CHARACTER*40 FFCOM,FFVAR
      CHARACTER*40 FFFCOM,FFFVAR
      NSTR=0
      NCMENU=0
      NA='VRMAIN.MNU'
      CALL MENURD(NA)
C SET UP THE SECOND BATCH FILE
      OPEN(UNIT=1,FILE='ROB2.BAT')
```

```
      WRITE(1,100)
100   FORMAT('ECHO OFF')
        IF(NSTR.EQ.1)THEN
C***********************************************
        WRITE(1,110)
110     FORMAT('RAWDD.EXE')
        CLOSE(1)
      ENDIF
C************************************
        IF(NSTR.EQ.2)THEN
C***********************************************
        WRITE(1,120)
120     FORMAT('STGEN.EXE')
      ENDIF
C************************************
        IF(NSTR.EQ.3)THEN
C***********************************************
        WRITE(1,130)
130     FORMAT('STDIS.EXE')
      ENDIF
C************************************
        IF(NSTR.EQ.4)THEN
C***********************************************
        WRITE(1,140)
140     FORMAT('BKGFIT.EXE')
      ENDIF
C************************************
        IF(NSTR.EQ.5)THEN
C***********************************************
        WRITE(1,150)
150     FORMAT('FSPFIT.EXE')
      ENDIF
C************************************
        IF(NSTR.EQ.6)THEN
C***********************************************
        WRITE(1,160)
160     FORMAT('FSPDIS.EXE')
      ENDIF
C************************************
        IF(NSTR.EQ.7)THEN
C***********************************************
        WRITE(1,170)
170     FORMAT('XCALIBER.EXE')
      ENDIF
C************************************
        IF(NSTR.EQ.8)THEN
C***********************************************
        WRITE(1,180)
180     FORMAT('EXEDIT.EXE')
      ENDIF
C************************************
        IF(NSTR.EQ.9)THEN
C***********************************************
        WRITE(1,190)
190     FORMAT('REM STOPPING ROBFIT')
      ENDIF
C************************************
```

```
      STOP
      END
C*********************************************************************
      SUBROUTINE MENURD(NA)
C *** The routine MENURD reads and writes to the *.MNU files
C *** which contain the option information for the ROBFIT modes.
C *** MENURD is used by VRMAIN and also by each individual mode.
      DIMENSION IR(10),FFCOM(160),FFVAR(160)
      CHARACTER*64 NA
      CHARACTER*40 FFCOM,FFVAR
      COMMON/USPAGE/FFVAR
      COMMON/FCALL/NSTR,NCMENU
      COMMON/BATCFL/IBATCH
      IBATCH=1
      OPEN(UNIT=17,FILE=NA)
      IRN=0
      DO 10 I=1,10
        READ(17,*)IR(I)
        IF(IR(I).EQ.100) GOTO 20
        IRN=IRN+1
10    CONTINUE
20    CONTINUE
      DO 30 I=1,160
        FFCOM(I)=' '
        FFVAR(I)=' '
30    CONTINUE
      DO 100 I=1,IRN
        IFF=(I-1)*20
        DO 200 J=1,IR(I)
          IFF=IFF+1
          READ(17,5000)K,FFCOM(IFF),FFVAR(IFF)
5000      FORMAT(I2,1X,A40,A37)
200     CONTINUE
100   CONTINUE
1000   CONTINUE
      DO 300 I=1,IR(1)
      IF(NCMENU.EQ.0)THEN
        FFVAR(I)=' '
        WRITE(5,5000)I,FFCOM(I),FFVAR(I)
      ELSE
        IF(I.GT.2)THEN
        WRITE(5,5000)I-2,FFCOM(I),FFVAR(I)
        ELSE
        WRITE(5,5010)FFCOM(I)
5010   FORMAT(3X,A40)
       ENDIF
      ENDIF
300   CONTINUE
      IF(NCMENU.EQ.0)THEN
        WRITE(5,5001)
5001    FORMAT(' ENTER THE LINE NUMBER TO RUN THE ROBFIT ROUTINE')
      ELSE
      WRITE(5,5100)
5100  FORMAT(' ENTER LINE NUMBER TO VIEW DATA OR 100 TO RUN'/
     *'OR 99 TO STOP EXECUTION')
      ENDIF
```

```
      READ(5,*)LLINE
      IF(LLINE.NE.99.AND.LLINE.NE.100)THEN
        IF(LLINE.LT.1.OR.LLINE.GT.IRN-1)GOTO 1000
      ENDIF
      IF(NCMENU.EQ.0)THEN
        NCMENU=1
        DO 2002 I=1,IR(1)
          FFVAR(I)='NO'
2002    CONTINUE
        FFVAR(LLINE)='YES'
        REWIND(17)
        DO 2005 I=1,IRN+1
          WRITE(17,*)IR(I)
2005    CONTINUE
        DO 2003 I=1,IR(1)
          WRITE(17,5000)I,FFCOM(I),FFVAR(I)
2003    CONTINUE
        DO 2004 I=1,IR(1)
          WRITE(17,5000)I,'DUMMY','DUMMY'
2004    CONTINUE
        NSTR=LLINE
        CLOSE(17)
        RETURN
      ENDIF
      IF(LLINE.EQ.99) STOP
      IF(LLINE.EQ.100)GOTO 3000
2000  CONTINUE
      IOFF=LLINE*20
      DO 400 I=1,IR(LLINE+1)
        IOFF=IOFF+1
      IF(I.GT.2)THEN
        WRITE(5,5000)I-2,FFCOM(IOFF),FFVAR(IOFF)
      ELSE
       WRITE(5,5010)FFCOM(IOFF)
      ENDIF
400   CONTINUE
      IOFF=LLINE*20
      WRITE(5,5300)
5300  FORMAT('ENTER LINE NUMBER TO CHANGE A LINE'/' 0,0 TO RETURN TO
THE
     1 MAIN MENU'
     1/'      AND ,1 TO DELETE THE LINE'/'           ,10 TO CHANGE THE
     1VARIABLE'/'           ,100 TO ENTER A COMPLETE NEW LINE
     1')
      READ(5,*)LSLINE,CHNGE
      LSLINE=LSLINE+2
      IF(LSLINE.LT.1.OR.LSLINE.GT.(IR(LLINE+1)+1))GOTO 1000
      IF(CHNGE.EQ.0)GOTO 1000
      IF(CHNGE.EQ.1)THEN
        NP=0
        DO 500 I=1,20
          NP=NP+1
          IOFF=IOFF+1
          IF(NP.GE.LSLINE)THEN
            LTEST=LLINE*20+20
            IF(NP.EQ.LTEST)THEN
              FFCOM(IOFF)=' '
```

```
                FFVAR(IOFF)=' '
              ELSE
                FFCOM(IOFF)=FFCOM(IOFF+1)
                FFVAR(IOFF)=FFVAR(IOFF+1)
              ENDIF
            ENDIF
500       CONTINUE
          IR(LLINE+1)=IR(LLINE+1)-1
        ENDIF
        IF(CHNGE.EQ.10)THEN
          IOFF=LLINE*20+LSLINE
          WRITE(5,5301)
5301      FORMAT('ENTER THE NEW VARIABLE')
          READ(5,5400)FFVAR(IOFF)
5400      FORMAT(A40)
        ENDIF
        IF(CHNGE.EQ.100) THEN
          IOFF=LLINE*20+LSLINE
          WRITE(5,5500)
5500      FORMAT('ENTER THE NEW COMMENT')
          READ(5,5400)FFCOM(IOFF)
          WRITE(5,5301)
          READ(5,5400)FFVAR(IOFF)
          IF(LSLINE.GT.IR(LLINE+1))THEN
            IR(LLINE+1)=IR(LLINE+1)+1
          ENDIF
        ENDIF
        GOTO 2000
3000    CONTINUE
        REWIND(17)
        DO 600 I=1,IRN+1
          WRITE(17,*)IR(I)
600     CONTINUE
        DO 700 I=1,IRN
          IFF=(I-1)*20
          DO 800 J=1,IR(I)
            IFF=IFF+1
            WRITE(17,5000)J,FFCOM(IFF),FFVAR(IFF)
800       CONTINUE
700     CONTINUE
        CLOSE(17)
        RETURN
        END
```

The following is a listing of the VRMAIN.MNU file; this file is read by the MENURD subroutine and the variables are passed to the VRMAIN program to decide which ROBFIT mode is to be run.

```
          9
          1
          1
          1
          1
          1
          1
          1
```

```
          1
          1
        100
 1 DISPLAY RAW DATA (RAWDD)                    NO
 2 GENERATE STANDARD (STGEN)                   NO
 3 DIPLAY STANDARD  (STDIS)                    NO
 4 FIT BACKGROUND (BKGFIT)                     NO
 5 FIT SPECTRUM (FSPFIT)                       NO
 6 DISPLAY SPECTRAL FIT (FSPDIS)               NO
 7 CALIBRATE SPECTRA (XCALIBER)                NO
 8 ENTER THE WATFOR77 EDITOR                   NO
 9      STOP EXECUTION                         YES
 1                                        DUMMY
DUMMY
 2                                        DUMMY
DUMMY
 3                                        DUMMY
DUMMY
 4                                        DUMMY
DUMMY
 5                                        DUMMY
DUMMY
 6                                        DUMMY
DUMMY
 7                                        DUMMY
DUMMY
 8                                        DUMMY
DUMMY
 9                                        DUMMY
DUMMY
```

The following is a listing of the code RAWDD which displays the data prior to any fitting. It can handle two data sets at a time which allows data subtraction while saving the proper weight file to go with the subtracted data so that susequent fits can be properly made. In addition there is a correction section which allows data dropouts or
bad electronic signals to be linearly interpolated out.

```
      PROGRAM RAWDD
C*********************************************************************
C     Raw data display program (RAWDD)
C*********************************************************************
      DIMENSION F(4100,2),FR(4100),FP(4100),NT(2),ACOUNTS(2)
C *** LINK IN THE MENU SELECTED QUANTITIES
      DIMENSION UFILE(2),CFILE(2),FBEGC2(2),FENDC2(2),IBEGC2(2),
     #IENDC2(2)
      CHARACTER*17 UFILE,CFILE,XYFILE,FPNAME,FMNAME,TITLE
      COMMON/GLFILE/UFILE,CFILE,XYFILE,FPNAME,FMNAME,TITLE
      COMMON/GLINK/CORR,DIFF,DIFPLT,RAT1,RAT2,FBEGC2,FENDC2,
     #IMIN2,IMAX2,LFLAG2,
     #IBEGC2,IENDC2,
     #IBEGC1,IENDC1,
     #IMIN1,IMAX1,
     #LFLAG1
C *** END THE MENU LINK
         COMMON/AXISP/BHORI,BVERT,SF,SVERT,EHORI,EVERT
```

```
      COMMON/BATCFL/IBATCH
        CHARACTER*1 ANS
         CHARACTER*64 NA,VALU
C
320   CONTINUE
      IBATCH=1
      CALL G4LINK
        DO 5 J=1,2
5       CALL REDAT(F(1,J),NT(J),UFILE(J))
        CLOSE(UNIT=1)
        N=MAX0(NT(1),NT(2))
        IX=31
302     IMC=N
        AMULT=1
55    CONTINUE
        WRITE(*,103)IX,IMC
103     FORMAT(' DATA HAS CHANNELS',2I5 /
     # ' INITIAL CHANNEL?, END CHANNEL?, <-1,0> FOR MENU')
      READ(*,*)IBEGC1,IENDC1
      IF(IBEGC1.EQ.-1)THEN
        GOTO 320
      ENDIF
C
C *** LINK IN MENU VALUES
C
      IBEGC=IBEGC1
      IENDC=IENDC1
C*************************
        IF(CORR.EQ.1.)CALL CORREC(F,NT)
        IF(CORR.EQ.1.)GOTO 320
      IBEGC=MAX0(1,IBEGC)
        IF(IENDC.NE.0)GOTO 330
        IBEGC=IX
        IENDC=IMC
330     IBEGC=MAX0(1,IBEGC)
        IENDC=MIN0(IMC,IENDC)
        IF(IBEGC.GE.IENDC)WRITE(*,104)
104     FORMAT(' THERE ARE NO CHANNELS IN THE
     #INTERVAL',F10.0,',',F10.0)
        IF(IBEGC.GE.IENDC)GOTO 320
        AMIN=2000000000
        AMAX=0
        AMULT=1
        DO 335 J=1,2
335     ACOUNTS(J)=0
        DO 341 J=1,2
        IF(NT(J).EQ.0)GOTO 341
        DO 340 I=IBEGC,IENDC
        IF(I.GT.NT(J))GOTO 340
        ACOUNTS(J)=ACOUNTS(J)+AMULT*F(I,J)
        AMIN=AMIN1(AMIN,AMULT*F(I,J))
340     AMAX=AMAX1(AMAX,AMULT*F(I,J))
341     CONTINUE
        AMIN=AMAX1(0.,AMIN)
        WRITE(*,108)AMIN,AMAX,ACOUNTS
108     FORMAT(' AMIN=',F10.0,'  AMAX=',F10.0,' TOTAL COUNTS
     #ARE',2F10.0
```

```
     # ,' ENTER DESIRED MIN ,DESIRED MAX')
      READ(*,*)IMIN1,IMAX1
C
C *** LINK IN MENU VALUES
C
      IMIN=IMIN1
      IMAX=IMAX1
C*************************
        IF(IMAX.EQ.0)GOTO 37
        AMIN=IMIN
        AMAX=IMAX
37      CONTINUE
        LFLAG=0
      WRITE(*,555)
555   FORMAT(' DO YOU WANT A LOG Y SCALE (Y/N)')
      READ(*,'(A)')ANS
      LFLAG1=0
      IF(ANS.EQ.'Y')LFLAG1=1
      CALL RSETG
C
C *** LINK IN MENU VALUES
C
      LFLAG=LFLAG1
C*************************
        IF(LFLAG.NE.1)GOTO 400
        AMIN=AMAX1(.01,AMIN)
        AMAX=ALOG(AMAX)
        AMIN=ALOG(AMIN)
400     IRES=0
115      FORMAT(A1)
        CALL CLEARS
         ABEGC=1.*IBEGC
         AENDC=1.*IENDC
         CHI=0.
        CALL AXIS(AMIN,AMAX,LFLAG,IRES,ABEGC,AENDC,TITLE,CHI)
        SFE=SF/4
        ISF=SF/2
        ISKIP=2/SF
        ISKIP=MAX0(1,ISKIP)
         DO 505 JO=1,2
        IF(NT(JO).EQ.0)GOTO 505
        IXO=BHORI
        DO 500 I=IBEGC,IENDC,ISKIP
        XP=BHORI+SF*(I-IBEGC)
        IXP=XP
        ADAT=AMULT*F(I,JO)
        IF(LFLAG.EQ.1)ADAT=ALOG(AMAX1(1.,ADAT))
        IYDAT=SVERT*(ADAT-AMIN)+BVERT
C *** THE FIRST PLOT CALL
        IF(I.EQ.IBEGC)CALL STPL(IXO,IYDAT)
        IF(ISKIP.GT.1)GOTO 495
        IF(I.GT.IBEGC)CALL PLOT(IXO,IYDAT)
        IXO=IXP+ISF
        CALL PLOT(IXO,IYDAT)
        GOTO 500
495     IDMAX=0
        IDMIN=1000000
```

```
      DO 497 J=1,ISKIP
      IARG=I+J-1
     IF(IARG.GT.IENDC)GOTO 498
      ICOMP=F(IARG,JO)
      IDMAX=MAX0(ICOMP,IDMAX)
497   IDMIN=MIN0(IDMIN,ICOMP)
498   ADMIN=AMULT*IDMIN
      ADMAX=AMULT*IDMAX
      IF(LFLAG.EQ.1)ADMIN=ALOG(AMAX1(1.,ADMIN))
      IF(LFLAG.EQ.1)ADMAX=ALOG(AMAX1(1.,ADMAX))
      IDMIN=SVERT*(ADMIN-AMIN)+BVERT
      IDMAX=SVERT*(ADMAX-AMIN)+BVERT
      CALL PLOT(IXO,IDMIN)
      IXO=IXP+ISF
      CALL PLOT(IXO,IDMAX)
500   CONTINUE
505   CONTINUE
C
C *** LINK IN MENU DIFFERENCE DECISION
C
      IF(DIFF.NE.1.)GOTO 950
      AMAX=-100000000
      AMINT=100000000
      ACOUNTS(1)=0
       RAT=RAT1/RAT2
       PRINT*,' F(30,1),F(30,2),RAT',F(30,1),F(30,2),RAT
      DO 740 I=IBEGC,IENDC
      FR(I)=F(I,1)-F(I,2) *RAT
       FP(I)=F(I,1)+F(I,2)*RAT*RAT
     FP(I)=AMAX1(1.,FP(I))
     IF(DIFPLT.EQ.1.)FR(I)=FR(I)/SQRT(FP(I))
      IF(I.GE.NT(1).OR.I.GE.NT(2))GOTO 740
      ACOUNTS(1)=ACOUNTS(1)+AMULT*FR(I)
      AMINT=AMIN1(AMINT,AMULT*FR(I))
740   AMAX=AMAX1(AMAX,AMULT*FR(I))
      AMIN=AMINT
C
C *** LINK IN MENU IMIN,IMAX
C
     IMIN=IMIN2
     IMAX=IMAX2
C***************************
      IF(IMAX.EQ.0)GOTO 750
      AMIN=IMIN
      AMAX=IMAX
750   CONTINUE
      LFLAG=0
C
C *** LINK IN MENU LOG FLAG
C
     LFLAG=LFLAG2
C**************************
      IF(LFLAG.NE.1)GOTO 760
      LFLAG=1
      AMIN=AMAX1(.01,AMIN)
      AMAX=ALOG(AMAX)
      AMIN=ALOG(AMIN)
```

```
760     IRES=0
         IW4FL=0
         CALL CLEARS
        CALL AXIS(AMIN,AMAX,LFLAG,IRES,ABEGC,AENDC,TITLE,CHI)
782     ISKIP=2/SF
        ISKIP=MAX0(1,ISKIP)
        IXO=BHORI
        DO 900 I=IBEGC,IENDC,ISKIP
        XP=BHORI+SF*(I-IBEGC)
        IXP=XP
        ADAT=AMULT*FR(I)
        IF(LFLAG.EQ.1)ADAT=ALOG(AMAX1(1.,ADAT))
        IYDAT=SVERT*(ADAT-AMIN)+BVERT
        IF(I.EQ.IBEGC)CALL STPL(IXO,IYDAT)
        IF(ISKIP.GT.1)GOTO 895
        IF(I.GT.IBEGC)CALL PLOT(IXO,IYDAT)
        IXO=IXP+ISF
        CALL PLOT(IXO,IYDAT)
        GOTO 900
895     IDMAX=0
        IDMIN=1000000
        DO 897 J=1,ISKIP
        IARG=I+J-1
      IF(IARG.GT.IENDC)GOTO 899
        ICOMP=FR(IARG)
        IDMAX=MAX0(ICOMP,IDMAX)
897     IDMIN=MIN0(IDMIN,ICOMP)
899     ADMIN=AMULT*IDMIN
        ADMAX=AMULT*IDMAX
        IF(LFLAG.EQ.1)ADMIN=ALOG(AMAX1(1.,ADMIN))
        IF(LFLAG.EQ.1)ADMAX=ALOG(AMAX1(1.,ADMAX))
        IDMIN=SVERT*(ADMIN-AMIN)+BVERT
        IDMAX=SVERT*(ADMAX-AMIN)+BVERT
        IXO=IXP+ISF
        CALL PLOT(IXO,IDMIN)
        CALL PLOT(IXO,IDMAX)
900     CONTINUE
950     CONTINUE
        WRITE(*,1947)
1947    FORMAT(' SEEN ENOUGH?  ( enter Y or N )')
        READ(*,115)ANS
        CALL ANSI
        IF(ANS.NE.'Y')GOTO 55
         IF(FPNAME.EQ.'NULL')GOTO 980
         NA=FPNAME
         OPEN(UNIT=2,FILE=NA,FORM='UNFORMATTED')
         XT=0.
        PRINT*,' F(30,1),F(30,2)=',F(30,1),F(30,2)
        PRINT*,' FP(30)=',FP(30),'FR(30)',FR(30)
         WRITE(2)N,XT
         WRITE(2)(FP(I),I=1,N)
         CLOSE(2)
         NA=FMNAME
         OPEN(UNIT=3,FILE=NA,FORM='UNFORMATTED')
         XT=IBEGC-1
         NOUT=IENDC-IBEGC+1
         WRITE(3)NOUT,XT
```

```
          WRITE(3)(FR(I),I=IBEGC,IENDC)
          CLOSE(3)
980      CONTINUE
         STOP
         END
C*******************************************************************
         SUBROUTINE REDAT(F,NIR,UFILE)
C *** This routine used by RAWDD to read data files is a subset of
C *** the routine BREAD, used by most other modes, BREAD
C *** adds information about weighting the data points.
         CHARACTER*64 NA
         CHARACTER*17 UFILE
         CHARACTER*4 HDAT,Z4DAT,IDAT,LCDAT,II,IRBKG,IBKG
         DIMENSION F(4100),II(40)
         DIMENSION IHD(4100)
       COMMON/BATCFL/IBATCH
       DATA LCDAT/'LCDA'/,XVB,XVE/2*0.D0/,HDAT/'HDAT'/IDAT/'IDAT'/
       DATA Z4DAT/'Z4DA'/IPW/0/,PCON/1.D6/,IRBKG/'RBKG'/,IBKG/'    '/
         N=0
       NIR=0
         NMULT=1
5      CONTINUE
       NA=UFILE
         IF(NA.EQ.'\'.OR.NA.EQ.'STOP'.OR.NA.EQ.'NULL'.OR.NA.EQ.
     # 'END')RETURN
         IF(IBATCH.EQ.0)IT=INDEX(NA,' ')
         IF(IBATCH.EQ.1)IT=INDEX(NA,'.')
         IRF=0
          IF(NA(IT+1 :IT+3).EQ.'UF')GOTO 1100
         IF(NA(IT+1:IT+3).EQ.'LEO')IRF=1
        IF(IRF.EQ.1)THEN
         OPEN(UNIT=1,FILE=NA,FORM='UNFORMATTED')
         CALL G4READ(F,IHD,NIR)
         GOTO 220
       ENDIF
         OPEN(UNIT=1,FILE=NA,FORM='FORMATTED')
         READ(1,105)II
105     FORMAT(10A4)
         IF(II(40).EQ.IBKG)GOTO 300
         IF(II(40).EQ.LCDAT)CALL GLCHEX(F,NIR,XE)
         IF(II(40).EQ.LCDAT)GOTO 220
         IF(II(40).NE.Z4DAT)READ(1,*)NXB,NEND
         IF(II(40).EQ.Z4DAT)READ(1,*)NXB,NEND,NMULT
1923     FORMAT(' NMULT=',I5)
         NXB=1
         IF(II(40).EQ.Z4DAT)GOTO 180
         DO 160 J=1,818
         JB=10*(J-1)
         IF(II(40).EQ.HDAT)READ(1,156,END=162)(IHD(JB+I),I=1,10)
         IF(II(40).EQ.IDAT)READ(1,157,END=162)(IHD(JB+I),I=1,10)
         NIR=J*10
156      FORMAT(Z7,9Z8)
157      FORMAT(10I8)
160      CONTINUE
         GOTO 162
180      DO 185 J=1,408
         JB=20*(J-1)
```

```
         READ(1,158,END=162)(IHD(JB+I),I=1,20)
158      FORMAT(20Z4)
         NIR=20*J
185      CONTINUE
162      CONTINUE
         DO 190 I=1,NIR
190      F(I)=IHD(I)*NMULT
220      CLOSE(1)
         RETURN
300       REWIND 1
          N=0
310       READ(1,*,END=320)XT,F(N+1)
          IF(N.GT.0)GOTO 315
          N=XT-1
          IF(N.LT.1)GOTO 315
          N=MIN0(N,8191)
          F(N+1)=F(1)
          DO 313 I=1,N
313       F(I)=0
315       N=N+1
          IF(N.LT.8192)GOTO 310
320       CONTINUE
          NIR=N
          RETURN
1100      CONTINUE
          OPEN(UNIT=1,FILE=NA,FORM='UNFORMATTED')
          READ(1)N,XE
          READ(1)(F(I),I=1,N)
          N1=AMAX1(0.,XE)
          NIR=N1+N
          DO 1110 I=1,N
          NT=N+1-I
1110      F(N1+NT)=F(NT)
          DO 1120 I=1,N1
1120      F(I)=.1
       IF(IBATCH.EQ.0)THEN
          CLOSE(1)
       ELSE
          ENDFILE(UNIT=1)
       ENDIF
          RETURN
        END
C*******************************************************************
        SUBROUTINE G4READ(F,IIDAT,N)
C *** This routine is used to read in a specific file format.
        DIMENSION F(1),IIDAT(1)
        DO 416 III=1,8064,128
        READ(1,END=418)(IIDAT(I),I=III,III+127)
416     N=III+127
418     DO 400 I=1,N
        F(I)=IIDAT(I)
400     CONTINUE
        RETURN
      END
C*******************************************************************
        SUBROUTINE GLCHEX(F,NF,XE)
```

```
C *** This is a somewhat obsolete routine used to convert characters
C *** to hexadecimal numbers on machines for which fortran is not
C *** extended to include hexadecimal formatting.
        DIMENSION F(1),IH(6)
        CHARACTER*1 HC(16),HI(64)
        DATA HC/'0','1','2','3','4','5','6','7','8',
     # '9','A','B','C','D','E','F'/
        READ(1,*)NMIN,NMAX
        NMIN=1
        NMAX=8190
C       WRITE(*,109)NMIN,NMAX
109     FORMAT(' NMIN, NMAX',2I6)
        NOPE=0
        NF=1
        N1=1
        N2=32
        DO 185 ISKIP=1,8
185     READ(1,1994)
10      READ(1,1994,END=50)(HI(I),I=N1,N2)
        N1=1
1994    FORMAT(9X,32A1)
11      IF(NOPE.LT.NMIN)GOTO 17
        DO 15 L=1,6
        J=L+N1-1
        DO 12 K=1,16
        IF(HI(J).EQ.' ')GOTO 15
        IF(HI(J).EQ.HC(K))GOTO 15
12      CONTINUE
15      IH(L)=K-1
        IF(NF.GE.1)F(NF)=IH(2)+16*IH(1)+256*(IH(4)+16*IH(3)+256*
     # (IH(6)+16*IH(5)))
        NF=NF+1
17      NOPE=NOPE+1
        IF(NOPE.GT.NMAX)GOTO 50
        N1=N1+6
        IF(N1+5.LT.N2)GOTO 11
        NT=N2-N1+1
        DO 20 I=1,NT
20      HI(I)=HI(I+N1-1)
        N1=NT+1
        N2=N1+31
        GOTO 10
50      CONTINUE
        XE=NOPE-1
60      CONTINUE
C0      WRITE(*,112)XE,NOPE
112     FORMAT(' XE, NOPE',F7.0,I6)
        IF(F(NF-1).NE.0)RETURN
        NOPE=NOPE-1
        NF=NF-1
        XE=XE-1
        GOTO 60
        END
C*********************************************************************
        SUBROUTINE CORREC(F,NT)
C *** This routine corrects corrupted data by linearly interpolating
C *** from the last low channel good point to the first high channel
```

```
C *** good point.
      CHARACTER*64 NA
      DIMENSION F(4100,2),NT(2),IHDAT(10)
      DIMENSION UFILE(2),CFILE(2),
     #FBEGC2(2),FENDC2(2),IBEGC2(2),IENDC2(2)
      CHARACTER*17 UFILE,CFILE,XYFILE,FPNAME,FMNAME,TITLE
      COMMON/GLFILE/UFILE,CFILE,XYFILE,FPNAME,FMNAME,TITLE
      COMMON/GLINK/CORR,DIFF,DIFPLT,RAT1,RAT2,FBEGC2,FENDC2,
     #IMIN2,IMAX2,LFLAG2,
     #IBEGC2,IENDC2,
     #IBEGC1,IENDC1,
     #IMIN1,IMAX1,
     #LFLAG1
      COMMON/BATCFL/IBATCH
      DO 10 I=1,2
      IF(NT(I).EQ.0)GOTO 10
C
C *** LINK IN MENU SELECTED IBEG,IEND
C
      IBEG=IBEGC2(I)
      IEND=IENDC2(I)
C************************************
      IF(IBEG.GT.IEND)GOTO 10
      IBEGM=IBEG-1
      IENDP=IEND+1
C
C *** LINK IN MENU SELECTION
C
      IF(FBEGC2(I).NE.0.)THEN
      PRINT*,' FBEG,FEND',FBEGC2(I),FENDC2(I)
        F(IBEGM,I)=FBEGC2(I)
        F(IENDP,I)=FENDC2(I)
      ENDIF
C**************************
      DF=F(IENDP,I)-F(IBEGM,I)
      DF=DF/(IENDP-IBEGM)
      DO 5 J=IBEG,IEND
      F(J,I)=F(IBEGM,I)+(J-IBEGM)*DF
5     CONTINUE
C
C *** LINK IN MENU SELECTED FILE
C
      NA=CFILE(I)
C************************
      IF(NA.EQ.'NULL')GOTO 10
      OPEN(UNIT=4,FILE=NA,FORM='UNFORMATTED')
      XOFF=0.
      WRITE(4)NT(I),XOFF
      WRITE(4)(F(J,I),J=1,NT(I))
        CLOSE(4)
10    CONTINUE
      RETURN
      END
C****************************************************************
      SUBROUTINE G4LINK
C *** This routine links in the menu selected data for RAWDD
C *** operation
```

```
      DIMENSION FFCOM(80),FFVAR(80)
      CHARACTER*64 CSTR,VALU,NA
      DIMENSION UFILE(2),CFILE(2),FBEGC2(2),FENDC2(2),IBEGC2(2),
     #IENDC2(2)
      CHARACTER*17 UFILE,CFILE,XYFILE,FPNAME,FMNAME,TITLE
      COMMON/GLFILE/UFILE,CFILE,XYFILE,FPNAME,FMNAME,TITLE
      COMMON/GLINK/CORR,DIFF,DIFPLT,RAT1,RAT2,FBEGC2,FENDC2,
     #IMIN2,IMAX2,LFLAG2,
     #IBEGC2,IENDC2,
     #IBEGC1,IENDC1,
     #IMIN1,IMAX1,
     #LFLAG1
      CHARACTER*40 FFCOM,FFVAR
      COMMON/USPAGE/FFVAR
      COMMON/FCALL/NSTR,NCMENU
      NA='RAWDD.MNU'
      NCMENU=1
      CALL MENURD(NA)
      DO 10 J=1,3
        NS=J*20+2
        IT=INDEX(FFVAR(NS+1),',')
        IF(J.EQ.1) THEN
C1*******************************
          NS=NS+1
          UFILE(1)=FFVAR(NS)(1:IT-1)
          UFILE(2)=FFVAR(NS)(IT+1:IT+18)
          NS=NS+1
        IT=INDEX(FFVAR(NS),',')
        IF(FFVAR(NS)(IT+1:IT+3).EQ.'DEF') THEN
          IBEGC1=0
          IENDC1=0
        ELSE
          CSTR=FFVAR(NS)(1:IT-1)
          CALL CTON(CSTR,IMISS,ANUM)
          IF(IMISS.NE.0)THEN
C           WRITE(9,200)IMISS,J
200         FORMAT(' NOT AN INTEGER. MISSED CHARS=',2I8)
          ENDIF
          IBEGC1=ANUM
          CSTR=FFVAR(NS)(IT+1:IT+18)
          CALL CTON(CSTR,IMISS,ANUM)
          IF(IMISS.NE.0)THEN
C           WRITE(9,200)IMISS,J
          ENDIF
          IENDC1=ANUM
        ENDIF
          NS=NS+1
          IT=INDEX(FFVAR(NS),',')
        IF(FFVAR(NS)(IT+1:IT+3).EQ.'DEF') THEN
          IMIN1=0
          IMAX1=0
        ELSE
          CSTR=FFVAR(NS)(1:IT-1)
          CALL CTON(CSTR,IMISS,ANUM)
          IMIN1=ANUM
          CSTR=FFVAR(NS)(IT+1:IT+18)
          CALL CTON(CSTR,IMISS,ANUM)
```

```
         IF(IMISS.NE.0)THEN
         ENDIF
         IMAX1=ANUM
       ENDIF
         NS=NS+1
         CSTR=FFVAR(NS)(1:3)
         IF(CSTR.EQ.'LOG')THEN
           LFLAG1=1
         ELSE
           LFLAG1=0
         ENDIF
         NS=NS+1
         TITLE=FFVAR(NS)(1:17)
C2*****************************************************
       ELSE IF(J.EQ.2) THEN
C5********************************
         NS=NS+1
         CSTR=FFVAR(NS)(1:1)
         DIFF=0
         IF(CSTR.EQ.'Y')THEN
           DIFF=1
           CSTR=FFVAR(NS+1)(1:1)
           IF(CSTR.EQ.'Y')THEN
             DIFPLT=1
           ELSE
              DIFPLT=0
           ENDIF
           XYFILE='NULL'
           CSTR=FFVAR(NS+2)(1:17)
           FPNAME=CSTR
           CSTR=FFVAR(NS+3)(1:17)
           FMNAME=CSTR
       IT=INDEX(FFVAR(NS+4),',')
       IF(FFVAR(NS+4)(IT+1:IT+3).EQ.'DEF') THEN
         IMIN2=0
         IMAX2=0
       ELSE
         CSTR=FFVAR(NS+4)(1:IT-1)
         CALL CTON(CSTR,IMISS,ANUM)
         IMIN2=ANUM
         CSTR=FFVAR(NS+4)(IT+1:IT+18)
         CALL CTON(CSTR,IMISS,ANUM)
         IMAX2=ANUM
       ENDIF
         CSTR=FFVAR(NS+5)(1:3)
         IF(CSTR.EQ.'LOG')THEN
           LFLAG2=1
         ELSE
           LFLAG2=0
         ENDIF
       IT=INDEX(FFVAR(NS+6),',')
       IF(FFVAR(NS+6)(IT+1:IT+3).EQ.'DEF') THEN
         RAT1=1
         RAT2=1
       ELSE
         CSTR=FFVAR(NS+6)(1:IT-1)
         CALL CTON(CSTR,IMISS,ANUM)
```

```
            RAT1=ANUM
            CSTR=FFVAR(NS+6)(IT+1:IT+18)
            CALL CTON(CSTR,IMISS,ANUM)
             RAT2=ANUM
          ENDIF
              ELSE
                DIFPLT=0
                IMIN2=0
                IMAX2=0
                LFLAG2=0
                XYFILE='NULL'
                FPNAME='NULL'
                FMNAME='NULL'
                RAT1=1
                RAT2=1
              ENDIF
C2*******************************
          ELSE IF(J.EQ.3) THEN
C3********************************
            NS=NS+1
            CSTR=FFVAR(NS)(1:1)
            CORR=0
            IF(CSTR.EQ.'Y')THEN
              CORR=1
C*****************************************************************
C*****************************************************************
             DO 20 I=1,2
            IT=INDEX(FFVAR(NS+1),'/')
            IF(I.EQ.1)THEN
              VALU=FFVAR(NS+1)(1:IT-1)
            ELSE
              VALU=FFVAR(NS+1)(IT+1:37)
            ENDIF
          IT=INDEX(VALU,',')
          IF(VALU(IT+1:IT+3).EQ.'DEF') THEN
            IBEGC2(I)=0
            IENDC2(I)=0
          ELSE
            CSTR=VALU(1:IT-1)
            CALL CTON(CSTR,IMISS,ANUM)
            IBEGC2(I)=ANUM
            CSTR=VALU(IT+1:IT+18)
            CALL CTON(CSTR,IMISS,ANUM)
            IENDC2(I)=ANUM
          ENDIF
C***************************************************************
C***************************************************************
        IT=INDEX(FFVAR(NS+2),'/')
        IF(I.EQ.1)THEN
          VALU=FFVAR(NS+2)(1:IT-1)
        ELSE
          VALU=FFVAR(NS+2)(IT+1:37)
        ENDIF
             CSTR=VALU(1:4)
             IF(CSTR.NE.'NONE')THEN
               CFILE(I)=VALU
             ELSE
```

```
            CFILE(I)='NULL'
           ENDIF
          IT=INDEX(FFVAR(NS+3),'/')
          IF(I.EQ.1)THEN
            VALU=FFVAR(NS+3)(1:IT-1)
          ELSE
            VALU=FFVAR(NS+3)(IT+1:37)
          ENDIF
        IT=INDEX(VALU,',')
        IF(VALU(IT+1:IT+3).EQ.'DEF') THEN
          FBEGC2(I)=0
          FENDC2(I)=0
        ELSE
          CSTR=VALU(1:IT-1)
          CALL CTON(CSTR,IMISS,ANUM)
          FBEGC2(I)=ANUM
          CSTR=VALU(IT+1:IT+18)
          CALL CTON(CSTR,IMISS,ANUM)
          FENDC2(I)=ANUM
        ENDIF
20    CONTINUE
         ELSE
           DO 30 I=1,2
           IBEGC2(I)=0
           IENDC2(I)=0
           CFILE(I)='NULL'
           FBEGC2(I)=0
           FENDC2(I)=0
30         CONTINUE
         ENDIF
C3******************************
      ENDIF
10    CONTINUE
        RETURN
        END
```

This is a listing of the RAWDD.MNU file; this file is read by the MENURD subroutine and the variables are passed to the RAWDD program to give it its options.

```
          5
          7
          9
          6
        100
 1 PAGE RAWDD
 2
 3 GENERAL INFORMATION FOR THE DISPLAY
 4 DIFFERENCE PLOT INFORMATION
 5 CORRECTED PLOT INFORMATION
 1 PAGE RAWDD1 (GENERAL INFORMATION)
 2
 3 FILES TO DISPLAY F(1),F(2)              ROBPC.SP,\
 4 INITIAL AND FINAL CHANNELS              1,1000
 5 MINIMUM AND MAXIMUM PEAK HEIGHTS        1,DEF
 6 LOG OR LINEAR PLOT                      LIN
 7 TITLE FOR THE GRAPH                     PHOBOS
```

```
 1 PAGE RAWDD2 (DIFFERENCE INFORMATION)
 2
 3 DO YOU WANT A DIFFERENCE PLOT (Y/N)      NO
 4   DIVIDE BY THE ERROR (Y/N)              NO
 5   THE FILE NAME FOR SUM                  PHOSUM.UF
 6   THE FILE NAME FOR DIFFERENCE           PHODIF.UF
 7   MIN AND MAX PEAK HEIGHTS               1,DEFAULT
 8   LOG OR LINEAR Y SCALE                  LOG
 9   NORMALIZATION FOR FILES 1 AND 2        1,1
 1 PAGE RAWDD3 (CORRECTION INFORMATION)
 2
 3 DO YOU WANT TO CORRECT THE DATA (Y/N)    NO
 4   CHANNEL RANGE TO BE CORRECTED          488,493/1,DEF
 5   FILE NAME FOR CORRECTED DATA           CORR1.UF/NONE
 6   BEGINNING AND ENDING HEIGHTS           2.3,1.2/1,DEF
```

The following is a listing of the standard generating code. A standard consists of up to 10 back to back cubic splines with varying heights, widths and locations representing fits to Gaussian peaks, Lorentzian peaks, Voigt peaks, or experimentally determined standard peak shapes. The standard determined here is used by FSPFIT to fit the data in terms of these standards plus background and by FSPDIS to display the results of the fit.

```
      PROGRAM STGEN
C*********************************************************************
C     The standard generating program (STGEN)
C*********************************************************************
      EXTERNAL GAUSS,FLOREN,EXALG,ALICOS
      REAL*8 CHI
        CHARACTER*64 NA
        CHARACTER*1 IAN,NO
        CHARACTER*4 INT1,INT,IGAUSS,ILOREN,ZERO,BKG,VOIGT,LCDAT,HDAT
        CHARACTER*4 IDAT,Z4DAT
        COMMON/DATA/XI(512),FI(512),NP
        DIMENSION IHD(512)
        COMMON/KNOTS/W(10),XP(10),C(10),P(34),NSC,NB,CB(4),XB
        COMMON/PASS/ETA,XSUP(100),YSUP(100),NPSUP
        EQUIVALENCE (IHD(1),FI(1))
        CHARACTER*4 II(40)
C**************************************************
C
C   VSHAPE DATA
C
C**************************************************
      CHARACTER*17 STNAME,TYNAME
      COMMON/CHVSHA/STNAME,TYNAME
      COMMON/CVSHAP/NBB,NS
C**************************************************
C   VOIGHT DATA
C**************************************************
      COMMON/CVOIGH/ETAA
C**************************************************
        DATA IGAUSS,ILOREN,NO,ZERO/'GAUS','LORE','N','ZERO'/
        DATA BKG,VOIGHT/'BKGR','VOIG'/,LCDAT/'LCDA'/,HDAT/'HDAT'/
        DATA IDAT/'IDAT'/,Z4DAT/'Z4DA'/
C****************************
```

```
C  CALL THE MENU PAGES
C****************************
      CALL SHLINK
      NB=NBB
      ETA=ETAA/DSQRT(DLOG(2.D0))
C *** ETAA IS THE RATION OF THE DOPPLER PEAK TO THE LORENTZIAN
C *** FLOREN IS A LORENTZIAN, GAUSS IS A GAUSSIAN
C         CALL ERRSET(208,256,-1,1)
C     CALL ERRSET(212,256,-1,1)
106     FORMAT(A64)
C       WRITE(*,107)
107     FORMAT(' ENTER THE FILE FOR THE STANDARD PEAK PARAMETERS')
        NA=STNAME
        OPEN(UNIT=12,FILE=NA)
        INT1=TYNAME
101      FORMAT(A4)
        IF(NB.EQ.0)CB(1)=0
        IF(NB.LT.2)CB(2)=0
        IF(INT1.EQ.IGAUSS.OR.INT1.EQ.ILOREN.OR.INT1.EQ.VOIGHT)GOTO 60
C *** NB IS THE NUMBER OF COEFFICIENTS IN THE BACKGROUND POLY
C *** NS IS THE MAXIMUM NUMBER OF SPLINES USED TO FIT THE SHAPE
        CALL VBREAD(II)
60       II(1)=INT1
         II(2)='    '
        IF(INT1.EQ.VOIGHT)GOTO 90
        IF(INT1.EQ.IGAUSS)GOTO 80
        IF(INT1.EQ.ILOREN)GOTO 70
        GOTO 290
70      CALL BLI(XI,FI,-32.E0,32.E0,NP,200,FLOREN,1)
        GOTO 290
80      CALL BLI(XI,FI,-3.E0,3.E0,NP,200,GAUSS,1)
        GOTO 290
90      CONTINUE
CO      WRITE(*,123)
123     FORMAT(' WHAT VALUE FOR ETA?')
C       READ(*,*)ETA
        RANGE=(2/ETA**2)*(-1+SQRT(1+85*ETA**2))
        WRITE(*,124)RANGE
124     FORMAT(' THE RANGE OF X IS 0 TO',F10.5)
        CALL BLI(XSUP,YSUP,0.E0,RANGE,NPSUP,100,EXALG,0)
        CALL BLI(XI,FI,-32.E0,32.E0,NP,200,ALICOS,1)
290     XB=XI(1)
        NSC=1
300   CALL VQMIN(NS,CHI)
        CALL VSPLOT(II,CHI)
        CALL LMAX
        STOP
        END
C******************************************************************
        SUBROUTINE LCHEX(F,NF,XE)
C *** This converts character variables to hexadecimal and in
C *** addition corrects for the strange way Lecroy multichannel
C *** analyzers store bytes on tape.
        DIMENSION F(1),IH(6)
        CHARACTER*1 HC(16),HI(64)
        DATA HC/'0','1','2','3','4','5','6','7','8',
     # '9','A','B','C','D','E','F'/
```

```
          READ(8,*)NMIN,NMAX
C         WRITE(*,109)NMIN,NMAX
109       FORMAT(' NMIN, NMAX',2I6)
          NOPE=0
          NF=1
          N1=1
          N2=32
          DO 185 ISKIP=1,8
185       READ(8,1994)
10        READ(8,1994,END=50)(HI(I),I=N1,N2)
          N1=1
1994      FORMAT(9X,32A1)
11        IF(NOPE.LT.NMIN)GOTO 17
          DO 15 L=1,6
          J=L+N1-1
          DO 12 K=1,16
          IF(HI(J).EQ.HC(K))GOTO 15
12        CONTINUE
15        IH(L)=K-1
          IF(NF.GE.1)F(NF)=IH(2)+16*IH(1)+256*(IH(4)+16*IH(3)+256*
     #    (IH(6)+16*IH(5)))
          NF=NF+1
17        NOPE=NOPE+1
          IF(NOPE.GT.NMAX)GOTO 50
          N1=N1+6
          IF(N1+5.LT.N2)GOTO 11
          NT=N2-N1+1
          DO 20 I=1,NT
20        HI(I)=HI(I+N1-1)
          N1=NT+1
          N2=N1+31
          GOTO 10
50        CONTINUE
          XE=NOPE-1
60        WRITE(*,112)XE,NOPE
112       FORMAT(' XE, NOPE',F7.0,I6)
          IF(F(NF-1).NE.0)RETURN
          NOPE=NOPE-1
          NF=NF-1
          XE=XE-1
          GOTO 60
          END
C*********************************************************************
          SUBROUTINE VQMIN(NS,CHI)
C *** This routine performs the minimization process for the standard
C *** being fitted. It controls the values of the constants used in
C *** the representation of the standard (CONS).
          COMMON/DATA/XI(512),FI(512),NP
          REAL*8 CHI,CHB,CHL,PC,PPCC,A,FR,CONS,SM,AM
          DIMENSION CONS(34),SM(34),PC(34),PPCC(595),FRES(512)
     #    ,AM(595),EC(10),EXP(10),EW(10)
          COMMON/KNOTS/W(10),XP(10),C(10),P(34),NSC,NB,CB(4),XB
C *** ASSUME THE XI'S ARE IN INCREASING ORDER AND DETERMINE
C *** AN APPROPRIATE PENALTY TERM
          NIB=NP/4
          NIE=(3*NP)/4
          ASP=2*ABS(XI(NIB)-XI(NIE))/NP
```

```
      WRITE(*,1932)NIB,NIE,ASP
1932  FORMAT(' NIB,NIE,ASP',2I5,E15.6)
      IF(NSC.GT.1)GOTO 16
      DO 14 J=2,NB
14    CB(J)=0.E0
      CB(1)=AMIN1(FI(1),FI(NP))
      IF(NB.EQ.1)GOTO 15
      CB(1)=FI(1)
      CB(2)=(FI(NP)-FI(1))/NP
15    CONTINUE
      DO 1501 I=1,NP
1501  FRES(I)=(FI(I)-CB(1)-CB(2)*(XI(I)-XB))/SQRT(AMAX1(1.,
     # FI(I)))
      CALL RESIDL(XI,FI,FRES,NP,C(1),XP(1),W(1))
      W(1)=AMAX1(ASP,W(1))
16    IF(NSC.GT.1)NSC=NSC-1
      DO 18 J=1,NB
      CONS(J)=CB(J)
18    SM(J)=10.**(J-1)
23    CONTINUE
      IE=0
      CHB=1.E31
      CHL=1.E32
      FR=0.95
      NT=NB+3*NSC
      NPART=NB+1
      DO 24 J=1,NSC
      CONS(NPART)=C(J)
      SM(NPART)=1.E-2
      CONS(NPART+1)=XP(J)
      SM(NPART+1)=SM(NPART)*1.D6
      CONS(NPART+2)=W(J)
      SM(NPART+2)=SM(NPART)*1.D4
24    NPART=NPART+3
      NKIT=30
      DO 2500 KIT=1,NKIT
      IF(KIT.EQ.NKIT)CHB=CHL
      CHI=0.
      K=0
      DO 30 J=1,NT
      PC(J)=0.
      DO 30 I=1,J
      K=K+1
30    PPCC(K)=0.
      DO 200 I=1,NP
      FA=VPOLY(XI(I))
      FRES(I)=FI(I)-FA
      FRES(I)=FRES(I)/SQRT(AMAX1(1.E0,FI(I),FA))
      CHI=CHI+FRES(I)**2
      IF(1.E0.GT.FA.AND.1.E0.GT.FI(I))GOTO 60
      IF(FA.LT.FI(I))GOTO 50
      W1=(1.-(FI(I)/FA)**2)
      W2=2.*(FI(I)/FA)**2/FA
      GOTO 70
50    W1=2.*(FA/FI(I)-1.)
      W2=2./FI(I)
      GOTO 70
```

```
60     W1=2.E0*(FA-FI(I))
         W2=2.E0
70     CONTINUE
         L=0
         DO 80 J=1,NT
         PC(J)=PC(J)+W1*P(J)
         W2P=W2*P(J)
         DO 80 K=1,J
         L=L+1
80     PPCC(L)=PPCC(L)+P(K)*W2P
200    CONTINUE
         NPART=NB+3
         PEN=0.E0
         DO 210 J=1,NSC
         IF(W(J).GE.ASP)GOTO 210
         PT=(ASP-W(J))**3
         PEN=PEN+PT
         CHI=CHI+PT
         PC(NPART)=PC(NPART)-3.E0*(ASP-W(J))**2
         L=NPART*(NPART+1)/2
         PPCC(L)=PPCC(L)+6.*(ASP-W(J))
210    NPART=NPART+3
         IF(CHL-CHB.GT..1E-1)GOTO 248
         IF(KIT.LE.3)GOTO 248
         IF(IE.EQ.1)GOTO 2502
         IE=1
         IF(CHI.LT.CHL)GOTO 2502
248    CALL SMSQ(CHI,CHB,CHL,PC,PPCC,AM,FR,CONS,SM,NT,1)
         DO 250 J=1,NB
250    CB(J)=CONS(J)
         NPART=NB+1
         DO 260 J=1,NSC
         C(J)=CONS(NPART)
         XP(J)=CONS(NPART+1)
         W(J)=ABS(CONS(NPART+2))
260    NPART=NPART+3
2500    CONTINUE
2502   CONTINUE
         IF(NSC.GE.NS)GOTO 3000
         CALL RESIDL(XI,FI,FRES,NP,C(NSC+1),XP(NSC+1),W(NSC+1))
         W(NSC+1)=AMAX1(ASP,W(NSC+1))
3000   CONTINUE
         WRITE(*,104)NP
104    FORMAT(' WE ARE FITTING',I5,' POINTS')
         CHI=CHI-PEN
         WRITE(*,102)CHI,(I,CB(I),I=1,NB)
102    FORMAT(' CHISQUARE IS',E20.6/' THE BKGRD CONS ARE'/(I5,E20.6))
         IF(NSC.GT.0)WRITE(*,103)(I,C(I),XP(I),W(I),I=1,NSC)
103    FORMAT(' THE B SPLINE COEFFS'/'    #',8X,'C(I)',17X,'XP(I)',
     # 18X,'W(I)'/(I5,3E22.8))
         CHIT=CHI-NP+NB+3*NSC
       IF(NSC.GE.NS)GOTO 3200
       NSC=NSC+1
       IF(NSC.LE.NS)GOTO 23
3200   NSC=NS
         RETURN
         END
```

```
C***********************************************************************
        FUNCTION VPOLY(X)
C *** THIS ROUTINE CALCULATES POLY=SUM CI*SI ALONG WITH DPOLY/DCI
C *** THE FIRST NB SI'S ARE THE BACKGROUND POLYNOMIAL
C *** THE NEXT NS SI'S, WHICH HAVE DERIVATIVES WITH RESPECT TO
C *** C,XP,AND W ARE BSPLINES
        COMMON/KNOTS/W(10),XP(10),C(10),P(34),NS,NB,CB(4),XB
C       DIMENSION P(1)
        VPOLY=CB(1)
        IF(NB.EQ.0)VPOLY=0
        P(1)=1.
        IF(NB.LT.2)GOTO 120
        DO 100 J=2,NB
        P(J)=(X-XB)**(J-1)
100   VPOLY=VPOLY+CB(J)*P(J)
120   NPART=NB+1
        DO 200 J=1,NS
        P(NPART)=0.
        P(NPART+1)=0.
        P(NPART+2)=0.
        XM=X-XP(J)
        IF(XM.LE.-W(J).OR.XM.GE.W(J))GOTO 200
        XM=XM/W(J)
C  PARTIAL WRT C(J)
        P(NPART)=2.*C(J)*((1.+XM)*(1.-XM))**3
        FADD=.5*C(J)*P(NPART)
        VPOLY=VPOLY+FADD
C PARTIAL WRT XP(J)
        P(NPART+1)=-3.*(FADD/(1.+XM)-FADD/(1.-XM))/W(J)
C PARTIAL WRT W(J)
        P(NPART+2)=P(NPART+1)*XM
200   NPART=NPART+3
        RETURN
        END
C***********************************************************************
        SUBROUTINE LMAX
C *** Finds the maximum of the fitted peak and its full width at half
C *** maximum.  Then converts the fitted peak to a standard peak
C *** which has a maximum and full width at half maximum of one.
C *** Finally a crude log plot is made on the screen which if
C *** observed would inform the user of obvious errors.
        CHARACTER*1 IA(80),IBL,IP
        COMMON/KNOTS/W(10),XP(10),C(10),P(34),NS,NB,CB(4),XB
        DIMENSION CS(10),ISS(5)
        DATA IBL,IP/' ','+'/
C     FIND MAX X AND MIN X OF PEAK
        XPS=1.D32
        XPL=-1.D32
        DO 30 J=1,NS
        XPS=AMIN1(XPS,XP(J)-W(J))
30    XPL=AMAX1(XPL,XP(J)+W(J))
       PRINT*,' XPS,XPL,C,XP,W',XPS,XPL,(C(I),XP(I),W(I),I=1,NS)
31    II=0
        H=(XPL-XPS)/200.
        X=XPS+H/2.
33    CONTINUE
        FB=0
```

```
        DO 40 I=1,200
         F=POLYF(X)
         IF(FB.GE.F)GOTO 40
         FB=F
         XB=X
40      X=X+H
        II=II+1
        IF(II.GE.4)GOTO 50
        X=XB-H
        H=H*.01
        GOTO 33
50      CONTINUE
        DO 55 I=1,NS
        C(I)=C(I)/SQRT(FB)
        XP(I)=XP(I)-XB
        CS(I)=C(I)**2
55      CONTINUE
        XPS=XPS-XB
        XPL=XPL-XB
C FINDING THE HALF MAXIMUM POINTS
        XLH=-1
        CALL VXHMAX(XLH)
        XUH=1
        CALL VXHMAX(XUH)
        WRITE(*,105)XLH,XUH
105     FORMAT(' THE VALUES FOR WHICH F IS .5 ARE',2E20.12)
        AM=1./(XUH-XLH)
        XPS=XPS*AM
        XPL=XPL*AM
        DO 65 I=1,NS
        XP(I)=XP(I)*AM
65      W(I)=W(I)*AM
        WRITE(*,106)XPS,XPL
106     FORMAT(' THE STANDARD COVERS THE INTERVAL',2E20.12)
        WRITE(*,103)
103     FORMAT(' THE FOLLOWING ARE THE CONSTANTS READY FOR RLFIT')
        WRITE(12,101)(I,CS(I),XP(I),W(I),I=1,NS)
101     FORMAT(I5,3E20.7)
        WRITE(*,101)(I,CS(I),XP(I),W(I),I=1,NS)
C GRAPHING THE STANDARD PEAK
        H=(XPL-XPS)/50.
        X=XPS+H/2.
        AI=0.
        IS=-4
        DO 72 I=1,5
        ISS(I)=IS
72      IS=IS+1
        WRITE(*,108)ISS
108     FORMAT(5I10)
        DO 80 I=1,50
        F=POLYF(X)
        AI=AI+.02*F
        DO 76 J=1,80
76      IA(J)=IBL
        IC=0
        IF(F.GT.0.E0)       IC=20.*ALOG10(F)+80.0000001E0
        IC=MIN0(80,IC)
```

```
      IF(IC.GE.1)IA(IC)=IP
      WRITE(*,102)X,F,IA
80    X=X+H
102   FORMAT(2E12.4/80A1)
      WRITE(*,107)XB,XB+XLH,XB+XUH,AI
107   FORMAT(' XB=',E14.6,' HALF MAXS AT',2E14.6,' AI=',E14.6)
      RETURN
      END
C*********************************************************************
      FUNCTION POLYF(X)
C *** THIS ROUTINE CALCULATES POLYF=SUM CI*SI
      COMMON/KNOTS/W(10),XP(10),C(10),P(34),NS,NB,CB(4),XB
      POLYF=0.
      DO 200 J=1,NS
      XM=X-XP(J)
      IF(XM.LE.-W(J).OR.XM.GE.W(J))GOTO 200
      XM=XM/W(J)
      FADD=C(J)*C(J)*((1.+XM)*(1.-XM))**3
      POLYF=POLYF+FADD
200   CONTINUE
      RETURN
      END
C*********************************************************************
      SUBROUTINE VSPLOT(II,CHIS)
C *** This routine writes the graphics file SHAPE.GR that is used in
C *** the STDIS mode.
      REAL*8 CHIS
      COMMON/DATA/XI(512),FI(512),NP
      COMMON/KNOTS/W(10),XP(10),C(10),P(34),NSC,NB,CB(4),XB
      CHARACTER*4 II(20)
C1    DIMENSION P(34)
      INTEGER*2 IDIV,IXMUL,IX,IRES,IDAT,IFA,IFB,ICH(6)
      DATA IFA,IFB,ICH/8*0/
C FIND HIGHEST VALUE TO BE PLOTTED
      FMAX=-1.E32
      DO 100 J=1,NP
      FMAX=AMAX1(FI(J),FMAX)
100   CONTINUE
      IDIV=1+FMAX/32767.
      IFA=FMAX/IDIV
      IXMUL=32767/XI(NP)
      WRITE(*,1234)IDIV,IFA,IXMUL
1234  FORMAT(' IDIV,IFA,IXMUL',3I10)
      OPEN(UNIT=3,FILE='SHAPE.GR',FORM='UNFORMATTED')
      WRITE(3)IDIV,IXMUL,(II(I),I=1,2),CHIS
      DO 1090 J=1,NP
      FA=VPOLY(XI(J))
      IRES=AMIN1(32767.,100*(FI(J)-FA)/SQRT(AMAX1(1.,FA,FI(J))))
      IFA=AMIN1(32767.,FA/IDIV)
      IX=AMIN1(32767.,XI(J)*IXMUL)
      IDAT=AMIN1(32767.,FI(J)/IDIV)
      IF(NSC.EQ.0)GOTO 1085
C *** NOW FOR THE PEAKS
      DO 1070 I=1,6
1070  ICH(I)=1
      KP=0
      DO 1080 K=1,NSC
```

```
      KP=KP+1
      IF(KP.EQ.7)KP=1
      ARG=.5*C(K)*P(NB+1+3*(K-1))
      IF(ARG.LT..001*C(K)*C(K))GOTO 1080
      ICH(KP)=AMIN1(32767.,ARG/IDIV)
1080  CONTINUE
1085  CONTINUE
1902  FORMAT(10I6)
      WRITE(3)IRES,IDAT,IFA,IX,ICH
1090  CONTINUE
      RETURN
      END
C*********************************************************************
      SUBROUTINE RESIDL(XI,FI,FR,NP,C,XP,W)
C *** Finds the location at which the next back to back cubic spline
C *** should be started for fitting purposes. Calls VRESL
      DIMENSION XI(1),FI(1),FR(1)
      CALL VRESL(FR,ALR,JB,NP)
       PRINT*,' ALR,JB,NP',ALR,JB,NP
       XP=XI(JB)
      C=ALR*SQRT(ABS(FI(JB)))
      C=SQRT(ABS(C))
      JS=MAX0(1,JB-2)
      JL=MIN0(NP,JB+2)
50    W=XI(JL)-XI(JS)
      WRITE(*,101)C,XP,W
101   FORMAT(' IN ZEROTH APPROX C,X,W'/3D12.5)
      RETURN
      END
C*********************************************************************
      SUBROUTINE VRESL(FRES,ALR,ILR,N)
C *** This routine calculates the position of the largest residual.
C *** This is the position at which an additional bspline will be
C *** added if a closer match to the peak shape is required.
      DIMENSION FRES(512)
20    ALR=0
      NM2=N-2
      FSP2=FRES(2)
      FSP1=FRES(1)
      FS=0
      FSM1=0
      FSM2=0
      FSM3=0
      SUM=FSP1+FSP2
     DO 40 I=1,N
      FSM3=FSM2
      FSM2=FSM1
      FSM1=FS
      FS=FSP1
      FSP1=FSP2
      FSP2=0
      IF(I.LT.NM2)FSP2=FRES(I+2)
      SUM=SUM+FSP2-FSM3
     IF(SUM.LT.ALR)GOTO 40
     ILR=I
     ALR=SUM
40   CONTINUE
```

```
        ALR=ALR/2.236
      RETURN
      END
C*******************************************************************
      SUBROUTINE BLI(XI,FI,B,E,NP,N,FLOREN,NW)
C *** The best linear interpolator.  Finds the f(x(i))-interpolation
C ***  (between f(xi-1) and f(xi+1)).  Finds the largest of this
C *** multiplied by x(i+1)-x(i-1).  Places new points between x(i-1)
C *** and x(i) and also between x(i) and x(i+1).  The result is a set
C *** of values of x and f which are spaced so as to achieve the
C *** "best" possible linear interpolation.
      DIMENSION D(200),XI(512),FI(512)
101     FORMAT(' IN BLI WITH B=',F8.3,' E=',F8.3,' AND N=',I5)
        NP=3
        XI(1)=B
        FI(1)=FLOREN(XI(1))
        XI(2)=(B+E)/2
        FI(2)=FLOREN(XI(2))
        XI(3)=E
        FI(3)=FLOREN(XI(3))
        D(1)=-1
        D(3)=-1
        IM=2
90      IB=MAX0(2,IM-1)
        IE=MIN0(NP-1,IM+3)
        DO 100 I=IB,IE
        D(I)=ABS(FI(I)-FI(I-1)-((XI(I)-XI(I-1))/(XI(I+1)-XI(I-1)
     #))*(FI(I+1)-FI(I-1)))
        D(I)=ABS(D(I)*(XI(I+1)-XI(I-1)))
100      CONTINUE
C *** FINDING THE NEW IM
        IM=2
        DM=D(2)
        DO 120 I=3,NP
        IF(DM.GT.D(I))GOTO 120
        DM=D(I)
        IM=I
120     CONTINUE
102     FORMAT(' IN BLI DM, IM',E20.6,I5)
        FSAVE=FI(IM)
        XSAVE=XI(IM)
        XP=.5*(XI(IM)+XI(IM+1))
        XM=.5*(XI(IM)+XI(IM-1))
C *** SHOVE THE STACK UP
        NMOVE=NP-IM
        J=NP
        DO 160 I=1,NMOVE
        XI(J+2)=XI(J)
        FI(J+2)=FI(J)
        D(J+2)=D(J)
160      J=NP-I
        XI(IM)=XM
        FI(IM)=FLOREN(XM)
        XI(IM+1)=XSAVE
        FI(IM+1)=FSAVE
        XI(IM+2)=XP
        FI(IM+2)=FLOREN(XP)
```

```
      NP=NP+2
      IF(NP.LT.N-2.AND.NP.LT.198)GOTO 90
      IF(NW.NE.1)RETURN
      WRITE(*,1985)NP
1985  FORMAT(I5)
      WRITE(*,1986)
1986    FORMAT(18X,'XI',18X,'FI')
      DO 170 I=1,NP
170   WRITE(*,1987)XI(I),FI(I)
1987 FORMAT(2E20.6)
      RETURN
      END
C*********************************************************************
      FUNCTION GAUSS(X)
C *** Used to generate a Gaussian peak shape
      GAUSS=0.E0
      X2=X*X
      IF(X2.GT.85.E0)RETURN
      GAUSS=1000.*EXP(-X2)
      RETURN
      END
C*********************************************************************
      FUNCTION FLOREN(X)
C *** Used to generate a Lorentzian peak shape
      FLOREN=1000./(X*X+1.E0)
      RETURN
      END
C*********************************************************************
      FUNCTION EXALG(X)
C *** Used in finding a Voigt peak
      COMMON/PASS/ETA,XSUP(200),NPSUP
      EXALG=1000.*(1+ETA)*EXP(-X-(ETA*X/2)**2)
      RETURN
      END
C*********************************************************************
      FUNCTION ALICOS(ALPHA)
C *** Used in finding a Voigt peak (evaluates the convolution
C *** integral) CALCULATES VM INTEGRAL OF DCOS(ALPHA*X)*F(X)
      IMPLICIT REAL*8 (A-H,O-Z)
      COMMON/PASS/ETA,XI(100),F(100),NP
       REAL*4 ALICOS,ALPHA,ETA,XI,F
      AIFUN(EF0,EF1,EF2,EF3,X)=DCOS(X)*(EF3*(3*X*X-6)+EF2*2*X
     # +EF1)+DSIN(X)*(EF3*X*(X*X-6)+EF2*(X*X-2)+EF1*X+EF0)
      ALPA=(1+ETA)*ALPHA
      IF(DABS(ALPA).LT.1.D-7)ALPA=1.D-7
      ALPAD=1/ALPA
      ALPAD2=ALPAD*ALPAD
      ALPAD3=ALPAD2*ALPAD
      ALPAD4=ALPAD3*ALPAD
      ALICOS=0
      NPMM=NP-2
      DO 100 I=2,NPMM
      X1=XI(I-1)
      X2=XI(I)
      X3=XI(I+1)
      X4=XI(I+2)
      F1=F(I-1)
```

```
      F1=F1/((X1-X2)*(X1-X3)*(X1-X4))
      F2=F(I)/((X2-X1)*(X2-X3)*(X2-X4))
      F3=F(I+1)/((X3-X1)*(X3-X2)*(X3-X4))
      F4=F(I+2)/((X4-X1)*(X4-X2)*(X4-X3))
      EF0=X2*X3*X4*F1+X1*X3*X4*F2+X1*X2*X4*F3+X1*X2*X3*F4
      EF0=-EF0*ALPAD
      EF1=(X2*X3+X2*X4+X3*X4)*F1+(X1*X3+X1*X4+X3*X4)*F2
     # +(X1*X2+X1*X4+X2*X4)*F3+(X1*X2+X1*X3+X2*X3)*F4
      EF1=EF1*ALPAD2
      EF2=(X2+X3+X4)*F1+(X1+X3+X4)*F2+(X1+X2+X4)*F3+(X1+X2+X3)*F4
      EF2=-EF2*ALPAD3
      EF3=(F1+F2+F3+F4)*ALPAD4
      XUB=XI(I+1)*ALPA
      IF(I.EQ.NPMM)XUB=XI(NP)*ALPA
      XLB=XI(I)*ALPA
      IF(I.EQ.2)XLB=XI(1)*ALPA
      FUB=AIFUN(EF0,EF1,EF2,EF3,XUB)
      FLB=AIFUN(EF0,EF1,EF2,EF3,XLB)
      IF(DABS((FUB-FLB)/FLB).LT.1.D-5.AND.DABS(XUB-
     #XLB).LT..5D0)GOTO90
      ALICOS=ALICOS+FUB-FLB
      GOTO 100
90      XA=.5*(XUB+XLB)
        ALICOS=ALICOS+DCOS(XA)*(EF0+XA*(EF1+XA*(EF2+XA*EF3)))*(XUB-
     #XLB)
100     CONTINUE
        RETURN
        END
C*********************************************************************
        SUBROUTINE VBREAD(II)
C *** This routine reads in the data files used when generating a
C *** standard fron the raw data.  This routine is also a subset of
C *** BREAD
        COMMON/DATA/XI(512),F(512),NP
        DIMENSION IIDAT(512)
        EQUIVALENCE (IIDAT(1),F(1))
        CHARACTER*4 II(40)
        CHARACTER*64 NA
C**************************************************
C   BREAD DATA
C**************************************************
      CHARACTER*17 DATAFN
      COMMON/CHBREA/DATAFN
      COMMON/CBREAD/IBEGC,IENDC
C**************************************************
C *** F(N) IS THE SPECTRUM CURRENTLY BEING FITTED.
13      WRITE(*,109)
109     FORMAT(' ENTER THE NAME OF THE DATA FILE')
C       READ(*,1943)NA
        NA=DATAFN
        II(1)=NA(:4)
        II(2)=NA(5:8)
1943    FORMAT(A)
        WRITE(*,1001)
1001    FORMAT(' ENTER BEG CHANNEL, END CHANNEL!')
C       READ(*,*)N1,N2
        N1=IBEGC
```

```
          N2=IENDC
          WRITE(*,1178)N1,N2
1178      FORMAT(' IN BREAD N1,N2',2I5)
          IT=INDEX(NA,'.')
          IF(NA(IT+1:IT+3).EQ.'UF')GOTO 1100
C        PRINT*,' TESTING FOR SALLY ',NA(IT+1:IT+5)
         IF(NA(IT+1 :IT+5).EQ.'SALLY')GOTO 2000
          IRF=0
          IF(NA(IT+1:).EQ.'LEO')IRF=1
          WRITE(*,*)IRF
        IF(IRF.EQ.0)THEN
C         CALL CMS('FILEDEF 8 DISK '//NA,IRT)
          OPEN(UNIT=8,FILE=NA,ERR=13)
        ENDIF
        IF(IRF.EQ.1)THEN
C         CALL CMS('FILEDEF 8 DISK '//NA,IRT)
          OPEN(UNIT=8,FILE=NA,FORM='UNFORMATTED',ERR=13)
        ENDIF
          IF(IRF.NE.1)GOTO 420
          NREC=(N2-1)/128+1
          NU=128*NREC
          DO 416 III=1,NU,128
416       READ(8)(IIDAT(I),I=III,III+127)
          N=N2-N1+2
          XE=N2-N
          DO 400 I=1,N
          XI(I)=I+XE
          F(I)=IIDAT(I+N1-1)
400       CONTINUE
          NP=N
          CLOSE(8)
          RETURN
420     CONTINUE
          READ(8,101)II
C          PRINT*,'II(40)=',II(40)
101       FORMAT(10A4)
        IF(II(40).NE.'HDAT'.AND.II(40).NE.'IDAT'.AND.II(40).NE.'Z4DA')
       # GOTO 195
          IF(II(40).NE.'Z4DA')READ(8,*)NXB,NEND
          NMULT=1
          IF(II(40).EQ.'Z4DA')READ(8,*)NXB,NEND,NMULT
          NXB=N1
          NEND=N2
          I10=10
          IF(II(40).EQ.'Z4DA')I10=20
          NXBS=(NXB-1)/I10
          IF(NXBS.EQ.0)GOTO 145
          DO 144 I=1,NXBS
144       READ(8,156)
145       NIR=NEND-I10*NXBS
          NIR=MIN0(NIR,8192)
          IF(II(40).EQ.'HDAT')READ(8,156)(IIDAT(I),I=1,NIR)
          IF(II(40).EQ.'IDAT')READ(8,157)(IIDAT(I),I=1,NIR)
          IF(II(40).EQ.'Z4DA')READ(8,158)(IIDAT(I),I=1,NIR)
156       FORMAT(10Z8)
157       FORMAT(10I8)
158       FORMAT(20Z4)
```

```
      CLOSE(8)
      NP=NEND-NXB+1
      NADD=NXB-1-I10*NXBS
      XE=NEND-NP
      DO 190 I=1,NP
      XI(I)=I+XE
190   F(I)=IIDAT(I+NADD)*NMULT
      RETURN
195    N=0
200   N=N+1
      IF(N.GT.8192)GOTO 220
      READ(8,*,END=220)XI(N),F(N)
      GOTO 200
220   NP=N-1
      CLOSE(8)
      RETURN
1100   CONTINUE
       OPEN(UNIT=8,FILE=NA,FORM='UNFORMATTED')
       READ(8)N,XE
       READ(8)(F(I),I=1,N)
       N1=N1-XE
       NT=N+XE
       NC=MIN0(NT,N2)
       N2=N2-XE
       N1=MAX0(1,N1)
      N=N2-N1+1
      XE=N2-N
      DO 1120 I=1,N
      F(I)=F(I+N1-1)
      XI(I)=I+XE
1120  CONTINUE
       NP=N
       CLOSE(8)
       RETURN
2000  CONTINUE
       OPEN(UNIT=8,FILE=NA,FORM='UNFORMATTED')
      READ(8)(IIDAT(I),I=1,128)
      I1=0
      DO 2010 I=1,300
      READ(8,END=2015)(IIDAT(I1+J),J=1,128)
      I1=I1+128
2010  CONTINUE
2015  N=I1+J-1
      NXB=N1
      NEND=N2
      NP=NEND-NXB+1
      NADD=NXB-1
      XE=NEND-NP
      DO 2190 I=1,NP
      XI(I)=I+XE
2190  F(I)=IIDAT(I+NADD)
      END
C*******************************************************************
      SUBROUTINE SHLINK
C *** This routine links the menu selected data into the standard
C *** generating routine STGEN.
      DIMENSION FFCOM(200),FFVAR(200)
```

```
      CHARACTER*64 CSTR,VALU,NA
      COMMON/USPAGE/FFVAR
      COMMON/FCALL/NSTR,NCMENU
      CHARACTER*40 FFCOM,FFVAR
C************************************************
C
C   DATA TO MAIN ROUTINE VSHAPE
C
C************************************************
C
C   VSHAPE DATA
C
C************************************************
      CHARACTER*17 STNAME,TYNAME
      COMMON/CHVSHA/STNAME,TYNAME
      COMMON/CVSHAP/NBB,NS
C************************************************
C   BREAD DATA
C************************************************
      CHARACTER*17 DATAFN
      COMMON/CHBREA/DATAFN
      COMMON/CBREAD/IBEGC,IENDC
C************************************************
C   VOIGHT DATA
C************************************************
      COMMON/CVOIGH/ETAA
C************************************************
      NA='STGEN.MNU'
      NCMENU=1
      CALL MENURD(NA)
      DO 10 J=1,10
        NSS=J*20+2
        IF(J.EQ.1)THEN
C************************************************
      NSS=NSS+1
      STNAME=FFVAR(NSS)
      PRINT*,'STNAME=',STNAME
      NSS=NSS+1
C  NOT USED AT PRESENT BUT SET IN THE PROGRAM
      NSS=NSS+1
      TYNAME=FFVAR(NSS)
      NSS=NSS+1
      IT=INDEX(FFVAR(NSS),',')
      CSTR=FFVAR(NSS)(1:IT-1)
      CALL CTON(CSTR,IMISS,ANUM)
      NBB=ANUM
      CSTR=FFVAR(NSS)(IT+1:37)
      CALL CTON(CSTR,IMISS,ANUM)
      NS=ANUM
C*******************************
        ENDIF
        IF(J.EQ.2)THEN
C*******************************
      NSS=NSS+1
      CSTR=FFVAR(NSS)
      CALL CTON(CSTR,IMISS,ANUM)
      ETAA=ANUM
```

```
      NSS=NSS+1
        DATAFN=FFVAR(NSS)
        NSS=NSS+1
        IT=INDEX(FFVAR(NSS),',')
        IF(FFVAR(NSS)(IT+1:IT+3).EQ.'DEF')THEN
          IBEGC=0
          IENDC=0
        ELSE
          CSTR=FFVAR(NSS)(1:IT-1)
          CALL CTON(CSTR,IMISS,ANUM)
          IBEGC=ANUM
          CSTR=FFVAR(NSS)(IT+1:37)
          CALL CTON(CSTR,IMISS,ANUM)
          IENDC=ANUM
        ENDIF
C********************************
        ENDIF
10    CONTINUE
        RETURN
        END
C********************************************************************
      SUBROUTINE VXHMAX(XT)
C *** IF XT INITIALLY < 0 THE XT RETURNED WILL BE THE LOWER VALUE
C *** IF XT INITIALLY > O THE XT RETURNED WILL BE THE UPPER VALUE
      NLOOP=0
      FA=.5
      XA=0
      XB=XT
5     FB=POLYF(XB)-.5
      IF(FB.LT.0.)GOTO 10
      XB=2*XB
      GOTO 5
10      XT=XA-FA*(XA-XB)/(FA-FB)
      FT=POLYF(XT)-.5
      NLOOP=NLOOP+1
      IF(NLOOP.GT.100)GOTO 50
      IF(ABS(FT).LT.1.E-6.OR.ABS(XA-XB).LT.1.E-6)RETURN
      IF(FT.LT.0)GOTO 20
      FA=FT
      XA=XT
      GOTO 10
20      FB=FT
      XB=XT
      GOTO 10
50      WRITE(*,100)NLOOP,XA,XB,XT,FA,FB,FT
100     FORMAT(' XHMAX IN A LOOP, NLOOP,XA,XB,XT,FA,FB,FT'/I5,6E
     # 20.6)
      RETURN
      END
```

The following is a listing of the STGEN.MNU file; this file is read by the MENURD subroutine and the variables are passed to the STGEN
program to give it the option information.

4

```
            6
            5
          100
 1 PAGE STGEN
 2
 3 GENERAL DATA ON THE GENERATION
 4 SELECTION OF THE FIT TYPE DETAILS
 1 PAGE STGEN1 (GENERATION DATA)
 2
 3 NAME OF THE OUTPUT STANDARDS FILE       LOREN4.ST
 4 GRAPHICAL OUTPUT TO BE PLACED IN FILE   SHAPE.GR
 5 FIT TYPE (GAUS,LORE,VOIG OR FITD)       LORE
 6 NUMBER OF BACKGRND COEFFS AND SPLINES   0,4
 1 PAGE STGEN2 (TYPE OPTIONS)
 2
 3 VOIGHT FIT ETA PARAMETER                1
 4 NAME OF INPUT DATA FILE                 OFFMARS.UF
 5 BEGINING POINT,END POINT                20,30
```

The mode STDIS reads the unformatted graph file produces by STGEN and displays the standard graphically along with its constituent splines. It is both menu and interactivelly driven allowing the user
to observe the details of the standard. Users should always look at their standards before proceeding!!!!

```
      PROGRAM STDIS
C*********************************************************************
C     The standard display program (STDIS)
C*********************************************************************
        REAL*8 CHIS
        COMMON/AXISP/BHORI,BVERT,SF,SVERT,EHORI,EVERT
       INTEGER*2 IMULT,IXDIV,IX,IFA,ICH,NRES,IDAT,IFMA
       DIMENSION NRES(4096),IDAT(4096),IFA(4096),IX(4096),
     # ICH(4096,6)
       DIMENSION FORIG(4),XORIG(4),FINT(4),XINT(4)
       CHARACTER*8 II
       CHARACTER*17 TITLE
       CHARACTER*40 NA
       CHARACTER*1 ANS
       CHARACTER*64 VALU
C**********************************************
C
C   GRASHAPE DATA
C
C**********************************************
      CHARACTER*17 GDNAME
      CHARACTER*1 VCF,VSP,VBG
      COMMON/CHGRAS/GDNAME,VCF,VSP,VBG
      COMMON/CGRASH/ABEGC,AENDC,IIMIN,IIMAX,LFLAG
C**********************************************
       IMFLAG=0
5      CONTINUE
      IF(IMFLAG.EQ.0)CALL GSLINK
      CALL RSETG
        NA=GDNAME
102     FORMAT(A40)
```

```
      OPEN(UNIT=3,FILE=NA,FORM='UNFORMATTED',ERR=999)
      READ(3)IMULT,IXDIV,II,CHIS
      TITLE=II
       CHI=CHIS
      AXDIV=IXDIV
      DO 300 I=1,4096
      READ(3,END=302)NRES(I),IDAT(I),IFA(I),IX(I),
     # (ICH(I,J),J=1,6)
300   CONTINUE
302   CLOSE (3)
      IMC=I-1
      AMULT=IMULT
320  CONTINUE
     IF(IMFLAG.EQ.1)THEN
       WRITE(*,103)1.*IX(1)/AXDIV,1.*IX(IMC)/AXDIV
103    FORMAT(' DATA HAS ABSICCA FROM',F8.2,' TO',F8.2 /
     # ' INITIAL X?, END X?','<-1,0> FOR MENU <0,0> FOR DEFAULT')
       READ(*,*)ABEGC,AENDC
       IF(ABEGC.EQ.-1)THEN
        IMFLAG=0
        GOTO 5
       ENDIF
      ENDIF
      IF(AENDC.EQ.0.)ABEGC=1.*IX(1)/AXDIV
      IF(AENDC.EQ.0.)AENDC=1.*IX(IMC)/AXDIV
      IF(ABEGC.GE.AENDC)WRITE(*,104)
104   FORMAT(' THERE ARE NO CHANNELS IN THE
     #INTERVAL',F10.0,',',F10.0)
      IF(ABEGC.GE.AENDC)GOTO 320
      SF=700./(AENDC-ABEGC)
      AMIN=2000000000
      AMAX=0
      DO 340 I=1,IMC
      XP=IX(I)/IXDIV
      IF(XP.LT.ABEGC.OR.XP.GT.AENDC)GOTO 340
      AMIN=AMIN1(AMIN,AMULT*IDAT(I))
      AMAX=AMAX1(AMAX,AMULT*IDAT(I))
340   CONTINUE
      AMIN=AMAX1(0.,AMIN)
      IMIN=AMIN
      IXR=ALOG10(AMIN+1)-1
      IXRDIV=10**IXR
      IXRDIV=MAX0(IXRDIV,1)
      IMIN=IXRDIV*(IMIN/IXRDIV)
      AMIN=IMIN
      IF(IMFLAG.EQ.1)THEN
        WRITE(*,108)AMIN,AMAX
108   FORMAT(' AMIN=',F8.0,'  AMAX=',F8.0/' ENTER DESIRED MIN',
     # ',DESIRED MAX')
        READ(*,*)IMIN,IMAX
      ENDIF
     IMIN=IIMIN
     IMAX=IIMAX
      IF(IMIN.EQ.0.AND.IMAX.EQ.0)GOTO 314
      AMIN=IMIN
      AMAX=IMAX
314   CONTINUE
```

```
      IF(IMFLAG.EQ.1)THEN
       WRITE(*,120)
120    FORMAT(' DO YOU WANT A LOG SCALE?')
       LFLAG=0
115    FORMAT(A1)
       READ(*,115)ANS
       IF(ANS.NE.'Y'.OR.ANS.EQ.'y')GOTO 400
       LFLAG=1
      ENDIF
     IF(LFLAG.NE.1)GOTO 400
       AMIN=AMAX1(.01,AMIN)
       AMAX=ALOG(AMAX)
       AMIN=ALOG(AMAX1(AMIN,1.E-6))
400    IRES=1
       IF(IMFLAG.EQ.1)THEN
       WRITE(*,1245)
1245   FORMAT(' DO YOU WANT TO CLEAR THE SCREEN?')
       READ(*,115)ANS
       ELSE
       ANS='Y'
       ENDIF
       IF(ANS.EQ.'Y'.OR.ANS.EQ.'y')CALL CLEARS
       CALL AXIS(AMIN,AMAX,LFLAG,IRES,ABEGC,AENDC,II,CHI)
        IBHORI=BHORI
        IHRES=EVERT-49
       CHISL=0
       IMOVE=0
       DO 500 I=1,IMC
        XTEST=IX(I)/AXDIV
        IF(XTEST.LT.ABEGC.OR.XTEST.GT.AENDC)GOTO 500
       XP=BHORI+SF*(XTEST-ABEGC)
       IF(I.LT.IMC)ISF=SF*(IX(I+1)-IX(I))/(2*AXDIV)
       IXP=XP
       CHISL=CHISL+(NRES(I)/100.)**2
       ADAT=AMULT*IDAT(I)
       IF(LFLAG.EQ.1)ADAT=ALOG(AMAX1(1.,ADAT))
       IYDAT=SVERT*(ADAT-AMIN)+BVERT
C *** THE FIRST PLOT CALL
       IF(IMOVE.EQ.0)CALL STPL(MAX0(IBHORI,IXP-ISF),IYDAT)
       IF(IMOVE.NE.0)CALL PLOT(MAX0(IBHORI,IXO),IYDAT)
       IMOVE=1
       IXO=IXP+ISF
       CALL PLOT(IXO,IYDAT)
500    CONTINUE
        I1=.5*(EHORI-BHORI)
        I2=.5*(EVERT-BVERT)
       CALL NUMOUT(I1+40,I2,CHISL,1,0)
       VALU=' CHISL=\'
       CALL BCHART(I1-60,I2,VALU)
505    CONTINUE
C     CALL ANSI
C      WRITE(*,1546)
1546   FORMAT(' DO YOU WANT TO SEE THE CURVE FIT?')
C      READ(*,115)ANS
       ANS=VCF
       IF(ANS.NE.'Y'.AND.ANS.NE.'y')GOTO 650
       IMOVE=0
```

```
        DO 600 I=1,IMC
         XTEST=IX(I)/AXDIV
         IF(XTEST.LT.ABEGC.OR.XTEST.GT.AENDC)GOTO 600
         XP=BHORI+SF*(XTEST-ABEGC)
        IXP=XP
        AFRA=AMULT*IFA(I)
        IF(LFLAG.EQ.1)AFRA=ALOG(AMAX1(1.,AFRA))
        IY=SVERT*(AFRA-AMIN)+BVERT
        IF(IMOVE.EQ.0)CALL STPL(IXP-ISF,IY)
        IMOVE=1
        CALL PLOT(IXP,IY)
        IF(I.LT.2)GOTO 600
        IF(I.GT.IMC-3)GOTO 600
        SFE=SF*.25*(IX(I+1)-IX(I))/AXDIV
        IF(SFE.LT.1.)GOTO 600

C *** INTERPOLATION SECTION
580     IB=MAX0(4+I-IMC,2)
      IF(I.EQ.1)IB=1
        IF(IB.EQ.4)GOTO 600
        IEX=2
        DO 582 K=1,4
        FORIG(K)=AMULT*IFA(I-IB+K)
        IF(LFLAG.EQ.1)FORIG(K)=ALOG(AMAX1(FORIG(K),1.E-6))
582     XORIG(K)=IX(I-IB+K)/AXDIV
        CALL PLINTERP(FORIG,XORIG,FINT,XINT,IB,IEX)
        DO 583 K=1,IEX
        IXP=BHORI+SF*(XINT(K)-ABEGC)
        IY=SVERT*(FINT(K)-AMIN)+BVERT
583     CALL PLOT(IXP,IY)
C *** END OF INTERPOLATION SECTION

600     CONTINUE
650     CONTINUE
CC    CALL ANSI
C       WRITE(*,1987)
1987    FORMAT(' DO YOU WANT TO SEE THE INDIVIDUAL SPLINES?')
C       READ(*,115)ANS
        ANS=VSP
        IF(ANS.NE.'Y'.AND.ANS.NE.'y')GOTO 780
        DO 750 J=1,6
        IFPC=0
        ITEST=1
        DO 750 I=1,IMC
         XTEST=IX(I)/AXDIV
         IF(XTEST.LT.ABEGC.OR.XTEST.GT.AENDC)GOTO 750
         XP=BHORI+SF*(XTEST-ABEGC)
        IXP=XP
        IF(ITEST*ICH(I,J).EQ.1)GOTO 750
        APK=AMULT*ICH(I,J)
        IF(LFLAG.EQ.1)APK=ALOG(AMAX1(1.,APK))
        IF(APK.LT.AMIN.OR.ITEST.EQ.1)IFPC=0
        IF(APK.LT.AMIN)GOTO 750
        IYP=SVERT*(APK-AMIN)+BVERT
        IF(IFPC.NE.0)GOTO 740
        IFPC=1
        CALL STPL(IXP,IYP)
```

```
740     CALL PLOT(IXP,IYP)
        IF(I.LT.2)GOTO 750
        IF(I.GT.IMC-3)GOTO 750
        SFE=SF*.25*(IX(I+1)-IX(I))/AXDIV
        IF(SFE.LT.1.)GOTO 750
C *** INTERPOLATION SECTION
        IB=MAX0(4+I-IMC,2)
      IF(I.EQ.1)IB=1
        IF(IB.EQ.4)GOTO 750
        IEX=2
        DO 742 K=1,4
        FORIG(K)=AMULT*ICH(I-IB+K,J)
        IF(LFLAG.EQ.1)FORIG(K)=ALOG(AMAX1(1.E-6,FORIG(K)))
742     XORIG(K)=IX(I-IB+K)/AXDIV
        CALL PLINTERP(FORIG,XORIG,FINT,XINT,IB,IEX)
        DO 743 K=1,IEX
        IXP=BHORI+SF*(XINT(K)-ABEGC)
        IY=SVERT*(FINT(K)-AMIN)+BVERT
743     CALL PLOT(IXP,IY)
C *** END OF INTERPOLATION SECTION
750     ITEST=ICH(I,J)
780     IF(IRES.EQ.0)GOTO 900
C *** NRES IS BHORI TIMES THE RESIDUAL ACTUALLY FOUND
        DO 800 I=1,IMC
         XTEST=IX(I)/AXDIV
         IF(XTEST.LT.ABEGC.OR.XTEST.GT.AENDC)GOTO 800
         XP=BHORI+SF*(XTEST-ABEGC)
        IXP=XP
        IFIX=IHRES+.15*NRES(I)
        IF(SF.GT.3.)GOTO 795
        CALL PONT(IXP,IFIX)
        GOTO 800
795     CONTINUE
        VALU='*\'
        CALL BCHART(IXP,IFIX-7,VALU)
800     CONTINUE
900     CONTINUE
C      CALL ANSI
C       WRITE(*,1573)
1573    FORMAT(' DO YOU WANT TO SEE THE BACKGROUND')
C       READ(*,115)ANS
        ANS=VBG
        IF(ANS.NE.'Y'.AND.ANS.NE.'y')GOTO 1650
        IMOVE=0
        DO 1600 I=1,IMC
         XTEST=IX(I)/AXDIV
         IF(XTEST.LT.ABEGC.OR.XTEST.GT.AENDC)GOTO 1600
         XP=BHORI+SF*(XTEST-ABEGC)
        IXP=XP
        ISP=0
        DO 1595 J=1,6
        IF(ICH(I,J).LE.1)GOTO 1595
        ISP=ISP+ICH(I,J)
1595    CONTINUE
        AFRA=AMULT*(IFA(I)-ISP)
        IF(LFLAG.EQ.1)AFRA=ALOG(AMAX1(1.,AFRA))
        IY=SVERT*(AFRA-AMIN)+BVERT
```

```
        IF(IMOVE.EQ.0)CALL STPL(IXP-ISF,IY)
        IMOVE=1
        CALL PLOT(IXP,IY)
1600    CONTINUE
1650    CONTINUE
C     CALL ANSI
        PRINT*,' ARE YOU FINISHED WITH THE CODE?'
        READ(*,115)ANS
      CALL ANSI
        IF(ANS.EQ.'Y'.OR.ANS.EQ.'y')GOTO 990
        IMFLAG=1
        GOTO 5
990      CONTINUE
        STOP
999     WRITE(*,1945)
1945    FORMAT(' ERROR IN OPENING FILE ',A40,' TRY AGAIN')
        GOTO 5
        END
C*********************************************************************
      SUBROUTINE GSLINK
C *** This routine links the standard display menu selected
C *** quantities into the STDIS routine.
      DIMENSION FFCOM(600),FFVAR(600)
      CHARACTER*64 CSTR,VALU,NA
      COMMON/USPAGE/FFVAR
      COMMON/FCALL/NSTR,NCMENU
      CHARACTER*40 FFCOM,FFVAR
C**************************************************
C
C   DATA TO MAIN ROUTINE GRASHAPE
C
C**************************************************
C
C   GRASHAPE DATA
C
C**************************************************
      CHARACTER*17 GDNAME
      CHARACTER*1 VCF,VSP,VBG
      COMMON/CHGRAS/GDNAME,VCF,VSP,VBG
      COMMON/CGRASH/ABEGC,AENDC,IMIN,IMAX,LFLAG
C**************************************************
      NA='STDIS.MNU'
      NCMENU=1
      CALL MENURD(NA)
      DO 10 J=1,10
        NSS=J*20+2
        IF(J.EQ.1)THEN
C**************************************************
      NSS=NSS+1
      GDNAME=FFVAR(NSS)
      NSS=NSS+1
        IT=INDEX(FFVAR(NSS),',')
        IF(FFVAR(NSS)(IT+1:IT+3).EQ.'DEF')THEN
          ABEGC=0
          AENDC=0
        ELSE
          CSTR=FFVAR(NSS)(1:IT-1)
```

```
            CALL CTON(CSTR,IMISS,ANUM)
            ABEGC=ANUM
            CSTR=FFVAR(NSS)(IT+1:37)
            CALL CTON(CSTR,IMISS,ANUM)
            AENDC=ANUM
          ENDIF
          NSS=NSS+1
          IT=INDEX(FFVAR(NSS),',')
          IF(FFVAR(NSS)(IT+1:IT+3).EQ.'DEF')THEN
            IMIN=0
            IMAX=0
          ELSE
            CSTR=FFVAR(NSS)(1:IT-1)
            CALL CTON(CSTR,IMISS,ANUM)
            IMIN=ANUM
            CSTR=FFVAR(NSS)(IT+1:37)
            CALL CTON(CSTR,IMISS,ANUM)
            IMAX=ANUM
          ENDIF
        NSS=NSS+1
            CSTR=FFVAR(NSS)(1:3)
            IF(CSTR.EQ.'LOG')THEN
              LFLAG=1
            ELSE
              LFLAG=0
            ENDIF
        ENDIF
C*************************************
        IF(J.EQ.2) THEN
C*************************************
          NSS=NSS+1
          VCF=FFVAR(NSS)(1:1)
          NSS=NSS+1
          VSP=FFVAR(NSS)(1:1)
          NSS=NSS+1
          VBG=FFVAR(NSS)(1:1)
C******************************
        ENDIF
10      CONTINUE
          RETURN
          END
```

The following is a listing of the STDIS.MNU file; this file is read by the MENURD subroutine and the variables are passed to the STDIS program to pass it its options.

```
           4
           6
           5
         100
 1 PAGE STDIS
 2
 3 INFORMATION FOR THE STANDARD DISPLAY
 4 VIEWING INFORMATION FOR THE DISPLAY
 1 PAGE STDIS1 (GENERAL INFORMATION)
 2
```

```
 3 NAME OF THE INPUT GRAPHICS FILE          SHAPE.GR
 4 BEGINNING/ENDING CHANNELS                0,0
 5 DESIRED MIN AND MAX Y SCALE              0,0
 6 LOG OR LINEAR FIT Y SCALE                LIN
 1 PAGE STDIS2 (VIEWING OPTIONS)
 2
 3 VIEW THE CURVE FIT (Y/N)                 YES
 4 VIEW THE SPLINES (Y/N)                   YES
 5 VIEW THE BACKGROUND FIT (Y/N)            YES
```

The following is a listing of the backgound fitting codes which fit the background (data - peaks) to either a cubic spline or the exponential of a cubic spline. The ROB part of ROBFIT comes from the fact that this is a robust rather than a least squares fit. The data above the fit (which probably contains peaks) is systematically made less important than that near the fit while data below the fit is made more important. All of this without biasing the final fit in either direction. Note that VBKGFI does not fit the peaks. The peak information is suppressed by the robust nature of the fit, but is eliminated completely only when FSPFIT is run. FSPFIT (see below) fits peaks and background giving a much more accurate background than VBKGFI. With a broad peak and narrow background, however, FSPFIT can go wild. VBKGFI is much more stable and should usually be used to get a good starting estimate for the background before FSPFIT is invoked.

```
      PROGRAM VBKGFI
C*********************************************************************
C     This calls the background fitting routine (BKGFIT)
C*********************************************************************
        IMPLICIT REAL *8 (A-H,O-Z)
         REAL*4 X,F,WX,FA,CUT
         CHARACTER*1 ANS
         CHARACTER*4 BKGF
         CHARACTER*64 NA
         DIMENSION CONS(100)
         DIMENSION FA(4096)
         DIMENSION  F(4096),WX(4096)
         DIMENSION IFDT(4096)
C**********************************************************
C
C      VBKGFIT DATA
C
C***********************************************************
C     REAL*4 CUT
      COMMON/CVBKGF/NVM,CUT
      COMMON/COUNTR/ICSMIN
C*****************************************************************
C   BREAD DATA
C**********************************************************
      CHARACTER*17 DATAFN,NADIFF
      COMMON/CHBREA/DATAFN,NADIFF
      COMMON/CBREAD/IBEGC,IENDC,IEXTWT
      COMMON/BATCFL/IBATCH
C****************************************************************
         DATA CONS/100*0.D0/
         DATA NE/0/,ICONT/0/
```

```
      DO 3 I=1,4096
3     FA(I)=0
      ICSMIN=0
      NV=0
        BKGF='FITB'
      CALL BKLINK
20      CALL BREAD(X,F,IFDT,WX,N)
         NCALL=0
25      CALL BKGFIT(X,F,FA,WX,N,CONS,NV,BKGF,LFLAG,NCALL,CUT)
        OPEN(UNIT=2,FILE='BKGCONS.CN')
         WRITE(2,*)LFLAG
        WRITE(2,'(G20.12)')(CONS(I),I=1,NV)
        CLOSE(UNIT=2)
        CHIS=0
        DO 30 I=1,N
        CHIS=CHIS+(F(I)-FA(I))**2*WX(I)
30      CONTINUE
        PRINT*,' CHIS=',CHIS
        OPEN(UNIT=3,FILE='BKGCONS.GR')
C******** WRITE STANDARDS DUMMY INFORMATION
      NSTAN=0
      NSS=0
      WRITE(3,*)NSTAN
C******** WRITE THE RAW DATA
      NSCALE=1000
      WRITE(3,*)IBEGC,IENDC,IEXTWT
      WRITE(3,*)N,X
      WRITE(3,*)NSCALE
      DO 556 I=1,N
      IFDT(I)=F(I)*NSCALE
556   CONTINUE
      WRITE(3,555)(IFDT(I),I=1,N)
555   FORMAT(10Z8)
C****** WRITE THE BACKGROUND DATA
      WRITE(3,*)NV
      WRITE(3,*)LFLAG
      DO 557 I=1,NV
      WRITE(3,*)CONS(I)
557   CONTINUE
C****** WRITE THE DUMMY PEAK DATA
      NPEAKT=0
      WRITE(3,*)NPEAKT
      IP=0
      NPP=0
      WRITE(3,*)IP,NPP
C****** WRITE THE GRAPHICS INFORMATION
      BWSPW=1
      IX=0
      WRITE(3,*)CHIS
      WRITE(3,*)BWSPW
      WRITE(3,*)IX
C*************************************
         CLOSE(UNIT=3)
          IF(NV.LT.NVM)GOTO 25
         STOP
          END
C*********************************************************************
```

```
      SUBROUTINE BKLINK
C *** This routine links the user selected values from the BKGFIT.MNU
C *** file
      DIMENSION FFCOM(200),FFVAR(200)
      CHARACTER*64 CSTR,VALU,NA
      COMMON/USPAGE/FFVAR
      COMMON/FCALL/ NSTR,NCMENU
      CHARACTER*40 FFCOM,FFVAR
C***********************************************************
C
C DATA TO MAIN ROUTINE
C
C***********************************************************
C
C       VBKGFIT DATA
C
C***********************************************************
      REAL*4 CUT
      COMMON/CVBKGF/NVM,CUT
C*******************************************************************
C       BREAD DATA
C*******************************************************************
      CHARACTER*17 DATAFN,NADIFF
      COMMON/CHBREA/DATAFN,NADIFF
      COMMON/CBREAD/IBEGC,IENDC,IEXTWT
C*******************************************************************
C       BKGFIT DATA
C*******************************************************************
      CHARACTER*17 IBKGFN
      COMMON/CHBKGF/IBKGFN
      COMMON/CBKGFI/IBKGFL
C*******************************************************************
      NCMENU=1
      NA='BKGFIT.MNU'
      CALL MENURD(NA)
      DO 10 J=1,10
        NS=J*20+2
        IF(J.EQ.1) THEN
C1*****************************
        NS=NS+1
        CSTR=FFVAR(NS)
        CALL CTON(CSTR,IMISS,ANUM)
        NVM=ANUM
        NS=NS+1
        CSTR=FFVAR(NS)
        CALL CTON(CSTR,IMISS,ANUM)
        CUT=ANUM
        NS=NS+1
        DATAFN=FFVAR(NS)
        NS=NS+1
        IT=INDEX(FFVAR(NS),',')
        IF(FFVAR(NS)(IT+1:IT+3).EQ.'DEF')THEN
          IBEGC=0
          IENDC=0
        ELSE
          CSTR=FFVAR(NS)(1:IT-1)
          CALL CTON(CSTR,IMISS,ANUM)
```

```
             IBEGC=ANUM
             CSTR=FFVAR(NS)(IT+1:37)
             CALL CTON(CSTR,IMISS,ANUM)
             IENDC=ANUM
           ENDIF
           NS=NS+1
           IT=INDEX(FFVAR(NS),',')
           CSTR=FFVAR(NS)(1:IT-1)
           IBKGFN=CSTR
           CSTR=FFVAR(NS)(IT+1:IT+1)
           IF(CSTR.EQ.'Y')THEN
             IBKGFL=1
           ELSE
             IBKGFL=0
           ENDIF
           NS=NS+1
           CSTR=FFVAR(NS)
           CALL CTON(CSTR,IMISS,ANUM)
           IEXTWT=ANUM
           NS=NS+1
           NADIFF=FFVAR(NS)
C1**********************************
      ELSE IF(J.EQ.2) THEN
C2**********************************
      ENDIF
10    CONTINUE
      RETURN
      END
```

The following is a listing of the BKGFIT.MNU file; this file is read by the MENURD subroutine and the variables are passed to VBKGFI program to decide the options for BKGFIT.

```
         4
         9
         4
       100
 1 PAGE BKGFIT
 2
 3 INPUT DATA INFORMATION
 4 OUTPUT DATA INFORMATION
 1 PAGE BKGFIT1 (INPUT DETAILS)
 2
 3 MAXIMUM NUMBER OF BACKGROUND CONSTANTS. 8
 4 CUTOFF VALUE SET TO                     100
 5 INPUT DATA FILE NAME                    ROBPC.SP
 6 BEGINNING AND ENDING CHANNELS FOR FIT.  1,1000
 7 INPUT BACKGROUND CONSTANTS FILE.        NONE,YES
 8 WEIGHTS. 0=SQRT(DATA),1=CALC,2=FILE     0
 9 FILE CONTAINING THE WEIGHTS             NONE
 1 PAGE BKGFIT2 (OUTPUT DETAILS)
 2
 3 BACKGROUND CONSTANTS OUTPUT TO          BKGCONS.CN
 4 BACKGROUND FIT                          BKGCONS.GR
```

The following is a listing of FSPFIT (The full spectrum fitting program). Note that this routine also calls BKGFIT in its iterative approach to the final fit.

```
      PROGRAM FSPFIT
C***********************************************************************
C     The full spectral fit program (FSPFIT)
C***********************************************************************
       IMPLICIT REAL*8 (A-H,O-Z)
        INTEGER*2 I35
C***** TIMING COUNTER
      COMMON/COUNTR/ICSMIN
         COMMON/SCONS/DU(300),XPS(5),XPL(5),IP,NGPCAL
       REAL*4 F,FA,FB,SQRT,CUTOFF,WX,FT,SU,ALR,XE,CHIOLD
     # ,ATEST,AMDIST,XOFF
      REAL*4 BKGCUT
C       CHARACTER*64 NAGR,NAPK,NABKG
        CHARACTER*4 IFW,NOGP,IFIXW,IRBKG,IVFW
        CHARACTER*1 ANS
C *** WX=1/(EP*EP)
        CHARACTER*4 HDAT,IDAT,Z4DAT,BKGF
        DIMENSION F(4096),FB(4096),WX(4096),FA(4096)
C **** GRAPHICS COPY OF RAW DATA
      REAL*4 FG3,XG3
        DIMENSION FG3(4096)
        DIMENSION C(256 ),W(256 ),XP(256 ),PD(768 ),IPT(256)
        DIMENSION BCONS(100),ARAT(5),WRAT(5)
        DIMENSION IHD(4096),SC(256 ),SW(256 ),SXP(256 ),WF(256 )
        DIMENSION CHIOLD(3),FW(3,5),SFW(5,5)
C *** SET UP GRAPHICS ARRAY
      COMMON/G3PLG/CG3(20,5),XPG3(20,5),WG3(20,5),NST(5)
C*********************************************************
C
C DATA TO MAIN ROUTINE VROBFIT
C
C*********************************************************
C
C      VROBFIT DATA
C
C*********************************************************
      DIMENSION WID(20),ERRO(20),ICHAN(20)
      CHARACTER*17 NABKG,NAPK,NAGR
      COMMON/CHVROB/BKGF,IFW,NOGP,NABKG,NAPK,NAGR
      COMMON/CVROB/WID,ERRO,CUTA,CUTB,ICHAN,NPPMAX,IMREF,NBMAX,NBMIN,
     # NPEAKT
C******************************************************************
C      POLYG DATA
C******************************************************************
      DIMENSION STAN(5)
      CHARACTER*17 STAN
      COMMON/CHPOLY/STAN
      COMMON/CPOLYG/NSTAN
C******************************************************************
C      BREAD DATA
C******************************************************************
      CHARACTER*17 DATAFN,NADIFF
      COMMON/CHBREA/DATAFN,NADIFF
      COMMON/CBREAD/IBEGC,IENDC,IEXTWT
```

```
C*****************************************************************
C      BKGFIT DATA
C*****************************************************************
      CHARACTER*17 IBKGFN
      COMMON/CHBKGF/IBKGFN
      COMMON/CBKGFI/IBKGFL
      DIMENSION VNODE(200)
      COMMON/PNODE/ VNODE,NOTS
C*****************************************************************
C      ROPKS DATA
C*****************************************************************
      CHARACTER*17 IPEKFN
      COMMON/CHROPK/IPEKFN
      COMMON/CROPKS/IPEKFL
C*****************************************************************
        DATA XVB,XVE/2*0.D0/
      DATA IPW/0/,PCON/1.D6/,BWSPW/-9.D9/
         DATA ALPHA/1.D-12/BCONS/100*0.D0/
C *** F(N) IS THE SPECTRUM CURRENTLY BEING FITTED.
C *** FA(N) IS A FIT TO THE BACKGROUND
C *** FB(N) IS THE FIT TO THE PEAKS.
      ICSMIN=0
      I1=1
      NOTS=0
      DO 3 I=1,4096
      FA(I)=0
3     FB(I)=0
      DO 31 I=1,256
      SC(I)=0.
      SXP(I)=0.
      SW(I)=0.
31    CONTINUE
      IFIXW='FWID'
      CALL VRLINK
      IP=NPEAKT
C*****************************************************************
C**** INITIALIZING THE STANDARD BY CALLING POLYG
C*** NV = ZERO ADDED
      NV=0
        NCALLB=0
        NGPCAL=0
        XT=0
C *** CAUSES POLYG TO INPUT THE STANDARD PEAK TYPES
        XTT=POLYG(XT,1,ST)
        DO 5 I=1,3
5       CHIOLD(I)=1.E30
        IREFIT=0
        IREC4=1
      NPPMAX=MAX0(1,MIN0(NPPMAX,255))
      WMAX=0.
      DO 7 I=1,IP
        IV=2*I-1
        W1=WID(IV)
        EW1=ERRO(IV)
        NC1=ICHAN(IV)
        W2=WID(IV+1)
        EW2=ERRO(IV+1)
```

```
         NC2=ICHAN(IV+1)
         WMAX=DMAX1(WMAX,W1+EW1,W2+EW2)
        FW(2,I)=(W2**2-W1**2)/(NC2-NC1)
         FW(1,I)=W1**2-NC1*FW(2,I)
         EW12=(2*W1*EW1)**2
         EW22=(2*W2*EW2)**2
         SFW(2,I)=(EW22 -EW12)/(NC2-NC1)
         SFW(1,I)=EW12-NC1*SFW(2,I)
         DO 6 J=3,5
6        SFW(J,I)=0
        FW(3,I)=0
7      CONTINUE
          CUTOFF=CUTA
          XVB=0
          XVE=1
105     FORMAT(10(A4,1X))
       IF(IFW.EQ.'FWID')WRITE(*,102)
102    FORMAT(' THE FIXED WIDTH OPTION IS IN EFFECT')
         CALL BREAD(XOFF,F,IHD,WX,N,FA)
C *** MAKE A COPY OF THE RAW DATA
       NG3=N
       XG3=XOFF
       DO 56 I=1,N
       FG3(I)=F(I)
56     CONTINUE
       PRINT*,' BACK CONS WILL BE IN ', NABKG
       PRINT*,' THE NEW PEAK FILE IS ',NAPK
       PRINT*,' THE GRAPHICAL DATA IS IN ',NAGR
         NPP=0
         CALL ROPKS(C,SC,W,IPT,WF,SW,XP,SXP,N,NPP,FW,XOFF)
          IF(NPP.EQ.0)GOTO 90
          PRINT*,' ARE THE PEAK HEIGHTS ACCURATE?'
C***********************************
          IF(IPEKFL.EQ.1)GOTO 77
C*************************************
        BKGCUT=CUTA
        DO 65 I=1,50
         CALL BKGFIT(XOFF,F,FA,WX,N,BCONS,NV,BKGF,LFLAG,NCALLB,
      # BKGCUT)
        IF(NCALLB.EQ.30)GOTO 65
        IF(NV.GE.NBMIN)GOTO 68
65      CONTINUE
68      CONTINUE
       PRINT*,'JUST ABOUT TO CALL CFEST',NPP
          DO 75 I=1,NPP
          C(I)=CFEST(W(I),IPT(I),F,FA,WX,XP(I)-XOFF,N,ERRSQ)
       PRINT*,'W(I)=',I,W(I)
       PRINT*,'XP=',XP(I)
       PRINT*,'IN MAIN C=',I,C(I)
75        CONTINUE
77        CONTINUE
         CHIT=0
          DO 80 I=1,N
          XI=I+XOFF
          FB(I)=POLYA(XI,C,XP,W,IPT,NPP)
          F(I)=F(I)-FA(I)
          CHIT=CHIT+(F(I)-FB(I))**2*WX(I)
```

```
80       CONTINUE
90        CONTINUE
        WRITE(*,99)W1,W2
99       FORMAT(' FOR THE FIRST 50 PEAKS, PEAK WIDTHS ARE ASSUMED'/
     # ' LINEAR BEGINNING AT',F7.2,' AND ENDING AT',F7.2)
       WRITE(*,101)N,NPPMAX,XOFF
101    FORMAT(I5,' CHANNELS, AT MOST',I5,' PEAKS.'/
     #'  ZEROTH CHANNEL IS ',F8.0)
       NIT=0
       IWF=2
        IF(IFW.EQ.'FWID')THEN
         PCON=1.D6
         IWF=4
        ENDIF
        IF(IFW.EQ.'FIXD')IWF=9
290     CONTINUE
C*********************************************************
       WRITE(*,107)CUTOFF
107    FORMAT(' THE CODE WILL ATTEMPT TO ADD PEAKS UNTIL THE LARGEST',
     # ' RESIDUAL IS <',F10.2)
         XT=XOFF+N/2
         IF(NPP.GT.20)XT=.5*(XP(1)+XP(NPP))
         DO 295 I=1,N
         XI=I+XOFF
         FB(I)=POLYA(XI,C,XP,W,IPT,NPP)
          F(I)=F(I)+FA(I)-FB(I)
C LINE FOLLOWING MODIFIED FOR SOLAR WORK
       IF(IEXTWT.EQ.0)WX(I)=WX(I)/(1+.5*FB(I)*FB(I)*WX(I))
295       CONTINUE
297      BKGCUT=CUTOFF
       IF(NV.GE.NBMAX.AND.BKGF.EQ.'CONT')BKGF='NONK'
         IF(NV.GE.NBMAX.AND.BKGF.EQ.'FIXK')BKGF='FINK'
         CALL BKGFIT(XOFF,F,FA,WX,N,BCONS,NV,BKGF,LFLAG,NCALLB,
     # BKGCUT)
       IF(NV.LT.NBMIN)THEN
         BKGF='CONT'
         GOTO 297
       ENDIF
300    CHIS=0.D0
       DO 1119 I=1,N
         F(I)=F(I)-FA(I)+FB(I)
C LINE FOLLOWING MODIFIED FOR SOLAR WORK
        IF(IEXTWT.EQ.0)WX(I)=WX(I)/(1-.5*FB(I)*FB(I)*WX(I))
        IF(IEXTWT.EQ.0)WX(I)=1./AMAX1(1.,FA(I)+FB(I))
       IF(IEXTWT.EQ.1.AND.I.GT.5.AND.I.LT.N-5)THEN
         SUMERS=0.
         DO 5551 J=1,10
           SUMERS=SUMERS+(F(I-6+J)-(FA(I-6+J)+FB(I-6+J)))**2
5551     CONTINUE
         SUMERS=SUMERS/10.
         WX(I)=1./DMAX1(1.D0,SUMERS)
       ENDIF
       CHIS=CHIS+(F(I)-FB(I))**2*WX(I)
1119   CONTINUE
       WRITE(*,1120)NV,CHIS
1120   FORMAT(' AFTER BKGFIT WITH',I5,' COEFS, CHIS=',E20.12)
         IPOSPK=0
```

```
      IF(IWF.EQ.9)GOTO 315
        CHIAVE=CHIOLD(1)
        DO 310 I=1,2
        CHIAVE=CHIAVE+CHIOLD(I+1)
310     CHIOLD(I)=CHIOLD(I+1)
        CHIOLD(3)=CHIS
        IF(CHIAVE-3*CHIS.LT.300)GOTO 315
        IF(IREC4.GT.0)GOTO 390

C *** REFIT SECTION

315     IREC4=1
      IF(NPP.EQ.0)GOTO 550
        IF(NPP.GT.15.AND.IFW.NE.'FWID'.AND.IWF.NE.9)IWF=3
        IF(IFW.EQ.'VFWI'.AND.NPP.GT.20)IWF=4
        WRITE(*,1399)
1399    FORMAT(' REFITTING ALL THE PEAKS')
        IP1=1
        I=1
320     ILR=I
        KWF=5
        IF(IWF.EQ.1)KWF=6
        IF(IWF.EQ.4)KWF=7
        IF(IWF.EQ.9)KWF=9
        PRINT*,'IN ROBFIT IP1=',IP1
        CALL PFIT(N,ILR,F,FB,WX,C,SC,W,IPT,WF,SW,XP,SXP,NPP,CHIS,
     # 2,KWF,IP1,IPJ,XVB,XVE,FW,SFW,XOFF,IPW,PCON,WP)
        PRINT*,'AFTER PFIT'
        DO 330 K=IP1,IPJ
1857    FORMAT(I5,F8.2,F6.2,F7.1,F6.1,F6.2,F6.2,I2,F10.0)
330     WRITE(*,1857)K,XP(K),SXP(K),C(K),SC(K),W(K),SW(K),IPT(K),CHIS
        I=IPJ+1
        IF(I.LE.NPP)GOTO 320
        DO 340 I=1,3
340     CHIOLD(I)=1.E32
        WRITE(*,1832)CHIS
1832    FORMAT(' THE CHIS AFTER FIXING THE WIDTHS IS',E12.6)
        GOTO 550
C *** NORMAL PEAK ADDING SECTION
390     CONTINUE
        IQM=0
501     CONTINUE
        IF(NPP.GT.NPPMAX)GOTO 290
        IF(IQM.NE.3)GOTO 510
        ATEST=(ALR-CUTOFF)**2
        IF(ATEST.GT.25..OR.IWF.GT.3.OR.NPP.LT.10.OR.IFW.EQ.IFIXW)
     # GOTO 508
        IWF=3
508     IF(ATEST.LT.5..AND.IFW.EQ.'VFWI'.AND.IWF.EQ.3)IWF=4
        IF(ATEST.LT.1.0)IREC4=IREC4-1
510     NRESP=5*WMAX
        CALL RESL(F,FB,WX,FW,ALR,IQM,ILR,IFTC,NRESP,N,XOFF,
     # CUTOFF)
        IF(ALR.GT.0)IPOSPK=IPOSPK+1
        IF(IFTC.EQ.1.AND.IPOSPK.EQ.0)GOTO 315
        IF(IPOSPK.EQ.0.AND.IFTC.EQ.1.AND.IREC4.EQ.4)GOTO 315
        IF(IFTC.EQ.1.AND.IPOSPK.LT.3)IREC4=IREC4-1
```

```
          IF(IFTC.EQ.1)GOTO 290
            IGG=1
            IF(ALR.LT.0.D0)IGG=3
          NPPO=NPP
          CHISO=CHIS
          CALL PFIT(N,ILR,F,FB,WX,C,SC,W,IPT,WF,SW,XP,SXP,NPP,CHIS,
     # IGG,IWF,I1,JP,XVB,XVE,FW,SFW,XOFF,IPW,PCON,WP)
          TNPKS=.1*(ALR-CUTOFF)**2
          TNPKS=DMAX1(TNPKS,1.D0)
          XT=ILR+XOFF
        WRITE(*,119)NPP,IQM,IWF,XT,WP,ALR,CHIS
119     FORMAT(I5,I3,I2,4G14.6)
            IF(ALR.LT.0.D0)GOTO 501
          IF(IWF.EQ.9)GOTO 501
          IF(NPP.GT.NPPO)GOTO 501
          IF(CHISO-CHIS.GT.TNPKS)GOTO 501
          CALL PFIT(N,ILR,F,FB,WX,C,SC,W,IPT,WF,SW,XP,SXP,NPP,CHIS,
     # 0,IWF,I1,JP,XVB,XVE,FW,SFW,XOFF,IPW,PCON,WP)
        WRITE(*,119)NPP,IQM,IWF,XT,WP,ALR,CHIS
          GOTO 501
C *** NORMAL EXITING
550       CONTINUE
          CHIS=0
          DO 600 I=1,N
          CHIS=CHIS+(F(I)-FB(I))**2*WX(I)
          F(I)=F(I)+FA(I)
600       FB(I)=FB(I)+FA(I)
          WRITE(*,108)NPP,CHIS
108       FORMAT(' AT END OF MINIMIZATION',I4,' PEAKS, CHIS',E20.6)
          OPEN(UNIT=2,FILE=NABKG)
           WRITE(*,*)LFLAG
           WRITE(2,*)LFLAG
           WRITE(2,'(G20.12)' )(BCONS(I),I=1,NV)
          CLOSE(2)
C*********************************************************
C  OPEN THE GRAPHICS OUTPUT FILE
C*********************************************************
        PRINT*,'HELLO JUST OPENNED',NAGR
        OPEN(UNIT=3,FILE=NAGR,FORM='FORMATTED')
        WRITE(3,'(A,G20.12)')DATAFN(1:8),CHIS
        WRITE(3,'(2I5)')IP
        DO 357 NIP=1,NSTAN
        WRITE(3,'(A)')STAN(NIP)
357     CONTINUE
        WRITE(3,'(A)')'\'
C*********************************************************
C  WRITE THE RAW DATA PARAMETERS FOR FIT
C*********************************************************
        WRITE(3,'(I4/(A64))')IEXTWT,DATAFN,NADIFF,NABKG,NAPK
        WRITE(3,'(2I5)')IBEGC,IENDC
        DO 506 I=1,IP
        WRITE(3,'(2G20.12)')FW(1,I),FW(2,I)
506     CONTINUE
        CLOSE(3)
C*********************************************************
          WRITE(*,110)(XP(I),SXP(I),C(I),SC(I),W(I),SW(I),I=1,NPP)
110       FORMAT(F10.3,F7.3,F10.3,F7.3,F8.3,F6.3)
```

```
        OPEN(UNIT=4,FILE=NAPK)
        WRITE(4,'('' OUTPUT FROM ROBFIT''/'' IP='',I5)') IP
      WRITE(4,113)NPP,N,CHIS,NV,CUTOFF
      WRITE(*,113)NPP,N,CHIS,NV,CUTOFF
113   FORMAT(I5,' PEAKS',I5,' CHAN, CHIS=',F12.0,' NITB=',I5/
     # ' CUTOFF=',F7.2)
C *** THE BACKGROUND ERROR IS TAKEN TO BE THAT IN A ONE CONSTANT FIT
C *** TO ALL POINTS WITHIN THE RANGE OF ITS SPLINES ABOUT THE PEAK
C *** CENTER THUS THIS SOURCE OF ERROR IN C(I)**2 IS SQRT(SUM((FB-
C *** F))**2*WX/(SUM WX)))
        DO 800 I=1,NPP
         ITYPE=IPT(I)
        IRES=XP(I)-XOFF
        NBEG=IRES+W(I)*XPS(ITYPE)*4
         NBEG=MAX0(1,NBEG)
        NEND=IRES+W(I)*XPL(ITYPE)*4
        IF(NEND.LE.N)GOTO 710
        NEND=N
        NBEG=N-IRES
710     ANUM=0
        ADEN=0
        DO 720 J=NBEG,NEND
        ANUM=ANUM+(F(J)-FB(J))**2*WX(J)
720     ADEN=ADEN+WX(J)
      EB2=ANUM*NV/(ADEN*N)
C *** IN THIS VERSION OF THE CODE SC IS THE ERROR IN C**2*W
C *** AND IS CALCULATED IN QMIN
C *** THOUGH CHANGED HERE, NOTE THAT SC IS NOT USED ANYWHERE
C *** AND WILL BE RECALCULATED IN THE NEXT REFIT
800     SC(I)=DSQRT(SC(I)*SC(I)+EB2*W(I))
      CALL FWHM(NPP,ARAT,WRAT)
      WRITE(4,'(10F8.4)')(ARAT(I),WRAT(I),I=1,IP)
      DO 910 I=1,NPP
        STR=C(I)*C(I)*W(I)
        STR=ARAT(IPT(I))*STR
        SC(I)=ARAT(IPT(I))*SC(I)
        SXP(I)=DMIN1(SXP(I),99.999D0)
        WOUT=WRAT(IPT(I))*W(I)
        STTW=WRAT(IPT(I))*SW(I)
        SC(I)=DMIN1(SC(I),999999.D0)
      WRITE(*,114)XP(I),SXP(I),WOUT,STTW,STR,SC(I),IPT(I)
114   FORMAT(F9.3,F7.3,F11.3,F8.3,F12.0,F8.0,I3)
      WRITE(4,114)XP(I),SXP(I),WOUT,STTW,STR,SC(I),IPT(I)
910     CONTINUE
C*********************************************************
        CLOSE(4)
     IF(NOGP.EQ.'F SP')CALL FSPECT(1,N,FA,F,WX,C,XP,W,IPT,PD,
    # NPP,XOFF)
      IF(NOGP.NE.'GP')GOTO 920
C **** NOTE THAT PD IS USED BY RCPLOT FOR DERIVATIVE VALUES
920   CONTINUE
C *** RECALCULATING CUTOFF AND EITHER STOPPING OR CONTINUING
         CUTOFF=CUTOFF-(CUTA-CUTB)/IMREF
         IF(CUTOFF.GE.CUTB)THEN
         DO 940 I=1,N
         F(I)=F(I)-FA(I)
940      FB(I)=FB(I)-FA(I)
```

```
          GOTO 290
         ENDIF
         STOP
        END
C*******************************************************************
         SUBROUTINE PFIT(N,ILR,F,FB,
     # WX,C,SC,W,IPT,WF,SW,XP,SXP,NPP,CHIS,
     # IFTC,IWF,JB,NPB,XVB,XVE,FW,SFW,XOFF,IPW,PCON,WPNNT)
C *** THIS ROUTINE REFITS THE PEAKS CLOSE TO ILR,  IF NO PEAK IS
C *** CLOSE TO ILR OR IF IFTC=0 A NEW PEAK IS ADDED
C *** IFTC=2 IMPLIES A REFIT, IN WHICH CASE ILR IS THE PEAK BEING
C *** REFITTED
C *** IFTC=-1 IMPLIES A NEGATIVE RESIDUAL IN WHICH CASE NO PEAK IS
C *** ADDED FINDING BEGINNING AND ENDING FIT POINTS WHEN REFITTING
         IMPLICIT REAL*8 (A-H,O-Z)
        REAL*4 F,FB,SQRT,WX,FT,XOFF
         COMMON/SCONS/DUM(300),XPS(5),XPL(5),NTYPE
         DIMENSION F(1),FB(1),WX(1),C(1),SC(1),XP(1),SXP(1),W(1),
     # IPT(1),WF(1),SW(1),ECN(10),EXN(10),EWN(10)
     # ,CN(10),XN(10),WN(10),IPTN(10),WFN(10),FW(3,5),SFW(5,5)
         NPB=JB
          XT=ILR+XOFF
          IF(IFTC.LT.0)THEN
            XTS=XT
            XTL=XT
            DO 100 J=1,NTYPE
            WPG=DSQRT(DMAX1(3.D0,FW(1,J)+XT*(FW(2,J)+XT*FW(3,J))))
            XTS=DMIN1(XTS,XT+XPS(J)*WPG)
            XTL=DMAX1(XTL,XT+XPL(J)*WPG)
100         CONTINUE
            GOTO 400
          ENDIF
          IF(IFTC.NE.2)GOTO 200
          INNT=IPT(ILR)
          XT=XP(ILR)
          XTS=XT+DMAX1(XPS(INNT),-2.1D0)*W(ILR)
          XTL=XT+DMIN1(XPL(INNT),2.1D0)*W(ILR)
          GOTO 400

200       CONTINUE

C *** FINDING BEGINNING AND ENDING POINTS WHEN NO PEAK IS AT ILR
C *** WHICH IS NOW THE CHANNEL NUMBER WITH THE LARGEST RESIDUAL
C *** THIS IS WHERE WE FIND THE PEAK TYPE WITH THE MINIMUM INITIAL
C *** ERROR
           IF(IFTC.NE.3)THEN
         ERRC=1.D32
         DO 300 J=1,NTYPE
          XTT=ILR+XOFF
          XTT=DMAX1(1.D0+XOFF,DMIN1(1.D0*N+XOFF,XTT))
         CT=CQMIN(FB,F,WX,XOFF,N,XTT,WPG,J,FW,SFW,ERRSQ)
         IF(ERRSQ.GT.ERRC)GOTO 300
         CNNT=CT
         ERRC=ERRSQ
         WPNNT=WPG
          XT=DMAX1(1.D0+XOFF,DMIN1(1.D0*N+XOFF,XTT))
         INNT=J
```

```
 300      CONTINUE
          XTS=XT+WPNNT*DMAX1(XPS(INNT),-2.1D0)
          XTL=XT+WPNNT*DMIN1(XPL(INNT),2.1D0)
          ELSE
          XT=ILR+XOFF
          XTS=XT
          XTL=XTS
            DO 350 J=1,NTYPE
          WPG=DSQRT(DMAX1(3D0,FW(1,J)+XT*(FW(2,J)+XT*FW(3,J))))
            XTS=DMIN1(XTS,ILR+XOFF+WPG*XPS(J))
            XTL=DMAX1(XTL,ILR+XOFF+WPG*XPL(J))
 350        CONTINUE
          ENDIF
C *** NOW USING XTS AND XTL FIND PEAKS IN THE REGION OF INTEREST
 400      CONTINUE
          CALL LOCATE(XTS,XP,NPP,JB)
          JB=JB+1
          CALL LOCATE(XTL,XP,NPP,JE)
C *** INCLUDING ALL POSSIBLE OVERLAPPING PEAKS BEWARE TIME GOES AS
C *** NP**3
 405    JBM=JB-1
        DO 408 I=1,JBM
        ITYPE=IPT(I)
        IF(XTS.GT.XP(I)+W(I)*DMAX1(XPS(ITYPE),-2.1D0))GOTO 408
        JB=I
        XTS=DMIN1(XTS,XP(I)+W(I)*DMAX1(XPS(ITYPE),-2.1D0))
        GOTO 405
 408    CONTINUE
 415    JEP=JE+1
        DO 418 J=JEP,NPP
        I=NPP+JEP-J
        ITYPE=IPT(I)
        IF(XTL.LT.XP(I)+W(I)*DMAX1(XPS(ITYPE),-2.1D0))GOTO 418
        JE=I
        XTL=DMAX1(XTL,XP(I)+W(I)*DMIN1(XPL(ITYPE),2.1D0))
        GOTO 415
 418    CONTINUE
C *** CUTTING THE FITTED NUMBER OF PEAKS TO 10
 420      IF(JE-JB+1.LT.10)GOTO 500
           IF(JE-JB+1.EQ.10.AND.IFTC.NE.0)GOTO 500
           ITY1=IPT(JE)
           ITY2=IPT(JB)
         IF(ITY1.LE.0.OR.ITY1.GE.5)PRINT*,' ARRAY PROB ITY1=',ITY1
         IF(ITY2.LE.0.OR.ITY2.GE.5)THEN
            PRINT*,' ARRAY PROB ITY2=',ITY2
            PRINT*,' JB,JE',JB,JE
         ENDIF
         IF(JE.LE.0.OR.JE.GE.256)PRINT*,' ARRAY PROB JE=',JE
         IF(JB.LE.0.OR.JB.GE.256)PRINT*,' ARRAY PROB JB=',JB
           IF((XT-XP(JE))/XPS(ITY1).LT.(XT-XP(JB))/XPL(ITY2))GOTO 450
         IF(XP(JE-1).LT.XT)GOTO 450
 425       JE=JE-1
           GOTO 420
 450     CONTINUE
         IF(XP(JB+1).GT.XT)GOTO 425
           JB=JB+1
           GOTO 420
```

```
C *** NOW THAT WE HAVE FOUND JB AND JE WE NEED TO RESTABLISH XTS AND
C *** XTL AND ADD THE OLD PEAKS TO THE STARTING GUESSES FOR THE NEW
500     CONTINUE
         IM=1
         IF(JB.GT.JE)GOTO 610
         IF(IFTC.NE.0)THEN
            XTS=1.D32
            XTL=-XTS
         ENDIF
        DO 600 I=JB,JE
        ECN(IM)=SC(I)
        EXN(IM)=SXP(I)
        EWN(IM)=SW(I)
        CN(IM)=C(I)
        XN(IM)=XP(I)
        WN(IM)=W(I)
        IPTN(IM)=IPT(I)
         ITYPE=IPT(I)
        XTS=DMIN1(XTS,XN(IM)+DMAX1(XPS(ITYPE),-2.1D0)*WN(IM))
        XTL=DMAX1(XTL,XN(IM)+DMIN1(XPL(ITYPE),2.1D0)*WN(IM))
        IM=IM+1
600     CONTINUE
C *** REMOVE THE NEW PEAKS FROM THE FILE LIST
610      IM=IM-1
           IF(IFTC.EQ.3.AND.IM.EQ.0)RETURN
           IF(IFTC.LT.0.AND.IM.EQ.0)RETURN
         IF(IM.EQ.0)GOTO 650
        NPP=NPP-IM
        DO 620 I=JB,NPP
        SC(I)=SC(I+IM)
        SXP(I)=SXP(I+IM)
        SW(I)=SW(I+IM)
        C(I)=C(I+IM)
        XP(I)=XP(I+IM)
        W(I)=W(I+IM)
620      IPT(I)=IPT(I+IM)
C *** UPDATING CHIS AND THE FILE FOR THE REMOVAL OF THESE TERMS
650     IB=XTS-.2*(XTL-XTS)-XOFF
        IE=XTL+.2*(XTL-XTS)-XOFF
        IB=MAX0(1,IB)
        IE=MIN0(N,IE)
      DO 700 I=IB,IE
700   CHIS=CHIS-(F(I)-FB(I))**2*WX(I)
       DO 750 I=IB,IE
      XI=I+XOFF
750   FB(I)=FB(I)-POLYA(XI,CN,XN,WN,IPTN,IM)
C *** GETTING READY TO CALL QMIN
         IF(IFTC.EQ.0.OR.(IFTC.EQ.1.AND.IM.EQ.0))THEN
            IM=IM+1
            CN(IM)=CNNT
            WN(IM)=WPNNT
            IPTN(IM)=INNT
            XN(IM)=XT
        ENDIF
        IQFL=0
        IBO=IB
        IEO=IE
```

```
5173    CONTINUE
        IF(IWF.NE.9)
     # CALL QMIN(FB,F,WX,IB,IE,IM,CN,XN,WN,IPTN,WFN,IWF,PCON,
     # FW,SFW,XOFF,IPW,ECN,EXN,EWN,XVB,XVE)
        IF(IWF.EQ.9)
     # CALL FQMIN(FB,F,WX,IB,IE,IM,CN,XN,WN,IPTN,WFN,IWF,PCON,
     # FW,SFW,XOFF,IPW,ECN,EXN,EWN,XVB,XVE)
C *** CHECKING TO SEE THAT WE ARE IN THE CORRECT RANGE
        IF(IQFL.EQ.1)GOTO 5184
        DO 5180 I=1,IM
        IPTT=XN(I)-XOFF-2*WN(I)
        IPTT=MAX0(1,IPTT)
        IF(IPTT.GE.IB)GOTO 5175
        IPC=XN(I)-XOFF+WN(I)
        IF(IPC.LT.IB)CN(I)=.1*SQRT(ABS(WX(IPTT)))
        IF(IPC.LT.IB)GOTO 5175
        IQFL=1
        IB=IPTT
5175    IPTT=XN(I)-XOFF+WN(I)
        IPTT=MIN0(IPTT,N)
        IF(IPTT.LE.IE)GOTO 5180
        IPC=XN(I)-XOFF-WN(I)
        IF(IPC.GT.IE)CN(I)=.1*SQRT(ABS(WX(IPTT)))
        IF(IPC.GT.IE)GOTO 5180
        IE=IPTT
        IQFL=1
5180    CONTINUE
        IF(IQFL.NE.1)GOTO 5184
        IF(IB.GE.IBO)GOTO 5182
        IBOM=IBO-1
        DO 5181 I=IB,IBOM
5181    CHIS=CHIS-(F(I)-FB(I))**2*WX(I)
5182    IF(IE.LE.IEO)GOTO 5173
        IEOP=IEO+1
        DO 5183 I=IEOP,IE
5183    CHIS=CHIS-(F(I)-FB(I))**2*WX(I)
        GOTO 5173
5184    CONTINUE
        IF(IM.EQ.0)GOTO 541
C FROM HERE TO 540 XN IS BEING INSERTED INTO XP IN SUCH A WAY AS TO
C KEEP IN ASCENDING ORDER
      DO 540 I=1,IM
       CN(I)=DABS(CN(I))
        JPT=XN(I)-XOFF
        WN(I)=DMAX1(1.25D0,WN(I))
        JPT=MAX0(1,MIN0(N,JPT))
        IF(JPT.EQ.N.OR.JPT.EQ.1)CN(I)=0.D0
        IF(IWF.EQ.9)GOTO 519
         IF(CN(I)*CN(I).LT.1./DSQRT(DABS(WX(JPT)*WN(I))))CN(I)=0.D0
        IF(CN(I).LT.1.D-12)GOTO 540
519   IF(NPP.LT.1)GOTO 532
520   CALL LOCATE(XN(I),XP,NPP,J)
      IF(J.EQ.NPP)GOTO 535
      JU=NPP+1
      JL=NPP
530   XP(JU)=XP(JL)
        WF(JU)=WF(JL)
```

```
      W(JU)=W(JL)
        IPT(JU)=IPT(JL)
      C(JU)=C(JL)
        SXP(JU)=SXP(JL)
        SW(JU)=SW(JL)
        SC(JU)=SC(JL)
      JU=JU-1
      JL=JL-1
      IF(JL-J)535,535,530
532   J=0
535   JP=J+1
        NPB=JP
        SC(JP)=ECN(I)
        SW(JP)=EWN(I)
        SXP(JP)=EXN(I)
      XP(JP)=XN(I)
        WF(JP)=WFN(I)
        IPT(JP)=IPTN(I)
      W(JP)=WN(I)
      C(JP)=CN(I)
      NPP=NPP+1
540   CONTINUE
541   CONTINUE
C CORRECTING THE FB AND RESIDUALS FOR THE NEW PEAK
      IF(NPP.EQ.0)RETURN
      DO 543 I=IB,IE
      XI=I+XOFF
        FB(I)=POLYA(XI,C,XP,W,IPT,NPP)
      CHIS=CHIS+(F(I)-FB(I))**2*WX(I)
543    CONTINUE
        RETURN
        END
C*********************************************************************
      SUBROUTINE LOCATE(X,R,NMAX,J)
C *** This routine finds the position of a value X in an array R
      IMPLICIT REAL*8 (A-H,O-Z)
      DIMENSION R(1),IC(9)
      DATA IC/128,64,32,16,8,4,2,1,1/
      J=256
      DO 20 IL=1,9
      IF(J.GT.NMAX)GOTO 12
      IF(X.GT.R(J))GOTO 15
12    J=J-IC(IL)
      GOTO 20
15    J=J+IC(IL)
20    CONTINUE
      IF(J.EQ.0)RETURN
      IF(J.GT.NMAX)J=NMAX
      IF(X.LT.R(J))J=J-1
      RETURN
      END
C*********************************************************************
      FUNCTION CEST(W,SW,ITT,F,FB,WX,XT,N,ERRSQ)
C *** THIS CODE WAS DEVELOPED IN ANALOGY TO FITTING LOG(F**2) TO
C *** A GAUSSIAN, I.E. TO ALPHA + LAMBDA * (X-XC)**2
C *** WITH THE POLY USED INSTEAD OF THE GAUSSIAN (X-XC)**2 BECOMES
C *** ALOG(POLY)/LAMBDA**2, WHILE THE CONSTANT PARTIAL OF
```

```
C *** LAMBDA*(X-XC)**2 WITH RESPECT TO LAMBDA BECOMES (X-
C *** XC)*DPOLY/POLY
      IMPLICIT REAL*8 (A-H,O-Z)
      COMMON/SCONS/DU(300),XPS(5),XPL(5),NTYPE
      REAL*4 F(1),FB(1),WX(1)
         WMIN=W-2*SW
         WMAX=W+2*SW
         NIT=0
         IBG=XT-WMAX
         IBG=MAX0(1,IBG)
         IEND=XT+WMAX
         IEND=MIN0(N,IEND)
         ERRSQT=1.D32
         DO 70 ICEN=1,5
5        P1=0
         P2=0
         A11=0
         A12=0
         A21=0
         A22=0
         XCEN=XT+.20*(ICEN-3)
         DO 10 I=IBG,IEND
         X=(I-XCEN)/W
         DF=(F(I)-FB(I))**2
         WT=WX(I)*DF
         ALF=0.D0
         IF(DF.GT.0.D0)ALF=DLOG(DF)
         P1=P1+ALF*WT
         A11=A11+WT
         P=DMAX1(1.D-4,POLYG(X,ITT,SP) )
         ALP=DLOG(P)
         A12=A12+ALP*WT
         SPSP=SP/P
         P2=P2+ALF*WT*SPSP*X
         A21=A21+SPSP*WT*X
         A22=A22+ALP*SPSP*WT*X
10       CONTINUE
       ALA=0
         A12=2*W*W*A12
         A22=2*W*W*A22
         IF(A22*A11.NE.A12*A21)ALA=(P2*A11-P1*A21)/(A22*A11-A12*A21)
         WEST=DMIN1(WMAX,1.01*W)
         IF(ALA.GT.0.D0)WEST=DMIN1(WMAX,DMAX1(WMIN,1.D0/DSQRT(ALA)))
         IF(DABS(WEST-W).GT..05.AND.NIT.LT.5)THEN
           W=WEST
           NIT=NIT+1
           GOTO 5
         ENDIF
         W=WEST
         ALP=DMIN1((P1-ALA*A12)/A11,50.D0)
         CEST=DEXP(.5*ALP)
         ERRSQ=0
         DO 50 I=IBG,IEND
         X=(I-XCEN)/W
         FA=CEST*POLYG(X,ITT,SP)
         ERRSQ=(F(I)-FB(I)-FA)**2*WX(I)+ERRSQ
50       CONTINUE
```

```
          ERRSQ=ERRSQ/(IEND-IBG+1)
          IF(ERRSQ.LT.ERRSQT)THEN
            ERRSQT=ERRSQ
            WTEMP=WEST
            XTEMP=XCEN
          ENDIF
          W=WTEMP
70        CONTINUE
          XT=XTEMP
          ERRSQ=ERRSQT
          RETURN
          END
C***********************************************************************
      FUNCTION POLYA(X,C,XP,W,IPT,NS)
C *** THIS ROUTINE CALCULATES POLYA= SUM C(I)*STAN(I,X) USING POLYG
C *** FOR STAN NOTE THAT BY NOT HAVING XPS OR XPL LARGE, WE ASSUME
C *** THAT POLYB HAS BEEN CALLED BEFORE THE FIRST CALL TO POLYA
      IMPLICIT REAL*8(A-H,O-Z)
      COMMON/SCONS/DU(300),XPS(5),XPL(5)
      DIMENSION C(1),XP(1),W(1),IPT(1)
      POLYA=0
      IF(NS.EQ.0)RETURN
      DO 200 J=1,NS
      XM=X-XP(J)
        ITEST=IPT(J)
      IF(XM.LE.W(J)*XPS(ITEST).OR.XM.GE.W(J)*XPL(ITEST))GOTO 200
      XM=XM/W(J)
C  PARTIAL WRT C(J)
       FADD=C(J)*POLYG(XM,ITEST,SP)
      POLYA=POLYA+C(J)*FADD
200   CONTINUE
      RETURN
      END
C***********************************************************************
      SUBROUTINE QMIN(FB,F,WX,IB,IE,NSC,C,XP,W,IPT,WF,IWF,PCON,FW,SFW
     # ,XOFF,IPP,EC,EX,EW,XVB,XVE)
C *** Peak heights, locations and widths are varied to minimize a
C *** weighted least squares fit to data minus background.
C FB ARE THE BACKGROUND VALUES, F THE MEASURED VALUES, FW THE
C FIT TO THE WIDTHS, C THE CONSTANTS FOR THE PEAKS, XP THE
C PEAK LOCATIONS, W THE PEAK WIDTHS.  THESE LAST THREE ARE
C RETURNED BY QMIN, BUT INITIAL ESTIMATES MUST BE SUPPLIED.
      IMPLICIT REAL*8(A-H,O-Z)
      REAL*4 FB,F,WX,FMIN,FMAX,FA,SQRT,AMAX1,AMIN1,ALOG10,XOFF
      DIMENSION IO(10),FB(1),F(1),WX(1),C(1),XP(1),W(1),WF(1),FW(3,5)
      DIMENSION IPT(1),CONS(30),
     # EC(1),EW(1),EX(1),SFW(5,5),AEWF(10),EFT(10),IPTS(10)
     # ,AEWPC(10),CS(10),ECS(10),WS(10),EWS(10),XPS(10),EXS(10)
      COMMON/PTIALD/PPCC(666),AM(666),P(36),PC(36),SM(100)
C**************************************
C KNOT DATA
C**************************************
      DIMENSION VNODE(200)
      COMMON/PNODE/VNODE,NOTS
C**************************************
        DATA PD/12./
C**************************************
```

```
C CALC NO.. OF KNOTS IN IB/IE RANGE
C***************************************
      NOTINR=0
      DO 505 J=1,NOTS
        IF(VNODE(J).GT.IB.AND.VNODE(J).LT.IE)THEN
          NOTINR=NOTINR+1
        ENDIF
505   CONTINUE
      NCOEF=NOTINR*2+2
      IF(NCOEF.GT.6)THEN
        PRINT*,'WARNING COMPLEX BACKGROUND'
      ENDIF
      IF(NCOEF.GT.6)NCOEF=6
C*****************************************
C SELECT IE,IB ON KNOT LOCATION
C*****************************************
      IEI=0
      IBI=0
      IBFL=0
      IEFL=0
      IF(NOTS.GT.2)THEN
        IF(IB+XOFF.GE.VNODE(1))THEN
        XIB=IB+XOFF
          CALL LOCATE(XIB,VNODE,NOTS,IBFL)
          IBI=VNODE(IBFL)-XOFF
        ELSE
          IBI=IB
        ENDIF
        IF(IE+XOFF.LE.VNODE(NOTS))THEN
          XIE=IE+XOFF
          CALL LOCATE(XIE,VNODE,NOTS,IEFL)
          IEI=VNODE(IEFL+1)-XOFF
        ELSE
          IEI=IE
        ENDIF
      ENDIF
      IF(IEI.NE.IBI)THEN
C       IE=IEI
C       IB=IBI
        IXMID=(IEI+IBI)/2
      ELSE
        IXMID=(IE+IB)/2
      ENDIF
C     PRINT*,VNODE(IEFL+1),VNODE(IBFL)
      DO 555 KNT=1,NOTS
555   CONTINUE
C************************************************
        CHI1=1.E32
9     CHB=1.E31
      CHL=1.E32
      FR=0
      IEN=0
      NT=3*NSC
      NPART=1
       DO 24 J=1,NSC
      CONS(NPART)=C(J)
      SM(NPART)=1.D-2
```

```
      CONS(NPART+1)=XP(J)
      SM(NPART+1)=SM(NPART)*1.D6
      L=IPT(J)
      WPP=DSQRT(DMAX1(3D0,FW(1,L)+XP(J)*(FW(2,L)+
     # XP(J)*FW(3,L))))
      SW=DSQRT(DABS(SFW(1,L)+XP(J)*(SFW(2,L)+XP(J)*(SFW(3,L)
     # +XP(J)*(SFW(4,L)+XP(J)*SFW(5,L))))))
      SW=SW/(2*WPP)
      W(J)=DMAX1(1.5D0,WPP-2*SW,DMIN1(W(J),WPP+2*SW))
      CONS(NPART+2)=W(J)
      SM(NPART+2)=SM(NPART)*1.D4
24    NPART=NPART+3
        NDT=(NT+NCOEF)*((NT+NCOEF)+1)/2
      NKIT=30
      KIT=1
28      CHI=0
         PEN=0
      DO 30 J=1,NT+NCOEF
30    PC(J)=0.
      DO 35 J=1,NDT
35    PPCC(J)=0.
      DO 200 I=IB,IE
      X=I+XOFF
      FA=POLYB(X,P,C,XP,W,IPT,NSC,IXMID)+FB(I)
        RES=AMIN1(1.E12,AMAX1(-1.E12,F(I)-FA))
      CHI=CHI+RES**2*WX(I)
      W1=2*(FA-F(I))*WX(I)
      W2=2*WX(I)
        KJ=0
      DO 80 J=1,NT+NCOEF
      PC(J)=PC(J)+W1*P(J)
      W2P=W2*P(J)
      DO 80 K=1,J
        KJ=KJ+1
80    PPCC(KJ)=PPCC(KJ)+P(K)*W2P
200   CONTINUE
C     PRINT*,'AFTER 200 LOOP. NDT=',NDT
       CALL WADJ(FW,SFW,PC,PPCC,AM,
     #  PEN,W,XP,IPT,NSC,KIT)

          CHI=CHI+PEN
      IF(CHI.LE.1.01*CHL.AND.KIT.GE.NKIT)GOTO 2502
      IF(CHI.LE.1.01*CHL.AND.IEN.EQ.1)GOTO 2502
      IF(CHL-CHB.GT..1D0)GOTO 248
      IF(KIT.LE.3)GOTO 248
        IEN=1
      IF(CHI.LT.CHL)GOTO 2502
248   CALL SMSQ(CHI,CHB,CHL,PC,PPCC,AM,FR,CONS,SM,NT,IPP)
250   NPART=1
      DO 260 J=1,NSC
      C(J)=DMAX1(.1D0,CONS(NPART))
      XP(J)=DMAX1(1.D0*(IB+XOFF),DMIN1(1.D0*(IE+XOFF),CONS(NPART+1)))
      W(J)=DABS(CONS(NPART+2))
      L=IPT(J)
      WPP=DSQRT(DMAX1(3D0,FW(1,L)+XP(J)*(FW(2,L)+
     # XP(J)*FW(3,L))))
      SW=DSQRT(DABS(SFW(1,L)+XP(J)*(SFW(2,L)+XP(J)*(SFW(3,L)
```

```
     # +XP(J)*(SFW(4,L)+XP(J)*SFW(5,L))))))
      SW=SW/(2*WPP)
      W(J)=DMAX1(1.5D0,WPP-2*SW,DMIN1(W(J),WPP+2*SW))
 260  NPART=NPART+3
      KIT=KIT+1
      IF(KIT.GT.2*NKIT)GOTO 280
      GOTO 28
 2502 CONTINUE
        CHI=CHI-PEN
        DO 2503 I=1,NDT
 2503   AM(I)=PPCC(I)
      NTSMIN=NT+NCOEF
        CALL SMINV(AM,NTSMIN,IFL)
        DO 2510 I=1,NDT
 2510   AM(I)=DMIN1(1.D9,DMAX1(-1.D9,AM(I)))
        CHIP=DMAX1(1.D0,CHI/MAX0(1,IE-IB+1-NT))
        DO 2520 J=1,NSC
        JC=1+3*(J-1)
        JX=JC+1
        JW=JX+1
        JCC=JC*(JC+1)/2
        JWW=JW*(JW+1)/2
        ICT=KIJ(JC,JC+2)
        ESTR2=(2*C(J)*W(J))**2*AM(JCC)
        ESTR2=ESTR2+4*C(J)**3*W(J)*AM(ICT)
        ESTR2=ESTR2+C(J)**4*AM(JWW)
        EC(J)=DSQRT(ESTR2*CHIP*2)
        EX(J)=DSQRT(DMAX1(1.D-30,2*AM(JX*(JX+1)/2)*CHIP))
        EX(J)=DMIN1(DMAX1(10.D0,2*W(J)),EX(J))
        EW(J)=DSQRT(DMAX1(1.D-30,2*AM(JW*(JW+1)/2)*CHIP))
        WF(J)=W(J)
 2520   CONTINUE
         IF(IPP.EQ.0)RETURN
        NP=IE-IB+1
      WRITE(*,104)NP
 104  FORMAT(' WE ARE FITTING',I5,' POINTS')
      WRITE(*,102)CHI,PEN
 102  FORMAT(' CHISQUARE IS',E20.7,' PEN IS',E20.7)
      IF(NSC.GT.0)WRITE(*,103)(I,C(I),EC(I),XP(I),EX(I),W(I),EW(
     # I),I=1,NSC)
 103  FORMAT(' THE B SPLINE COEFFS'/'    #',8X,'C(I)',17X,'XP(I)',
     # 18X,'W(I)'/(I5,3(F12.4,F10.6)))
       RETURN
 280  WRITE(*,105)CHI,CHB,CHL,(CONS(I),I=1,NT)
 105  FORMAT(' LOOPING IN QMIN, CHI,CHB,CHL/CONS'/(3E20.12))
      NPART=1
      DO 290 J=1,NSC
      C(NPART)=0
 290  NPART=NPART+3
      RETURN
       END
C*******************************************************************
      SUBROUTINE FQMIN(FB,F,WX,IB,
     # IE,NSC,C,XP,W,IPT,WF,IWF,PCON,FW,SFW
     # ,XOFF,IPP,EC,EX,EW,XVB,XVE)
C *** In FQMIN the peak locations and widths are not allowed to vary
C FB ARE THE BACKGROUND VALUES, F THE MEASURED VALUES, FW THE
```

```
C FIT TO THE WIDTHS, C THE CONSTANTS FOR THE PEAKS, XP THE
C PEAK LOCATIONS, W THE PEAK WIDTHS.  THESE LAST THREE ARE
C RETURNED BY QMIN, BUT INITIAL ESTIMATES MUST BE SUPPLIED.
      IMPLICIT REAL*8(A-H,O-Z)
      REAL*4 FB,F,WX,FMIN,FMAX,FA,SQRT,AMAX1,AMIN1,ALOG10,XOFF
      DIMENSION IO(10),FB(1),F(1),WX(1),C(1),XP(1),W(1),WF(1),FW(3,5)
      DIMENSION IPT(1),CONS(30),
     # EC(1),EW(1),EX(1),SFW(5,5),AEWF(10),EFT(10),IPTS(10)
     # ,AEWPC(10),CS(10),ECS(10),WS(10),EWS(10),XPS(10),EXS(10)
      COMMON/PTIALD/PPCC(666),AM(666),P(36),PC(36),SM(100)
C****************************************
C KNOT DATA
C****************************************
      DIMENSION VNODE(200)
      COMMON/PNODE/VNODE,NOTS
C****************************************
        DATA PD/12./
C****************************************
C CALC NO.. OF KNOTS IN IB/IE RANGE
C****************************************
      NOTINR=0
      DO 505 J=1,NOTS
        IF(VNODE(J).GT.IB.AND.VNODE(J).LT.IE)THEN
          NOTINR=NOTINR+1
        ENDIF
505   CONTINUE
      NCOEF=NOTINR*2+2
      IF(NCOEF.GT.6)THEN
        PRINT*,'WARNING COMPLEX BACKGROUND'
      ENDIF
      IF(NCOEF.GT.6)NCOEF=6
C****************************************
        CHI1=1.E32
9     CHB=1.E31
      CHL=1.E32
      FR=0
      IEN=0
      NT=NSC
      PRINT*,' INSIDE FQMIN',NSC,NT
      NPART=1
       DO 24 J=1,NSC
      CONS(NPART)=C(J)
      SM(NPART)=1.D-2
24    NPART=NPART+1
        NDT=(NT+NCOEF)*((NT+NCOEF)+1)/2
      NKIT=30
      KIT=1
28      CHI=0
         PEN=0
      DO 30 J=1,NT+NCOEF
30    PC(J)=0.
      DO 35 J=1,NDT
35    PPCC(J)=0.
C*******************************************
C SELECT IE,IB ON KNOT LOCATION
C*******************************************
      IEI=0
```

```
      IBI=0
      IBFL=0
      IEFL=0
      IF(NOTS.GT.2)THEN
        IF(IB+XOFF.GE.VNODE(1))THEN
        XIB=IB+XOFF
          CALL LOCATE(XIB,VNODE,NOTS,IBFL)
          IBI=VNODE(IBFL)-XOFF
        ELSE
          IBI=IB
        ENDIF
        IF(IE+XOFF.LE.VNODE(NOTS))THEN
          XIE=IE+XOFF
          CALL LOCATE(XIE,VNODE,NOTS,IEFL)
          IEI=VNODE(IEFL+1)-XOFF
        ELSE
          IEI=IE
        ENDIF
      ENDIF
      IF(IEI.NE.IBI)THEN
        IXMID=(IBI+IEI)/2
      ELSE
        IXMID=(IB+IE)/2
      ENDIF
      PRINT*,'IEI,IBI=',IEI,IBI
      PRINT*,'AFTER',IE,IB
      PRINT*,'IEFL,IBFL=',IEFL,IBFL
      DO 555 KNT=1,NOTS
        PRINT*,'VNODE=',KNT,VNODE(KNT)
555   CONTINUE
      PRINT*,'IXMID=',IXMID
C**************************************************
      DO 200 I=IB,IE
      X=I+XOFF
      FA=POLYB(X,P,C,XP,W,IPT,NSC,IXMID)+FB(I)
        RES=AMIN1(1.E12,AMAX1(-1.E12,F(I)-FA))
      CHI=CHI+RES**2*WX(I)
      W1=2*(FA-F(I))*WX(I)
      W2=2*WX(I)
        KJ=0
      DO 80 J=1,NT+NCOEF
      JC=3*(J-1)+1
      PC(J)=PC(J)+W1*P(JC)
      W2P=W2*P(JC)
      DO 80 K=1,J
        KC=3*(K-1)+1
        KJ=KJ+1
80    PPCC(KJ)=PPCC(KJ)+P(KC)*W2P
200   CONTINUE
      IF(CHI.LE.1.01*CHL.AND.KIT.GE.NKIT)GOTO 2502
      IF(CHI.LE.1.01*CHL.AND.IEN.EQ.1)GOTO 2502
      IF(CHL-CHB.GT..1D0)GOTO 248
      IF(KIT.LE.3)GOTO 248
        IEN=1
      IF(CHI.LT.CHL)GOTO 2502
248   CALL SMSQ(CHI,CHB,CHL,PC,PPCC,AM,FR,CONS,SM,NT,IPP)
      DO 260 J=1,NSC
```

```
260   C(J)=DMAX1(.1D0,CONS(J))
      KIT=KIT+1
      IF(KIT.GT.2*NKIT)GOTO 280
      GOTO 28
2502  CONTINUE
        CHI=CHI-PEN
        DO 2503 I=1,NDT
2503    AM(I)=PPCC(I)
      NTSMIN=NT+NCOEF
        CALL SMINV(AM,NTSMIN,IFL)
        DO 2510 I=1,NDT
2510    AM(I)=DMIN1(1.D9,DMAX1(-1.D9,AM(I)))
        CHIP=DMAX1(1.D0,CHI/MAX0(1,IE-IB+1-NT))
        DO 2520 J=1,NSC
        EC(J)=2*(2*C(J)*W(J))**2*AM(J*(J+1)/2)*CHIP
      EC(J)=DSQRT(EC(J))
        EW(J)=0
        EX(J)=0
        WF(J)=W(J)
2520    CONTINUE
         IF(IPP.EQ.0)RETURN
        NP=IE-IB+1
      WRITE(*,104)NP
104   FORMAT(' WE ARE FITTING',I5,' POINTS')
      WRITE(*,102)CHI,PEN
102   FORMAT(' CHISQUARE IS',E20.7,' PEN IS',E20.7)
      IF(NSC.GT.0)WRITE(*,103)(I,C(I),EC(I),XP(I),EX(I),W(I),EW(
     # I),I=1,NSC)
103   FORMAT(' THE B SPLINE COEFFS'/'     #',8X,'C(I)',17X,'XP(I)',
     # 18X,'W(I)'/(I5,3(F12.4,F10.6)))
       RETURN
280   WRITE(*,105)CHI,CHB,CHL,(CONS(I),I=1,NT)
105   FORMAT(' LOOPING IN QMIN, CHI,CHB,CHL/CONS'/(3E20.12))
      NPART=1
      DO 290 J=1,NSC
290   C(J)=0
      RETURN
      END
C*********************************************************************
         FUNCTION CFEST(W,ITT,F,FB,WX,XT,N,ERRSQ)
C *** Provides a crude first estimate for peak heights and the
C *** reduction in chi to be found by adding them.  Used currently
C *** only for peak height estimation.
         IMPLICIT REAL*8 (A-H,O-Z)
         REAL*4 F(1),FB(1),WX(1)
         NIT=0
         IBG=XT-2*W
         IBG=MAX0(1,IBG)
         IEND=XT+W
         IEND=MIN0(N,IEND)
         ERRSQT=1.D32
         ANUM=0
         ADEN=0
         ERRT=0
         ERRSQ=0
         DO 10 I=IBG,IEND
         X=(I-XT)/W
```

```
          DF=(F(I)-FB(I))
          ERRSQ=ERRSQ+DF*DF*WX(I)
          FA=DMAX1(1.D-4,POLYG(X,ITT,SP) )
          ANUM=ANUM+DF*FA*WX(I)
          ADEN=ADEN+FA*FA*WX(I)
10        CONTINUE
          CFEST=0
          IF(ANUM.LT.0..OR.ADEN.LE.0.)RETURN
          CFEST=ANUM/ADEN
          ERRSQ=ERRSQ-CFEST*ANUM
          ERRSQ=ERRSQ/(IEND+1-IBG)
          CFEST=DSQRT(CFEST)
          RETURN
          END
C*******************************************************************
         SUBROUTINE WADJ(FW,SFW,PC,PPCC,AI,
     #   PEN,W,XP,IPT,NSC,KIT)
C *** This routine adjusts the widths of the fitted peaks to conform
C *** to the user specified limits. Used in conjunction with QMIN to
C *** keep fitted peaks close to user specified bounds.  This causes
C *** wide regions to leave residuals, which are eventually added as
C *** new peaks, giving the routine its power to break up multiple
C *** peak groupings when properly used.
         IMPLICIT REAL*8 (A-H,O-Z)
         DIMENSION PC(1),PPCC(1),FW(3,5),SFW(5,5),AI(1),
     # XP(1),IPT(1),W(1)
       NPART=3
          PEN=0.D0
          DO 210 J=1,NSC
         L=IPT(J)
       NPM22=(NPART-2)*(NPART-1)/2
       T11=DMAX1(1.D-12,DMIN1(1.D12,PPCC(NPM22)))
       NPM2NP=NPART-2+NPART*(NPART-1)/2
       T12=DMAX1(-1.D12,DMIN1(1.D12,PPCC(NPM2NP)))
       NPNP=NPART*(NPART+1)/2
       T22=DMAX1(1.D-12,DMIN1(1.D12,PPCC(NPNP)))
       WPP=DSQRT(DMAX1(3D0,FW(1,L)+XP(J)*(FW(2,L)+
     # XP(J)*FW(3,L))))
       SW=DSQRT(DABS(SFW(1,L)+XP(J)*(SFW(2,L)+XP(J)*(SFW(3,L)
     # +XP(J)*(SFW(4,L)+XP(J)*SFW(5,L))))))
       SW=SW/(2*WPP)
       SW=DMAX1(.1D-2,SW)
       SW2=SW*SW
       DELTAW=(W(J)-WPP)
208    PT=DMAX1(0.D0,(2*T11/SW2+T12*T12-T11*T22)/(2*T11))
       PC(NPART)=2*PT*DELTAW+PC(NPART)
       PPCC(NPNP)=2*PT+PPCC(NPNP)
       PEN=PEN+PT*DELTAW*DELTAW
210    NPART=NPART+3
       RETURN
         END
C*******************************************************************
       SUBROUTINE FWHM(NSC,AI,WRAT)
C *** Calculates full width at half max of a peak standard.  This is
C *** usually very close to 1., but may be different if the user has
C *** removed some of the splines from the list created by STGEN in
C *** response to looking at the fit using STDIS.
```

```
        IMPLICIT REAL*8 (A-H,O-Z)
      DIMENSION AI(1),WRAT(1)
        COMMON/SCONS/DU(300),XPS(5),XPL(5),NTYPE
C     TRAP RULE INTEGRATION AND HALF IFW FINDING
      DO 80 ITY=1,NTYPE
        XLH=-1.D0
      CALL XHMAX(XLH,ITY)
      XUH=1.D0
      CALL XHMAX(XUH,ITY)
      WRAT(ITY) =XUH-XLH
      H=WRAT(ITY)/50
        X=XPS(ITY)-H/2
        AI(ITY)=0.D0
25      X=X+H
        F=POLYG(X,ITY,SP)
        AI(ITY)=AI(ITY)+F
      IF(X.LT.XPL(ITY))GOTO 25
        AI(ITY)=AI(ITY)*H
        WRITE(*,101)XLH,XUH,AI(ITY)
101     FORMAT(' FWHM --- XLH=',E16.7,' XUH=',E16.7,' AI=',E16.7)
80      CONTINUE
        RETURN
        END
C*******************************************************************
      SUBROUTINE XHMAX(XT,ITYPE)
C *** RETURNS XT FOR WHICH THE ITYPE STANDARD IS HALF ITS MAXIMUM
C *** IF XT INITIALLY < 0 THE XT RETURNED WILL BE THE LOWER VALUE
C *** IF XT INITIALLY > O THE XT RETURNED WILL BE THE UPPER VALUE
      IMPLICIT REAL*8(A-H,O-Z)
      NLOOP=0
      FA=.5
      XA=0
      XB=XT
5     FB=POLYG(XB,ITYPE,SP)-.5
      IF(FB.LT.0.)GOTO 10
      XB=2*XB
      GOTO 5
10      XT=XA-FA*(XA-XB)/(FA-FB)
      FT=POLYG(XT,ITYPE,SF)-.5
      NLOOP=NLOOP+1
      IF(NLOOP.GT.100)GOTO 50
      IF(DABS(FT).LT.1.D-6.OR.DABS(XA-XB).LT.1.D-6)RETURN
      IF(FT.LT.0)GOTO 20
      FA=FT
      XA=XT
      GOTO 10
20      FB=FT
      XB=XT
      GOTO 10
50      WRITE(*,100)NLOOP,XA,XB,XT,FA,FB,FT
100     FORMAT(' XHMAX IN A LOOP, NLOOP,XA,XB,XT,FA,FB,FT'/I5,6E
     # 20.6)
      RETURN
      END
C*******************************************************************
      SUBROUTINE VRLINK
C *** This routine links in the menu selected quantities into FSPFIT.
```

```
      DIMENSION FFVAR(160)
      CHARACTER*64 CSTR,VALU,NA
      COMMON/USPAGE/FFVAR
      COMMON/FCALL/NSTR,NCMENU
      CHARACTER*40 FFVAR
C*****************************************************
C
C DATA TO MAIN ROUTINE VROBFIT
C
C*****************************************************
C
C       VROBFIT DATA
C
C*****************************************************
      REAL*8 WID,ERRO,CUTA,CUTB
      DIMENSION WID(20),ERRO(20),ICHAN(20)
      CHARACTER*17 NABKG,NAPK,NAGR
      CHARACTER*4 IFW,BKGF,NOGP
      COMMON/CHVROB/BKGF,IFW,NOGP,NABKG,NAPK,NAGR

      COMMON/CVROB/WID,ERRO,CUTA,CUTB,ICHAN,NPPMAX,IMREF,NBMAX,NBMIN,
     # NPEAKT
C************************************************************
C       POLYG DATA
C************************************************************
      DIMENSION STAN(5)
      CHARACTER*17 STAN
      COMMON/CHPOLY/STAN
      COMMON/CPOLYG/NSTAN
C*************************************************************
C       BREAD DATA
C*************************************************************
      CHARACTER*17 DATAFN,NADIFF
      COMMON/CHBREA/DATAFN,NADIFF
      COMMON/CBREAD/IBEGC,IENDC,IEXTWT
C*************************************************************
C       BKGFIT DATA
C*************************************************************
      CHARACTER*17 IBKGFN
      COMMON/CHBKGF/IBKGFN
      COMMON/CBKGFI/IBKGFL
C*************************************************************
C       ROPKS DATA
C*************************************************************
      CHARACTER*17 IPEKFN
      COMMON/CHROPK/IPEKFN
      COMMON/CROPKS/IPEKFL
C************************************************************
      NCMENU=1
      NA='ROBFIT.MNU'
      CALL MENURD(NA)
      DO 10 J=1,10
        NS=J*20+2
        IF(J.EQ.1) THEN
C1*****************************
        NS=NS+1
        CSTR=FFVAR(NS)
```

```
      CALL CTON(CSTR,IMISS,ANUM)
      NPPMAX=ANUM
      NS=NS+1
      IT=INDEX(FFVAR(NS),',')
      CSTR=FFVAR(NS)(1:IT-1)
      CALL CTON(CSTR,IMISS,ANUM)
      CUTA=ANUM
      VALU=FFVAR(NS)(IT+1:37)
      IT=INDEX(VALU,',')
      CSTR=VALU(1:IT-1)
      CALL CTON(CSTR,IMISS,ANUM)
      CUTB=ANUM
      CSTR=VALU(IT+1:37)
      CALL CTON(CSTR,IMISS,ANUM)
      IMREF=ANUM
      NS=NS+1
      IT=INDEX(FFVAR(NS),',')
      CSTR=FFVAR(NS)(1:IT-1)
      IFW=CSTR
      CSTR=FFVAR(NS)(IT+1:37)
      BKGF=CSTR
      NS=NS+1
        IT=INDEX(FFVAR(NS),',')
      CSTR=FFVAR(NS)(1:IT-1)
      CALL CTON(CSTR,IMISS,ANUM)
      NBMIN=ANUM
        CSTR=FFVAR(NS)(IT+1:37)
        CALL CTON(CSTR,IMISS,ANUM)
     NBMAX=ANUM
      NOGP='GP'
      NS=NS+1
      CSTR=FFVAR(NS)
      CALL CTON(CSTR,IMISS,ANUM)
      IEXTWT=ANUM
      NS=NS+1
      NADIFF=FFVAR(NS)
C1*******************************
      ELSE IF(J.EQ.2) THEN
C2*******************************
      NS=NS+1
      DATAFN=FFVAR(NS)
      NS=NS+1
      IT=INDEX(FFVAR(NS),',')
      IF(FFVAR(NS)(IT+1:IT+3).EQ.'DEF')THEN
        IBEGC=0
        IENDC=0
      ELSE
        CSTR=FFVAR(NS)(1:IT-1)
        CALL CTON(CSTR,IMISS,ANUM)
        IBEGC=ANUM
        CSTR=FFVAR(NS)(IT+1:37)
        CALL CTON(CSTR,IMISS,ANUM)
        IENDC=ANUM
      ENDIF
      NS=NS+1
      IT=INDEX(FFVAR(NS),',')
      CSTR=FFVAR(NS)(1:IT-1)
```

```
        IPEKFN=CSTR
        CSTR=FFVAR(NS)(IT+1:IT+1)
        IF(CSTR.EQ.'Y')THEN
          IPEKFL=1
        ELSE
          IPEKFL=0
        ENDIF
        NS=NS+1
        IT=INDEX(FFVAR(NS),',')
        CSTR=FFVAR(NS)(1:IT-1)
        IBKGFN=CSTR
        CSTR=FFVAR(NS)(IT+1:IT+1)
        IF(CSTR.EQ.'Y')THEN
          IBKGFL=1
        ELSE
          IBKGFL=0
        ENDIF
C2*********************************
      ELSE IF(J.EQ.3) THEN
C3*********************************
        NS=NS+1
        NABKG=FFVAR(NS)
        NS=NS+1
        NAPK=FFVAR(NS)
        NS=NS+1
        NAGR=FFVAR(NS)
C3*********************************
      ELSE IF(J.EQ.4) THEN
C4*********************************
           NSTAN=0
           DO 20 I=1,5
             NS=NS+1
             IF(FFVAR(NS)(1:1).EQ.' ')GOTO 20
             STAN(I)=FFVAR(NS)
             NSTAN=NSTAN+1
20         CONTINUE
C4*********************************
      ELSE IF(J.EQ.5) THEN
C5*********************************
        NPEAKT=0
        DO 40 K=1,5
        KV=2*K+1
        KVV=100+KV
        CSTR=FFVAR(KVV)
        IF(CSTR.EQ.' ')GOTO 40
        IV=KV-3
        DO 30 I=1,2
          IV=IV+1
          NS=NS+1
        IT=INDEX(FFVAR(NS),',')
        CSTR=FFVAR(NS)(1:IT-1)
        CALL CTON(CSTR,IMISS,ANUM)
        WID(IV)=ANUM
        VALU=FFVAR(NS)(IT+1:37)
        IT=INDEX(VALU,',')
        CSTR=VALU(1:IT-1)
        CALL CTON(CSTR,IMISS,ANUM)
```

```
            ERRO(IV)=ANUM
            CSTR=VALU(IT+1:37)
            CALL CTON(CSTR,IMISS,ANUM)
            ICHAN(IV)=ANUM
30          CONTINUE
            NPEAKT=NPEAKT+1
40           CONTINUE
C5***********************************
         ENDIF
10       CONTINUE
         RETURN
         END
C*******************************************************************
            FUNCTION CQMIN(FB,F,WX,XOFF,N,XPG,WG,L,FW,SFW,ERRSQ)
C *** Performs a preliminary fit to a peak.  This routine is used by
C *** PFIT to determine which standard to use for a newly introduced
C *** peak.  It is by no means able to unambiguously determine this.
C *** The user should always consult his own experience and will
C *** frequently want to change the type determined here by changing
C *** the type listed in the peak file and  running the code again
C *** with the modified file as input.
C FB ARE THE BACKGROUND VALUES, F THE MEASURED VALUES, FW THE
C FIT TO THE WIDTHS,  XP THE PEAK LOCATIONS, W THE PEAK WIDTHS.
C THESE LAST THREE ARE RETURNED, BUT INITIAL ESTIMATES MUST BE
C SUPPLIED.
         IMPLICIT REAL*8(A-H,O-Z)
         REAL*4 FB,F,WX,SQRT,ABS,XOFF,FA
         DIMENSION FB(1),F(1),WX(1),FW(3,5),IPT(2),SFW(5,5)
         DIMENSION P(3),CONS(3),SM(3),PC(3),
        #PPCC(6),AM(6),C(2),W(2),XP(2)
 9       CHB=1.E31
         CHL=1.E32
         IPT(1)=L
         FR=0
         IEN=0
         NT=3
         IX=XPG-XOFF
           PRINT*,' IN CQMIN XPG,XOFF,IX',XPG,XOFF,IX
         C(1)=SQRT(ABS(F(IX)-FB(IX)))
         CONS(1)=C(1)
         SM(1)=1.D-2
         XP(1)=XPG
         CONS(2)=XPG
         SM(2)=SM(1)*1.D6
         WPP=DSQRT(DMAX1(2D0,FW(1,L)+XPG*(FW(2,L)+
        # XPG*FW(3,L))))
         W(1)=WPP
         SW=DSQRT(DABS(SFW(1,L)+XPG*(SFW(2,L)+XPG*(SFW(3,L)
        # +XPG*(SFW(4,L)+XPG*SFW(5,L))))))
         SW=SW/(2*WPP)
         CONS(3)=W(1)
         SM(3)=SM(2)*1.D4
           NDT=NT*(NT+1)/2
         IB=XPG-2*W(1)-2*SW-XOFF
         IB=MAX0(1,IB)
         IE=XPG+2*W(1)+2*SW-XOFF
         IE=MIN0(N,IE)
```

```
      NKIT=30
      KIT=1
28      CHI=0
          PEN=0
      DO 30 J=1,NT
30    PC(J)=0.
      DO 35 J=1,NDT
35    PPCC(J)=0.
      DO 200 I=IB,IE
      X=I+XOFF
      FA=POLYB(X,P,C,XP,W,IPT,1,0)+FB(I)
        RES=AMIN1(1.E12,AMAX1(-1.E12,F(I)-FA))
      CHI=CHI+RES**2*WX(I)
      W1=2*(FA-F(I))*WX(I)
      W2=2*WX(I)
        KJ=0
      DO 80 J=1,NT
      PC(J)=PC(J)+W1*P(J)
      W2P=W2*P(J)
      DO 80 K=1,J
        KJ=KJ+1
80    PPCC(KJ)=PPCC(KJ)+P(K)*W2P
200   CONTINUE
       CALL WADJ(FW,SFW,PC,PPCC,AM,
     #  PEN,W,XP,IPT,1,KIT)
         CHI=CHI+PEN
      IF(CHI.LE.1.01*CHL.AND.KIT.GE.NKIT)GOTO 2502
      IF(CHI.LE.1.01*CHL.AND.IEN.EQ.1)GOTO 2502
      IF(CHL-CHB.GT..1D0)GOTO 248
      IF(KIT.LE.3)GOTO 248
        IEN=1
      IF(CHI.LT.CHL)GOTO 2502
248   CALL SMSQ(CHI,CHB,CHL,PC,PPCC,AM,FR,CONS,SM,NT,0)
250   NPART=1
      C(1)=DMAX1(.1D0,CONS(NPART))
      XP(1)=DMAX1(1.D0*(IB+XOFF),DMIN1(1.D0*(IE+XOFF),CONS(2)))
      W(1)=DMAX1(1.5D0,WPP-2*SW,DMIN1(CONS(3),WPP+2*SW))
      KIT=KIT+1
      IF(KIT.GT.2*NKIT)GOTO 280
      GOTO 28
2502  CONTINUE
        CHI=CHI-PEN
      ERRSQ=CHI/(IE+1-IB)
      CQMIN=C(1)
      WG=W(1)
      XPG=XP(1)
       RETURN
280   WRITE(*,105)CHI,CHB,CHL,(CONS(I),I=1,NT)
105   FORMAT(' LOOPING IN CQMIN, CHI,CHB,CHL/CONS'/(3E20.12))
      NPART=1
      CQMIN=0
      ERRSQ=1.D32
      RETURN
       END
C*******************************************************************
      SUBROUTINE RESL(F,FB,WX,FW,ALR,IQM,ILR,IFTC,IC,N,XOFF,CUTOFF)
C *** LOCATES THE LARGEST RESIDUAL IN THE FIT TO DETERMINE THE
```

```
C *** NEXT POTENTIAL PEAK LOCATION
      DIMENSION ILRO(50),ISRO(50),F(1),FB(1),WX(1),IWT(5)
         COMMON/SCONS/DU(300),XPS(5),XPL(5),IP,NGPCAL
      REAL*8 FW(3,5),XPS,XPL,DU
         SAVE ILRO,N10,ISRO
        DATA N10/50/
      IFTC=0
      IF(IQM.LE.N10)GOTO 20
15    IFTC=1
      RETURN
20    ALR=0
      XT=XOFF+N/2
      DO 90 J=1,IP
        WPG=DSQRT(DMAX1(3D0,FW(1,J)+XT*(FW(2,J)+XT*FW(3,J))))
      IW=WPG
          IW=MAX0(4,IW)
      IW=MIN0(100,IW)
      IWT(J)=IW
      JM=J-1
      DO 30 K=1,JM
30    IF(IW.EQ.IWT(K))GOTO 90
        CALL SCAN(F,FB,WX,N,ILRTE,ALRT,IW,ILRO,ISRO,IQM,IC)
      IF(ABS(ALR).GT.ABS(ALRT))GOTO 40
      ILR=ILRTE
      ILRT=J
      ALR=ALRT
40    CONTINUE
90    CONTINUE
      PRINT*,' IN RESL ILR,ILRT,ALR,CUTOFF',ILR,ILRT,ALR,CUTOFF
      IF(ABS(ALR).LT.CUTOFF)GOTO 15
        IQM=IQM+1
        IF(IQM.LE.50)THEN
              ILRO(IQM)=ILR
              ISRO(IQM)=1
               IF(ALR.LT.0.)ISRO(IQM)=-1
            ENDIF
      RETURN
      END
C*********************************************************************
         SUBROUTINE SCAN(F,FB,WX,N,ILR,ALR,IW,ILRO,ISRO,IQM,IC)
C *** Used with RESL in finding potentila new peak locations
      DIMENSION FSM(100),ILRO(50),ISRO(50),F(1),FB(1),WX(1)
      IWM=IW-1
      IWS2=IW/2
      DO 10 I=1,IW
10    FSM(I)=0
      SUM=0
      ALR=0
      DO 100 I=1,N
      FSM(IW)=(F(I)-FB(I))*SQRT(ABS(WX(I)))
      SUM=SUM+FSM(IW)-FSM(1)
      DO 20 J=1,IWM
20    FSM(J)=FSM(J+1)
      IF(ABS(ALR).GT.ABS(SUM))GOTO 40
      ILRT=I-IW
      FMAX=-1.E32
      DO 25 J=2,IW
```

```
      IF(FSM(J).LT.FMAX)GOTO 25
      FMAX=FSM(J)
      ILRA=ILRT+J
25    CONTINUE
      ILRT=MAX0(1,MIN0(N,ILRA))
         IF(IQM.EQ.0)GOTO 35
      DO 30 J=1,IQM
      IF(ILRT.EQ.ILRO(J))GOTO 40
      IF(SUM*ISRO(J).GT.0.AND.IABS(ILRT-ILRO(J)).LE.IC)GOTO 40
30    CONTINUE
35    CONTINUE
      ALR=SUM
      ILR=ILRT
40    CONTINUE
100   CONTINUE
      ALR=ALR/SQRT(1.*IW)
      RETURN
      END
```

The following is a listing of the FSPFIT.MNU file; this file is read by the MENURD subroutine and the variables are passed to the FSPFIT program to give it its options.

```
          7
          8
          6
          5
          3
          4
        100
 1 PAGE FSPFIT
 2
 3 GENERAL DATA ON THE SPECTRAL FIT
 4 INPUT DATA INFORMATION
 5 OUTPUT DATA INFORMATION
 6 STANDARDS FILES
 7 PEAK TYPE INFORMATION
 1 PAGE FSPFIT1 (GENERAL FIT INFORMATION)
 2
 3 MAXIMUM NUMBER OF PEAKS.                    255
 4 STARTING,ENDING CUTOFFS IN N STEPS          40,30,1
 5 VWID,FIXD AND CONT,NONK,NOBF,FIXK           VWID,CONT
 6 MAXIMUM NUMBER OF BACKGROUND CONSTANTS.     4,4
 7 WEIGHTS. 0=SQRT(DATA),1=CALC,2=FILE         0
 8 FILE CONTAINING THE WEIGHTS                 NONE
 1 PAGE FSPFIT2 (INPUT DATA)
 2
 3 INPUT DATA FILE NAME                        BTEST.UF
 4 BEGINNING AND ENDING CHANNELS FOR FIT.      1,1000
 5 INPUT PEAK FILE.                            NONE,NO
 6 INPUT BACKGROUND CONSTANTS FILE.            NONE,YES
 1 PAGE FSPFIT3 (OUTPUT DATA)
 2
 3 OUTPUT FILE FOR BACKGROUND CONSTANTS.       BTEST.CN
 4 OUTPUT FILE FOR PEAK DATA.                  BTEST.PK
 5 OUTPUT FILE FOR GRAPH DATA.                 BTEST.GR
```

```
1 PAGE FSPFIT4 (STANDARDS DATA)
2
3 GAUSSIAN STANDARD                         LOREN4.ST
1 PAGE FSPFIT5 (STANDARDS WIDTH VARIATION)
2
3 TYPE 1 PEAK WIDTH,ERROR AND CHANNEL NO. 10,5,10
4 TYPE 1 PEAK WIDTH,ERROR AND CHANNEL NO. 10,5,1000
```

The following is a listing of FSPDIS, the full spectral display program. This routine needs only the standard files, the background constants, the peak file and the data file to display the complete spectrum. The names of these files are contained in the *.GR file created by FSPFIT. The fittery is CPU intensive, but the display is not. It is therefore quite possible to fit on a main frame and then download the necessary files to a smaller machine to display the fit.

```
      PROGRAM FSPDIS
C*********************************************************************
C     The full spectral display program (FSPDIS)
C*********************************************************************
      INTEGER*2 IX,IFB
      REAL*8 POLYG3,POLYB3,SP
      REAL*8 XT,XTT,ST
      REAL*8 CONS,W1,W2,FW
      REAL*8 C,SC,W,WF,SW,XP,SXP
      REAL*8 XDP,P,PD,FDP
      REAL*8 ERRSQ,CHIS,BWSPW
      REAL*8 DU,XPS,XPL,XG3,X
      DIMENSION FAA(4096)
      DIMENSION FORIG(4),XORIG(4),FINT(4),XINT(4)
      DIMENSION F(4096),WX(4096),IHD(4096)
      EQUIVALENCE (IHD(1),F(1))
      DIMENSION C(256 ),W(256 ),PD(768 ),IPT(256 ),WF(256 )
      DIMENSION XP(256 ),SW(256 ),SXP(256 ),SC(256 )
      DIMENSION FW(3,5)
      DIMENSION CONS(200)
      DIMENSION P(756)
        CHARACTER*8 II,IFTEST
        CHARACTER*64 CLINE
        CHARACTER*40 NA
        CHARACTER*1 ANS
         COMMON/AXISP/BHORI,BVERT,SHORI,SVERT,EHORI,EVERT
C *** BEGIN OF SETWIN
        COMMON/PL/IGR,IGWIND,ITWIN
C**************************************************
C
C   GRAF3 DATA
C
C**************************************************
      CHARACTER*17 GDNAME
      CHARACTER*1 VCF,VPE,VBG,VRE
      COMMON/CHGRAS/GDNAME,VCF,VPE,VBG,VRE
      COMMON/CGRASH/IBEGC,IENDC,IMIN,IMAX,LFLAG
C *** GRAPHICS COMMON
      COMMON/BATCFL/IBATCH
C**********************************************
C  GRAPHICS COMMONS
```

```
C************************************************
      CHARACTER*17 NAPK
      COMMON/SCONS/DU(150),XPS(5),XPL(5),IP,NC
C*****RAW DATA
      CHARACTER*17 DATAFN,NADIFF
      COMMON/CHBREA/DATAFN,NADIFF
      COMMON/CBREAD/IB,IE,IEXTWT
C*****ROPKS STUFF
      CHARACTER*17 IPEKFN
      COMMON/CHROPK/IPEKFN
      COMMON/CROPKS/IPEKFL
C*****BKGFIT
      CHARACTER*17 IBKGFN
      COMMON/CHBKGF/IBKGFN
      COMMON/CBKGFI/IBKGFL
C*****************************************************
      DIMENSION STAN(5)
      CHARACTER*17 STAN
      COMMON/CHPOLY/STAN
      COMMON/CPOLYG/NSTAN
C*****************************************************
      IBATCH=0
5      CONTINUE
        CALL G3LINK
C       READ(*,102)NA
      NA=GDNAME
102     FORMAT(A40)
        IRF=0
          OPEN(UNIT=3,FILE=NA,ERR=999)
        II=NA
      READ(3,'(A,G020.12)')II,CHIS
      READ(3,*)IP
C *** INITIALISE THE STANDARD
      NSTAN=IP
      NC=0
      XT=0
      XTT=POLYG3(XT,1,ST)
      READ(3,'(I4/3(A64/),A64)')IEXTWT,DATAFN,NADIFF,IBKGFN,IPEKFN
      READ(3,'(2I5)')IB,IE
      READ(3,'(4G20.12)')(FW(1,I),FW(2,I),I=1,IP)
      DO 333 J=1,IP
333   FW(3,J)=0.
      CLOSE(3)
      CALL BREAD(XOFF,F,IHD,WX,N,FAA)
      IXOFF=XOFF
      IF(IBEGC.LT.IB.OR.IBEGC.EQ.0)IBEGC=IB
      IF(IENDC.GT.IE.OR.IENDC.EQ.0)IENDC=IE
      NDT=N+XOFF
C *** READ IN THE BACKROUND CONSTS
      OPEN(1,FILE=IBKGFN,STATUS='OLD')
      READ(1,*)IBKGFL
      NV=0
20    READ(1,*,END=25)CONS(NV+1)
      NV=NV+1
      GOTO 20
25    CLOSE(1)
C *** SECOND CALL ROPKS
```

```
      NPP=0
      CALL ROPKS(C,SC,W,IPT,WF,SW,XP,SXP,N,NPP,FW,XOFF)
        RCHIS=CHIS
        IXDUM=IXOFF+1
        IXDUM=MIN0(4096,MAX0(1,IXDUM))
        IX=IXDUM
301     CONTINUE
302     IMC=NDT-1
320     CONTINUE
        IREP=1
       IF(IREP.EQ.1)WRITE(*,103)IX,IMC
103     FORMAT(' DATA HAS CHANNELS',2I5,
     # ' INITIAL CHANNEL?, END CHANNEL?,<-1,0>FOR MENU')
        IF(IREP.EQ.1)READ(*,*)IBEGC,IENDC
        IF(IBEGC.EQ.-1)THEN
             GOTO 5
           ENDIF
        IF(IENDC.NE.0)GOTO 330
        IBEGC=IX
        IENDC=IMC
330     IXDUM=IX
        IBEGC=MAX0(IXDUM,IBEGC)
        IENDC=MIN0(IMC,IENDC)
        IF(IBEGC.GE.IENDC)THEN
        WRITE(*,104)IBEGC,IENDC
104     FORMAT(' THERE ARE NO CHANNELS IN THE
     #INTERVAL',F10.0,',',F10.0)
        STOP
        ENDIF
        AMIN=2000000000
        AMAX=-AMIN
        DO 340 I=IBEGC,IENDC
        AMIN=AMIN1(AMIN,F(I-IXOFF))
340     AMAX=AMAX1(AMAX,F(I-IXOFF))
        AMINT=AMIN
       IF(IREP.EQ.1)WRITE(*,108)AMINT,AMAX
108     FORMAT(' AMIN=',F8.0,'  AMAX=',F8.0,' ENTER DESIRED MIN',
     # ',DESIRED MAX')
        IF(IREP.EQ.1)READ(*,*)IMIN,IMAX
        IF(IMAX.EQ.0)GOTO 37
        AMIN=IMIN
        AMINT=AMIN
        AMAX=IMAX
37      CONTINUE
        IF(IREP.EQ.1)WRITE(*,120)
120     FORMAT(' DO YOU WANT A LOG SCALE?')
        IF(IREP.EQ.1)THEN
          READ(*,115)ANS
          LFLAG=0
          IF(ANS.EQ.'Y')LFLAG=1
        ENDIF
       CALL RSETG
       IF(LFLAG.NE.1)GOTO 400
        AMIN=AMAX1(.01,AMIN)
        AMAX=ALOG(AMAX)
        AMIN=ALOG(AMIN)
400     IRES=0
```

```
        IF(LFLAG.EQ.0)AMIN=AMINT
C       WRITE(*,110)
110     FORMAT(' DO YOU WANT TO SEE THE RESIDUALS?')
C       READ(*,115)ANS
      ANS=VRE
115     FORMAT(A1)
        IF(ANS.EQ.'Y')IRES=1
      ANS='Y'
      CALL STPL(0,0)
      IF(IBATCH.EQ.0)THEN
        IF(ANS.EQ.'Y'.OR.ANS.EQ.'y')CALL CLEARS
      ENDIF
        IAXIS=1
        CALL AXIS(AMIN,AMAX,LFLAG,IRES,1.*IBEGC,1.*IENDC,II,RCHIS)
        ISF=SHORI/2
        ISKIP=2/SHORI
        ISKIP=MAX0(1,ISKIP)
        IXO=BHORI
        DO 500 I=IBEGC,IENDC,ISKIP
420     XPP=BHORI+SHORI*(I-IBEGC)
        IXP=XPP
      IF(LFLAG.EQ.0)THEN
        ADAT=F(I-IXOFF)
      ELSE
        ADAT=ALOG(AMAX1(1E-3,F(I-IXOFF)))
      ENDIF
        IYDAT=SVERT*(ADAT-AMIN)+BVERT
C *** THE FIRST PLOT CALL
        IF(I.EQ.IBEGC)CALL STPL(IXO,IYDAT)
        IF(ISKIP.GT.1)GOTO 495
        IF(I.GT.IBEGC)CALL PLOT(IXO,IYDAT)
        IXO=IXP+ISF
        CALL PLOT(IXO,IYDAT)
        GOTO 500
495     ADMAX=-2.E9
        ADMIN=-ADMAX
        DO 497 J=1,ISKIP
        IARG=MIN0(IMC,I+J-1)
        ACOMP=F(IARG-IXOFF)
        ADMAX=AMAX1(ACOMP,ADMAX)
497     ADMIN=AMIN1(ADMIN,ACOMP)
      IF(LFLAG.EQ.1)THEN
        ADMIN=ALOG(AMAX1(1.E-3,ADMIN))
        ADMAX=ALOG(AMAX1(1.E-3,ADMAX))
      ENDIF
        IDMIN=SVERT*(ADMIN-AMIN)+BVERT
        IDMAX=SVERT*(ADMAX-AMIN)+BVERT
        CALL PLOT(IXO,IDMIN)
        IXO=IXP+ISF
        CALL PLOT(IXO,IDMAX)
500     CONTINUE
505     CONTINUE
C       CALL ANSI
C       WRITE(*,1546)
1546    FORMAT(' DO YOU WANT TO SEE THE CURVE FIT?')
C       READ(*,115)ANS
      ANS=VCF
```

```
          IF(ANS.NE.'Y')GOTO 650
          IIPKS=0
            INT=1+SHORI/2
            IMIN=INT/2+1
          CLINE='GREEN'
          CALL COLOR(CLINE)
          DO 600 I=IBEGC,IENDC
            DO 580 J=1,INT
            XG3=I+(J-IMIN)/(1.*INT)
          XPP=BHORI+SHORI*(XG3-IBEGC)
          IXP=XPP
            CALL G3POLY(XG3,P,NV,CONS,FDP)
        FA=POLYB3(XG3,P,C,XP,W,IPT,NPP,0)+FDP
          AFRA=FA
        IF(LFLAG.EQ.1)THEN
          AFRA=ALOG(AMAX1(1.E-3,FA))
        ENDIF
          IY=SVERT*(AFRA-AMIN)+BVERT
          IF(I.EQ.IBEGC.AND.J.EQ.1)THEN
           CALL STPL(IXP,IY)
           GOTO 580
           ENDIF
          CALL PLOT(IXP,IY)
580         CONTINUE
600       CONTINUE
650       CONTINUE
          CLINE='WHITE'
          CALL COLOR(CLINE)
C         CALL ANSI
C         WRITE(*,1556)
1556      FORMAT(' DO YOU WANT TO SEE THE BACKGROUND?')
C         READ(*,115)ANS
        ANS=VBG
          IF(ANS.NE.'Y')GOTO 700
          CLINE='BLUE'
          CALL COLOR(CLINE)
          DO 680 I=IBEGC,IENDC,ISKIP
          XPP=BHORI+SHORI*(I-IBEGC)
          IXP=XPP
          CALL G3POLY(1.D0*I,P,NV,CONS,FDP)
        IF(LFLAG.EQ.1)THEN
          FDP=DLOG(DMAX1(1.D-3,FDP))
        ENDIF
          IYB=SVERT*(FDP-AMIN)+BVERT
          IF(I.EQ.IBEGC)CALL STPL(IXP,IYB)
          CALL PLOT(IXP,IYB)
680       CONTINUE
          CLINE='WHITE'
          CALL COLOR(CLINE)
700       CONTINUE
C         CALL ANSI
C         WRITE(*,1566)
1566      FORMAT(' DO YOU WANT TO SEE THE INDIVIDUAL PEAKS?')
          IIPKS=1
C         READ(*,115)ANS
        ANS=VPE
          IF(ANS.NE.'Y')GOTO 780
```

```
         CLINE='RED'
         CALL COLOR(CLINE)
          INT=1+SHORI/2
          IMIN=INT/2+1
        DO 755 K=1,NPP
        ISM=W(K)*XPS(IPT(K))+XP(K)
        ILA=W(K)*XPL(IPT(K))+XP(K)
        I1=MAX0(ISM,IBEGC)
        I2=MIN0(ILA,IENDC)
        DO 750 I=I1,I2
          DO 745 J=1,INT
          XG3=I+(J-IMIN)/(1.*INT)
        XPP=BHORI+SHORI*(XG3-IBEGC)
        IXP=XPP
          CALL G3POLY(XG3,P,NV,CONS,FDP)
      FA=C(K)*C(K)*POLYG3((XG3-XP(K))/W(K),IPT(K),SP)+FDP
        AFRA=FA
      IF(LFLAG.EQ.1)THEN
        AFRA=ALOG(AMAX1(1.E-3,FA))
      ENDIF
        IY=SVERT*(AFRA-AMIN)+BVERT
        IF(I.EQ.I1.AND.J.EQ.1)THEN
         CALL STPL(IXP,IY)
         GOTO 745
         ENDIF
        CALL PLOT(IXP,IY)
745       CONTINUE
750     CONTINUE
755     CONTINUE
          CLINE='WHITE'
          CALL COLOR(CLINE)
780     IF(IRES.EQ.0)GOTO 900
        CHISL=0
        DO 800 I=IBEGC,IENDC
      XG3=I
        CALL G3POLY(XG3,P,NV,CONS,FDP)
      FA=POLYB3(XG3,P,C,XP,W,IPT,NPP,0)+FDP
          NRES=100*AMIN1(327.,(F(I-IXOFF)-FA)*SQRT(WX(I-IXOFF)))
        IF(NRES.GT.32767)NRES=NRES-65528
        XPP=BHORI+SHORI*(I-IBEGC)
        IXP=XPP
        IFIX=EVERT-49+.15*NRES
        CHISL=CHISL+(NRES/100.)**2
        IF(SHORI.GT.3)GOTO 795
        CALL PONT(IXP,IFIX)
        GOTO 800
795     CONTINUE
       DO 798 K=1,3
       DO 798 L=1,3
798    CALL PONT(IXP+(L-2),IFIX+(K-2))
800     CONTINUE
        CLINE='CHISL=\'
        CALL BCHART(100,200,CLINE)
        CALL NUMOUT(170,200,CHISL,0,0)
900     CONTINUE
C       CALL ANSI
        WRITE(*,1576)
```

```
1576    FORMAT(' ARE YOU FINISHED WITH THE CODE?')
        READ(*,115)ANS
      CALL ANSI
        IF(ANS.EQ.'Y')STOP
        IREP=1
       GOTO 320
999     WRITE(*,1945)
1945    FORMAT(' ERROR IN OPENING FILE ',A40,' TRY AGAIN')
        GOTO 5
        END
*********************************************************************
      SUBROUTINE G3LINK
C *** This routine links in the menu selected variables into the
C *** FSPDIS program.
      DIMENSION FFCOM(200),FFVAR(200)
      CHARACTER*64 CSTR,VALU,NA
      COMMON/USPAGE/FFVAR
      COMMON/FCALL/NSTR,NCMENU
      CHARACTER*40 FFCOM,FFVAR
C**************************************************
C
C   DATA TO MAIN ROUTINE GRAF3
C
C**************************************************
C
C   GRAF3 DATA
C
C**************************************************
      CHARACTER*17 GDNAME
      CHARACTER*1 VCF,VPE,VBG,VRE
      COMMON/CHGRAS/GDNAME,VCF,VPE,VBG,VRE
      COMMON/CGRASH/IBEGC,IENDC,IMIN,IMAX,LFLAG
      COMMON/CBREAD/IB,IE,IEXTWT
C**************************************************
      NCMENU=1
      NA='FSPDIS.MNU'
      CALL MENURD(NA)
      DO 10 J=1,10
        NSS=J*20+2
        IF(J.EQ.1)THEN
C**************************************************
      NSS=NSS+1
      GDNAME=FFVAR(NSS)
      NSS=NSS+1
        IT=INDEX(FFVAR(NSS),',')
        IF(FFVAR(NSS)(IT+1:IT+3).EQ.'DEF')THEN
          IBEGC=0
          IENDC=0
        ELSE
          CSTR=FFVAR(NSS)(1:IT-1)
          CALL CTON(CSTR,IMISS,ANUM)
          IBEGC=ANUM
          CSTR=FFVAR(NSS)(IT+1:37)
          CALL CTON(CSTR,IMISS,ANUM)
          IENDC=ANUM
        ENDIF
        NSS=NSS+1
```

```
        IT=INDEX(FFVAR(NSS),',')
        IF(FFVAR(NSS)(IT+1:IT+3).EQ.'DEF')THEN
          IMIN=0  .
          IMAX=0
        ELSE
          CSTR=FFVAR(NSS)(1:IT-1)
          CALL CTON(CSTR,IMISS,ANUM)
          IMIN=ANUM
          CSTR=FFVAR(NSS)(IT+1:37)
          CALL CTON(CSTR,IMISS,ANUM)
          IMAX=ANUM
        ENDIF
      NSS=NSS+1
          CSTR=FFVAR(NSS)(1:3)
          IF(CSTR.EQ.'LOG')THEN
            LFLAG=1
          ELSE
            LFLAG=0
          ENDIF
      ENDIF
      NSS=NSS+1
      CSTR=FFVAR(NSS)
C *** CALIBRATION NOT IMPLEMENTED IN THIS VERSION
C*************************************
      IF(J.EQ.2) THEN
C*************************************
        NSS=NSS+1
        VRE=FFVAR(NSS)(1:1)
        NSS=NSS+1
        VCF=FFVAR(NSS)(1:1)
        NSS=NSS+1
        VPE=FFVAR(NSS)(1:1)
        NSS=NSS+1
        VBG=FFVAR(NSS)(1:1)
        NSS=NSS+1
        CSTR=FFVAR(NSS)
C *** CALIBRATION NOT IMPLEMENTED IN THIS VERSION
C****************************
      ENDIF
10    CONTINUE
        RETURN
        END
C*******************************************************************
       FUNCTION POLYB3(X,P,C,XP,W,IPT,NPPS,IXMID)
C THIS ROUTINE CALCULATES POLYB=SUM CI*SI ALONG WITH DPOLYB/DCI
C THE NEXT NS SI'S, WHICH HAVE DERIVATIVES WITH RESPECT TO C,XP,AND W
C ARE BSPLINES
       IMPLICIT REAL*8(A-H,O-Z)
      COMMON/SCONS/DU(150),XPS(5),XPL(5)
       DIMENSION P(100),C(256),XP(256),W(256),IPT(256)
      NS=NPPS
      NC=1
5     NPART=1
      POLYB3=0
      DO 200 J=1,NS
        ITEST=IPT(J)
      P(NPART)=0.
```

```
      P(NPART+1)=0.
      P(NPART+2)=0.
      XM=X-XP(J)
      IF(XM.LE.W(J)*XPS(ITEST).OR.XM.GE.W(J)*XPL(ITEST))GOTO 200
      XM=XM/W(J)
C  PARTIAL WRT C(J)
       FADD=C(J)*POLYG3(XM,ITEST,SP)
      P(NPART)=2.*FADD
      POLYB3=POLYB3+C(J)*FADD
C PARTIAL WRT XP(J)
      P(NPART+1)=-C(J)*C(J)*SP/W(J)
C PARTIAL WRT W(J)
      P(NPART+2)=P(NPART+1)*XM
200   NPART=NPART+3
C*******************************
C EXTRA BACKGROUND CONTRIBUTION
C*******************************
      IF(NS.EQ.0)GOTO 181
      IF(IXMID.EQ.0) GOTO 181
      P(NPART)=1
      NPART=NPART+1
      P(NPART)=X-IXMID
      NPART=NPART+1
      P(NPART)=P(NPART-1)**2
      NPART=NPART+1
      P(NPART)=P(NPART-1)*P(NPART-2)
      NPART=NPART+1
      P(NPART)=P(NPART-2)*P(NPART-2)
      NPART=NPART+1
      P(NPART)=P(NPART-2)*P(NPART-3)
C********************************
181   CONTINUE
      RETURN
      END
C*********************************************************************
        SUBROUTINE G3POLY(X,P,NV,CONS,FA)
C *** THE FIRST FOUR COEFFICIENTS REPRESENT A CUBIC
C *** THE REST ARE IN THE FORM C(I)*(C(I+1)-X)+ **3
        IMPLICIT REAL*8 (A-H,O-Z)
         COMMON/CBKGFI/IBKGFL
        DIMENSION P(100),CONS(200)
        NVS=NV-4
        P(1)=1.D0
        FA=CONS(1)
        DO 5 I=2,4
        P(I)=X*P(I-1)
 5      FA=FA+CONS(I)*P(I)
        DO 20 I=5,NV,2
        P(I)=0.D0
        P(I+1)=0.D0
        IF(CONS(I+1).LE.X)GOTO 20
        DIFF=CONS(I+1)-X
        DIFF2=DIFF*DIFF
        P(I+1)=3*CONS(I)*DIFF2
        P(I)=DIFF2*DIFF
        FA=FA+CONS(I)*P(I)
 20     CONTINUE
```

```
      IF(IBKGFL.EQ.0)RETURN
      FA=DEXP(DMIN1(30D0,FA))
      DO 30 I=1,NV
30    P(I)=FA*P(I)
        RETURN
         END
C*********************************************************************
     FUNCTION POLYG3(X,IPT,SP)
C *** reads the standard files on its first call
C THIS ROUTINE CALCULATES POLYG=SUM CI*SI ALONG WITH DPOLY/DX
C IPT GIVES THE PEAK TYPE RANGING FROM 1 TO 5
       IMPLICIT REAL*8(A-H,O-Z)
        CHARACTER*64 NA
C*****************************************************
C
C  DATA FROM MAIN ROUTINE
C
C*****************************************************
      DIMENSION STAN(5)
      CHARACTER*17 STAN
      COMMON/CHPOLY/STAN
      COMMON/CPOLYG/NSTAN
      COMMON/SCONS/W(10,5),XP(10,5),C(10,5),XPS(5),XPL(5),IP,NC
      COMMON/PYG3/NST(5)
      IF(NC.EQ.1)GOTO 30
      NC=1
      NS=0
      IF(NSTAN.EQ.0)THEN
        POLYG3=0.
        NST(1)=0
        RETURN
      ENDIF
      DO 3 IP=1,NSTAN
      IF(IP.GT.5)GOTO 300
      READ(3,'(A)')STAN(IP)
      OPEN(11,FILE=STAN(IP),STATUS='OLD',ERR=3)
      NS=1
5     READ(11,101,END=20)C(NS,IP),XP(NS,IP),W(NS,IP)
101   FORMAT(5X,3E20.7)
      WRITE(*,103)NS,C(NS,IP),XP(NS,IP),W(NS,IP)
103   FORMAT(I5,3E20.7)
      NS=NS+1
      GOTO 5
20    NS=NS-1
      CLOSE(11)
      XPS(IP)=1.D32
      XPL(IP)=-1.D32
      DO 25 J=1,NS
      XPS(IP)=DMIN1(XPS(IP),XP(J,IP)-W(J,IP))
25    XPL(IP)=DMAX1(XPL(IP),XP(J,IP)+W(J,IP))
        NST(IP)=NS
3     CONTINUE
C *** READ THE BLANK CARD
      READ(3,'(A)')NA
30    CONTINUE
       POLYG3=0.
      SP=0.
```

```
        IF(NST(IPT).EQ.0)RETURN
        NS=NST(IPT)
      DO 200 J=1,NS
      XM=X-XP(J,IPT)
      IF(XM.LE.-W(J,IPT).OR.XM.GE.W(J,IPT))GOTO 200
      XM=XM/W(J,IPT)
      FADD=C(J,IPT)*((1.+XM)*(1.-XM))**3
      POLYG3=POLYG3+FADD
       SP=SP+3.*(FADD/(1.+XM)-FADD/(1.-XM))/W(J,IPT)
200   CONTINUE
      RETURN
300     PRINT*,' ATTEMPT TO DEFINE MORE THAN FIVE STANDARDS'
        STOP
      END
```

The following is a listing of the FSPDIS.MNU file; this file is read by the MENURD subroutine and the variables are passed to the FSPDIS program to give it its options.

```
          4
          7
          7
        100
 1 PAGE FSPDIS
 2
 3 INFORMATION ON THE SPECTRAL FIT DISPLAY
 4 VIEWING INFORMATION FOR THE FIT
 1 PAGE FSPDIS1 (GENERAL INFORMATION)
 2
 3 NAME OF THE GRAPHICAL FILE TO BE READ   FILENAME.GR
 4 BEGINNING/ENDING CHANNELS FOR DISPLAY   0,0
 5 DESIRED MIN AND MAX Y SCALE             0,0
 6 LOG OR LINEAR Y SCALE                   LOG
 7 XCALIBER FILE                           NONE
 1 PAGE FSPDIS2 (VIEWING OPTIONS)
 2
 3 VIEW THE RESIDUALS (Y/N)                YES
 4 VIEW THE CURVE FIT (Y/N)                YES
 5 VIEW THE PEAKS (Y/N)                    YES
 6 VIEW THE BACKGROUND FIT (Y/N)           YES
 7 CALIBRATE X-AXIS                        YES
```

The following is a listing of the calibration program XCALIBER. Most of this program is interactive in nature. The routines ask relevant questions at certain points in the calibration to which the user must respond. XCALIBER reads in a user supplied file of channel positions versus energies and efficiencies and creates a set of calibration constants. The code then uses these constants to calibrate peak files.

```
        PROGRAM XCALIBER
C*********************************************************************
```

```
C This is the calibration program
C*********************************************************************
        IMPLICIT REAL*8 (A-H,O-Z)
        DIMENSION XC(80),EXC(80),EC(80),AANS(40),
     2   EEC(80),STR(80),ESTR(80),STRC(80),ESTRC(80),
     3   ECON(40),EFFCON(40),
     4   P(40),PC(40),PPCC(820),EPPCCI(820),EFPPCI(820)
        CHARACTER*80 EFFTIT,DATTIT
        CHARACTER*64 NA
        CHARACTER*1 COM(30,80)
        CHARACTER*1 ANS,ANSE,CANS(80)
        CHARACTER*20 QUEST
C**************************************************
C
C   DATA TO MAIN ROUTINE XCALIBER
C
C**************************************************
      CHARACTER*17 NAENER,NAEFF,NACAL,NAPEAK
      COMMON/XCNAME/NAENER,NAEFF,NACAL,NPEAKS,NAPEAK(20)
C**************************************************
        EXTERNAL EPOLY,EFPOLY
        CALL XCLINK
        PRINT'(A17)',NAENER,NAEFF,NACAL
        PRINT'(A,I3/(A))',' NPEAKS ',NPEAKS,NAPEAK(1),NAPEAK(2)
        ANSE='N'
        IF(NAENER.EQ.'NONE'.AND.NACAL.EQ.'NONE')THEN
          PRINT*,' A FIT OF ENERGY VERSUS CHANNEL CANNOT BE MADE'
          PRINT*,' UNTIL YOU CONSTRUCT A FILE OF CHANNELS VERSUS',
     2 'ENERGY AND PLACE ITS NAME IN THE MENU'
          STOP
        ENDIF
        IF(NAENER.NE.'NONE'.AND.NAENER.NE.'MAKE')THEN
           READ(4,*)JM
           JMD=JM*(JM+1)/2
           READ(4,*)(ECON(I),I=1,JM)
           READ(4,*)(EPPCCI(I),I=1,JMD)
           PRINT'(A/(4G16.6))',' ECON ARE',(ECON(I),I=1,JM)
        ENDIF
        IF(NAEFF.NE.'NONE'.AND.NAEFF.NE.'MAKE')THEN
           READ(2,*)JME
           JMED=JME*(JME+1)/2
           READ(2,*)(EFFCON(I),I=1,JME)
           READ(2,*)(EFPPCI(I),I=1,JMED)
           PRINT'(A/(4G16.6))',' EFFCON ARE',(EFFCON(I),I=1,JME)
        ENDIF
        IF(NACAL.NE.'NONE')THEN
          IL=INDEX(NACAL,'.')-1
          NA=NACAL(:IL)//'.FCA'
          OPEN(1,FILE=NACAL)
          OPEN(3,FILE=NA)
          DO 13 JK=1,4
          READ(1,'(A80,A20)')EFFTIT,QUEST
          WRITE(*,'(1X,A80,A20)')EFFTIT,QUEST
          WRITE(3,'(1X,A80)')EFFTIT
13         CONTINUE
          N=1
5        CALL LINPUT(1,CANS,AANS,8,ICANS,IEND)
```

```
      IF(IEND.EQ.1)GOTO 10
      XC(N)=AANS(1)
      EXC(N)=AANS(2)
      IF(EXC(N).LE.0.D0)GOTO 5
      EC(N)=AANS(3)
      EEC(N)=AANS(4)
      STR(N)=AANS(5)
      ESTR(N)=AANS(6)
      STRC(N)=AANS(7)
      ESTRC(N)=AANS(8)
      DO 9 I=1,30
9     COM(I,N)=CANS(I)
        N=N+1
        IF(N.LE.80)GOTO 5
10      N=N-1
        CLOSE(1)
        IF(NAENER.EQ.'MAKE')THEN
          CALL EFIT(ECON,PC,EPPCCI,PPCC,P,XC,EXC,EC,EEC,N,JM)
          WRITE(*,'('' THE ENERGY CONSTANTS''/(4E15.6))')
     2     (ECON(I),I=1,JM)
          WRITE(3,'('' THE ENERGY CONSTANTS''/(4E15.6))')
     2     (ECON(I),I=1,JM)
          IL=INDEX(NACAL,'.')-1
          NA=NACAL(:IL)//'.XDA'
          OPEN(4,FILE=NA)
          WRITE(4,*)JM
          WRITE(4,'(2G16.6)')(ECON(I),I=1,JM)
          JMD=JM*(JM+1)/2
          WRITE(4,'(2G16.6)')(EPPCCI(I),I=1,JMD)
          WRITE(4,*)
          CLOSE(4)
C ***     DATA FOR GPLOT TO SHOW THE NON-LINEARITY
          H=1.2*XC(N)/100
          NA='EFIT.DAT'
          OPEN(4,FILE=NA)
          DO 8 I=1,100
          X=H*I
          CALL FAEFA(X,.01D0,ECON,EPPCCI,JM,P,EPOLY,E,DE,ERR)
          E=E-(ECON(1)+X*ECON(2))
8         WRITE(4,*)X,E
          CLOSE(4)
          NA='EFIT.POI'
          OPEN(4,FILE=NA)
          DO 12 I=1,N
12        WRITE(4,*)XC(I),EC(I)
          CLOSE(4)
C ***     DISPLAYING THE ENERGY FIT
          WRITE(3,'(7X,''CHAN'',11X,''ENERGY'',9X,''ECAL'',
     2     10X,''RES'')')
          WRITE(*,'(7X,''CHAN'',11X,''ENERGY'',9X,''ECAL'',
     2     10X,''RES'')')
          RESM=0
          DO 20 I=1,N
          CALL FAEFA(XC(I),EXC(I),ECON,EPPCCI,JM,P,EPOLY,E,DE,ERR)
          RES=(E-EC(I))/ERR
          IF(DABS(RES).GT.RESM)THEN
            IRESM=I
```

```
              RESM=DABS(RES)
             ENDIF
           WRITE(3,'(F8.2,'' +-'',F5.2,F8.2,'' +-'',F5.2,F8.2,'' +-'',
     2     F5.2,F8.2,5X,30A1)')XC(I),EXC(I),E,ERR,EC(I),EEC(I),RES,
     3      (COM(J,I),J=1,30)
           WRITE(*,'(F8.2,'' +-'',F5.2,F8.2,'' +-'',F5.2,F8.2,'' +-'',
     2     F5.2,F8.2,5X,30A1)')XC(I),EXC(I),E,ERR,EC(I),EEC(I),RES,
     3      (COM(J,I),J=1,30)
20         CONTINUE
           IF(RESM.GT.3.D0)THEN
            PRINT'('' NOTE THAT THE RESIDUAL
    1FOR THE'',I3,'' THE POINT'',
     2      '' IS'',F5.2,'' THIS MAY INDICATE AN ERROR'')',IRESM,RESM
             READ(*,'(A)')ANS
           ENDIF
         ENDIF
C ***    FITTING THE EFFICIENCIES
         IF(NAEFF.EQ.'NONE')GOTO 35
         IF(NAEFF.EQ.'MAKE')THEN
         PRINT*,' DO YOU HAVE EFFICIENCY DATA'
         READ(*,'(A1)')ANSE
         IF(ANSE.NE.'Y')GOTO 35
         CALL EFFFIT(EFFCON,PC,EFPPCI,PPCC,P,XC,EXC,STR,ESTR,
     2    STRC,ESTRC,N,JME)
            WRITE(*,'('' THE EFFICIENCY CONSTANTS''/(4E15.6))')
     2        (EFFCON(I),I=1,JME)
            WRITE(3,'('' THE EFFICIENCY CONSTANTS''/(4E15.6))')
     2        (EFFCON(I),I=1,JME)
            IL=INDEX(NACAL,'.')-1
            NA=NACAL(:IL)//'.EFF'
            OPEN(4,FILE=NA)
            WRITE(4,*)JME
            WRITE(4,'(2G16.6)')(EFFCON(I),I=1,JME)
            JMED=JME*(JME+1)/2
            WRITE(4,'(2G16.6)')(EFPPCI(I),I=1,JMED)
            WRITE(4,*)
            CLOSE(4)
            H=1.2*XC(N)/100
            NA='EFFIT.DAT'
            OPEN(4,FILE=NA)
            DO 22 I=1,100
            X=H*I
            CALL
FAEFA(X,.01,EFFCON,EFPPCI,JME,P,EFPOLY,EFFI,DEFFI,ERR)
22          WRITE(4,*)X,EFFI
            CLOSE(4)
            NA='EFFIT.POI'
            OPEN(4,FILE=NA)
            DO 24 I=1,N
24          WRITE(4,*)XC(I),STR(I)/STRC(I)
            CLOSE(4)
C *** DISPLAYING THE EFFICIENCY FIT
        WRITE(3,'(''     CHAN  ENERGY   CSTRENGTH  STRENGTH     RES'')')
        WRITE(*,'(''     CHAN  ENERGY   CSTRENGTH  STRENGTH     RES'')')
        RESM=0
         DO 30 I=1,N
         CALL FAEFA(XC(I),EXC(I),EFFCON,EFPPCI,JME,P,EFPOLY,EFFI,
```

```
     2    DEFFI,ERR)
         STRENG=STR(I)/EFFI
         ESTREN=DSQRT((STR(I)*ERR/EFFI**2)**2+(ESTR(I)/EFFI)**2)
         RES=(STRENG-STRC(I))/ESTREN
         IF(DABS(RES).GT.RESM)THEN
            IRESM=I
            RESM=DABS(RES)
         ENDIF
         WRITE(3,'(4F8.2,'' +-'',2F6.2,5X,30A1)')XC(I),EC(I),
     2 STRC(I),STRENG,ESTREN,RES,(COM(J,I),J=1,30)
         WRITE(*,'(4F8.2,'' +-'',2F6.2,5X,30A1)')XC(I),EC(I),
     2 STRC(I),STRENG,ESTREN,RES,(COM(J,I),J=1,30)
30      CONTINUE
         IF(RESM.GT.3.D0)THEN
            PRINT'('' NOTE THAT THE RESIDUAL
     1FOR THE'',I3,'' THE POINT'',
     2      '' IS'',F5.2,'' THIS MAY INDICATE AN ERROR'')',IRESM,RESM
         READ(*,*)ITEST
         ENDIF
         ENDIF
35      CONTINUE
        ENDIF
         DO 500 IPK=1,NPEAKS
          CLOSE(3)
         WRITE(*,'(1X,A)')NAPEAK(IPK)
         OPEN(1,FILE=NAPEAK(IPK),STATUS='OLD')
         IL=INDEX(NAPEAK(IPK),'.')-1
         NA=NAPEAK(IPK)(:IL)//'.PCA'
          OPEN(3,FILE=NA)
          WRITE(3,'(A)')' DATA FROM ',NAPEAK(IPK)
          WRITE(3,'(A)')' CALIBRATION FROM ',NACAL,NAENER,NAEFF
          WRITE(3,'('' THE ENERGY CONSTANTS''/(4E15.6))')
     2     (ECON(I),I=1,JM)
          IF(NAEFF.NE.'NONE'.OR.ANSE.EQ.'Y')WRITE(3,
     2      '('' THE EFF CONSTANTS''/(4E15.6))')(EFFCON(I),I=1,JME)
         READ(1,'(A80)')DATTIT
         WRITE(3,'(1X,A80)')DATTIT
         WRITE(*,'(1X,A80)')DATTIT
         READ(1,'(A4,I5)')DATTIT,IP
         WRITE(3,'(1X,A4,I3)')DATTIT,IP
         WRITE(*,'(1X,A4,I3)')DATTIT,IP
         DO 15 I=1,2
         READ(1,'(A80)')DATTIT
         WRITE(3,'(1X,A80)')DATTIT
15       WRITE(*,'(1X,A80)')DATTIT
         IF(IP.NE.0)READ(1,'(10F8.4)')(ARAT,WRAT,I=1,IP)
        IF(NAEFF.NE.'NONE')GOTO 327
           WRITE(3,'(7X,''CHANNEL'',11X,''ENERGY'',11X,''FWHM'',11X,
     2     ''STRENGTH'')')
           WRITE(*,'(7X,''CHANNEL'',11X,''ENERGY'',11X,''FWHM'',11X,
     2     ''STRENGTH'')')
295        READ(1,*,END=490)X,EX,FWHM,EFWHM,AREA,EAREA
           CALL FAEFA(X,EX,ECON,EPPCCI,JM,P,EPOLY,E,EDER,ERR)
           FWHM=EDER*FWHM
           EFWHM=EFWHM*EDER
           WRITE(3,'(3(F8.2,'' +-'',F6.2),F12.2,'' +-'',F8.2)')
     2       X,EX,E,ERR,FWHM,EFWHM,AREA,EAREA
```

```
        WRITE(*,'(3(F8.2,'' +-'',F6.2),F12.2,'' +-'',F8.2)')
    2     X,EX,E,ERR,FWHM,EFWHM,AREA,EAREA
        GOTO 295
327   CONTINUE
      WRITE(*,110)
110   FORMAT(' ENTER THE ENERGY RANGE FOR INTENSITY',
    # ' NORMALIZATION'/' OR 0,0 TO NORMALIZE USING THE STRONGEST'
    # ' PEAK')
      READ(*,*)ERB,ERE
      WRITE(*,*)ERB,ERE
298   ANORM=0
300   READ(1,*,END=400)X,EX,FWHM,EFWHM,AREA,EAREA
      IF(ERB.LT.1.D-5)GOTO 350
      CALL FAEFA(X,EX,ECON,EPPCCI,JM,P,EPOLY,E,DE,ERR)
      IF(E.GT.ERE)GOTO 400
      IF(E.LT.ERB)GOTO 300
      CALL FAEFA(X,EX,EFFCON,EFPPCI,JME,P,EFPOLY,EFFIC,DEFF,ERR)
      ANORM=ANORM+AREA/EFFIC
      GOTO 300
350   CALL FAEFA(X,EX,EFFCON,EFPPCI,JME,P,EFPOLY,EFFIC,DEFF,ERR)
      ANORM=DMAX1(ANORM,AREA/EFFIC)
      GOTO 300
400   REWIND 1
      READ(1,'(A80)')DATTIT
      READ(1,'(A4,I5)')DATTIT,IP
      READ(1,'(A80)')DATTIT
      READ(1,'(A80)')DATTIT
      IF(IP.NE.0)READ(1,'(10F8.4)')(ARAT,WRAT,I=1,IP)
      PRINT*,' ANORM=',ANORM
      WRITE(*,1909)
1909  FORMAT(' ENTER THE SIZE DESIRED FOR THE NORMALIZING PEAK')
      READ(*,*)ASIZE
      ANORM=ANORM/ASIZE
      PRINT*,' ANORM=',ANORM
      WRITE(*,1978)
      WRITE(3,1978)
1978  FORMAT('   CHANNEL',6X,'ENERGY',11X,'FWHM',10X,'AREA',10X,
    2 'INTENSITY')
410   READ(1,*,END=490)X,EX,FWHM,EFWHM,AREA,EAREA
      IF(X.LT.0.D0)GOTO 490
      CALL FAEFA(X,EX,ECON,EPPCCI,JM,P,EPOLY,E,DE,ERR)
      CALL FAEFA(X,EX,EFFCON,EFPPCI,JME,P,EFPOLY,EFFIC,DEFF,ERRF)
      AR=AREA/(EFFIC*ANORM)
      EARC=DSQRT((EAREA/(ANORM*EFFIC))**2+(AR*ERRF/EFFIC)**2)
      FWHM=DE*FWHM
      EFWHM=EFWHM*DE
      WRITE(*,460)X,E,ERR,FWHM,EFWHM,AREA,EAREA,AR,EARC
      WRITE(3,460)X,E,ERR,FWHM,EFWHM,AREA,EAREA,AR,EARC
460   FORMAT(2F8.2,' +-',2F6.2,' +-',F4.2,F10.2,' +-',2F8.2,
    2 ' +-',F6.2)
      GOTO 410
490   CLOSE (1)
500   CONTINUE
510   WRITE(*,115)
115   FORMAT(' DO YOU WISH TO CONVERT A COUNTS VS CHANNEL FILE TO',
    # ' A COUNTS VS ENERGY FILE?')
      READ(*,'(A1)')ANS
```

```
      IF(ANS.EQ.'Y')THEN
         CLOSE(3)
        CALL ECHAN(ECON,EPPCCI,JM)
        GOTO 510
      ENDIF
         CLOSE(3)
        STOP
        END
C*******************************************************************
      SUBROUTINE XCLINK
C *** This routine passes the menu variables through to the XCALIBER
C *** program.
      DIMENSION FFCOM(200),FFVAR(200)
      CHARACTER*64 CSTR,VALU,NA
      COMMON/USPAGE/FFVAR
      COMMON/FCALL/NSTR,NCMENU
      CHARACTER*40 FFCOM,FFVAR
C**********************************************
C
C   DATA TO MAIN ROUTINE XCALIBER
C
C**********************************************
      CHARACTER*17 NAENER,NAEFF,NACAL,NAPEAK
      COMMON/XCNAME/NAENER,NAEFF,NACAL,NPEAKS,NAPEAK(20)
C**********************************************
      NA='XCALIBER.MNU'
      NCMENU=1
      CALL MENURD(NA)
      DO 10 J=1,10
        NS=J*20+2
        IF(J.EQ.1)THEN
C**********************************************
        NS=NS+1
        NACAL=FFVAR(NS)
        NS=NS+1
        NAENER=FFVAR(NS)
        NS=NS+1
        NAEFF=FFVAR(NS)
C*****************************
        ENDIF
        IF(J.EQ.2)THEN
C*****************************
           NPEAKS=0
           DO 20 I=1,20
             NS=NS+1
             IF(FFVAR(NS)(1:1).EQ.' ')GOTO 20
             NAPEAK(I)=FFVAR(NS)
             NPEAKS=NPEAKS+1
20         CONTINUE
C*****************************
        ENDIF
10     CONTINUE
        RETURN
        END
C*******************************************************************
        SUBROUTINE ECHAN(ECON,EPPCCI,JM)
C *** Converts a histogram in channels to a histogram in energies
```

```
      IMPLICIT REAL*8 (A-H,O-Z)
      REAL*4 F1(8192),F2(8192),FTRUN
      DIMENSION ECON(1),EPPCCI(1),P(40),IHD(128),IHOUT(8192)
      CHARACTER*4 IHEAD(40),ITT(40)
      CHARACTER*78 HEAD
      CHARACTER*64 NA,NAS,NAOLD,NAOUT
      CHARACTER*1 ANS,ANSOLD
      EXTERNAL EPOLY
      EQUIVALENCE (E1,IHD(124)),(E2,IHD(126)),(NCHAN,IHD(128))
      DATA NAOLD/'NULL'/

      PRINT*,' FIRST THE FILE TO BE CONVERTED'
      READ(*,'(A)')NA
      CALL REDAT(F1,IMRD,NA,IHEAD)
      PRINT*,' IS THE DATA BEING ADDED TO THAT IN ANOTHER FILE'
      READ(*,*)ANSOLD
      IF(ANSOLD.NE.'Y'.AND.ANSOLD.NE.'y')GOTO 150
      PRINT*,' ENTER THE NAME OF THE OLD DATA FILE'
      READ(*,'(A)')NAOLD
      IT=INDEX(NAOLD,'.')
      IF(NAOLD(IT+1 :IT+5).EQ.'SALLY')GOTO 140
      OPEN(3,FILE=NAOLD,STATUS='OLD')
      READ(3,*)E1,E2,NCHAN
      READ(3,'(20A4)')IHEAD
      READ(3,'(A)')HEAD
      CLOSE(3)
      GOTO 170
140   CONTINUE
       OPEN(3,FILE=NAOLD,STATUS='OLD',FORM='UNFORMATTED')
       READ(3)(IHD(I),I=1,128)
       CLOSE(3)
      GOTO 170
150   PRINT*,' ENTER EBEG, EEND, # OF CHANNELS'
      READ(*,*)E1,E2,NCHAN
170   EB=E1
      PRINT*,' EB,E2,NCHAN',EB,E2,NCHAN
      DO 200 I=1,NCHAN
200   F2(I)=0
      NOUTC=0
      NADJ=0
      DE=(E2-E1)/(NCHAN-1)
      E1=E1-DE
      COLD=CHAN(E1,ECON,EPPCCI,JM,300D0)
      PRINT*,' AFTER CALL TO CHAN COLD=',COLD
      CG=COLD+2
10    NOUTC=NOUTC+1
1001  NCOLD=COLD-NADJ
12    E1=E1+DE
      F2(NOUTC)=(NCOLD+1+NADJ-COLD)*FTRUN(F1,(NCOLD+1),IMRD)
     # +F2(NOUTC)
      N=NCOLD+1
      C1=CHAN(E1,ECON,EPPCCI,JM,CG)
      NC1=C1-NADJ
15    IF(N.GE.NC1)GOTO 20
      F2(NOUTC)=F2(NOUTC)+FTRUN(F1,(N+1),IMRD)
      N=N+1
      GOTO 15
```

```
20      F2(NOUTC)=(C1-N-NADJ)*FTRUN(F1,(NC1+1),IMRD)+F2(NOUTC)
        CG=C1+(C1-COLD)
        COLD=C1
        IF(E1.LT.E2)GOTO 10
        E1=EB
        IF(ANSOLD.EQ.'Y'.OR.ANSOLD.EQ.'y')THEN
          CALL REDAT(F1,NIR,NAOLD,ITT)
          PRINT*,' IMRD=',IMRD,'NIR=',NIR,' NCHAN=',NCHAN
          DO 300 I=1,NCHAN
          F2(I)=F2(I)+FTRUN(F1,I,NIR)
300       CONTINUE
          ENDIF
        PRINT*,' ENTER THE OUTPUT FILE NAME'
        READ(*,'(A)')NAOUT
        IT=INDEX(NAOUT,'.')
        IF(NAOUT(IT+1 :IT+5).EQ.'SALLY')GOTO 640
        OPEN(4,FILE=NAOUT,STATUS='NEW')
        WRITE(4,*)E1,E2,NCHAN
        WRITE(4,'('' DATA FROM '',A20,'' PLUS '',A20)')NA,NAOLD
        IHEAD(30)='HDAT'
        WRITE(4,'(20A4)')IHEAD
        WRITE(4,'('' 1,'',I5,'' 1'')')NCHAN
        DO 400 I=1,NCHAN
400     IHOUT(I)=F2(I)+.5
        WRITE(4,'(Z7,9Z8)')(IHOUT(I),I=1,NCHAN)
        CLOSE(4)
        RETURN
640     CONTINUE
         OPEN(4,FILE=NAOUT,STATUS='NEW',FORM='UNFORMATTED')
         WRITE(4)(IHD(I),I=1,128)
         I1=0
1950    CONTINUE
        DO 2000 J=1,128
2000    IHOUT(J)=FTRUN(F2,I1+J,NCHAN)+.5
        WRITE(4)(IHOUT(J),J=1,128)
        I1=I1+128
        IF(I1.LT.NCHAN)GOTO 1950
        CLOSE(4)
        RETURN
        END
C*********************************************************************
        FUNCTION FTRUN(F1,JK,IM)
C *** Used by ECHAN
        DIMENSION F1(1)
        FTRUN=0
        IF(JK.GT.0.AND.JK.LT.IM)FTRUN=F1(JK)
        RETURN
        END
C*********************************************************************
        FUNCTION CHAN(E,ECON,EPPCCI,JM,CG)
C *** Finds the channel for a given energy
        IMPLICIT REAL*8 (A-H,O-Z)
        DIMENSION ECON(1),EPPCCI(1),P(40)
        EXTERNAL EPOLY
        EX=0
        CHAN=CG
        NLOOP=0
```

```
10      CALL FAEFA(CHAN,EX,ECON,EPPCCI,JM,P,EPOLY,EC,EDER,ERR)
        IF(DABS((E-EC)/DMAX1(E,1.D0)).LT.1.D-8)GOTO 30
        CHAN=CHAN-(EC-E)/EDER
        NLOOP=NLOOP+1
        IF(NLOOP.LT.100)GOTO 10
        WRITE(*,1935)NLOOP,E,EC,EDER
1935    FORMAT(' CHAN NOT SOLUBLE IN DILUTE MATH, E, EC, EDER',
     # I5,3E20.6)
        READ(*,*)ITEST
        STOP
30      CHAN=DMAX1(1D-7,CHAN)
        RETURN
        END
C*********************************************************************
        SUBROUTINE REDAT(F,NIR,NA,II)
C *** Reads in the data files
        CHARACTER*64 NA
        CHARACTER*4 HDAT,Z4DAT,IDAT,LCDAT,II,IRBKG,IBKG
        DIMENSION F(8192),II(40)
        DIMENSION IHD(8192)
      DATA LCDAT/'LCDA'/,XVB,XVE/2*0.D0/,HDAT/'HDAT'/IDAT/'IDAT'/
      DATA Z4DAT/'Z4DA'/IPW/0/,PCON/1.D6/,IRBKG/'RBKG'/,IBKG/'    '/
5        N=0
        DO 7 I=1,40
7       II(I)='    '
        NIR=0
        NMULT=1
        IF(NA.EQ.'\'.OR.NA.EQ.'STOP'.OR.NA.EQ.'NULL'.OR.NA.EQ.
     # 'END')RETURN
        IT=INDEX(NA,'.')
        IRF=0
         IF(NA(IT+1 :IT+3).EQ.'UF')GOTO 1100
        IF(NA(IT+1 :IT+5).EQ.'SALLY')GOTO 2000
        IF(NA(IT+1:).EQ.'LEO'.OR.NA(IT+1:).EQ.'leo')IRF=1
        IF(IRF.EQ.1)OPEN(8,FILE=NA,STATUS='OLD',
     # FORM='UNFORMATTED',ERR=999)
        IF(IRF.EQ.1)CALL BREAD(F,IHD,IHD,NIR)
        IF(IRF.EQ.1)GOTO 220
        OPEN(8,FILE=NA,STATUS='OLD',ERR=999)
        READ(8,105)II
105    FORMAT(10A4)
        WRITE(*,105)II
         IF(II(40).EQ.IBKG)GOTO 300
        IF(II(40).EQ.LCDAT)CALL LCHEX(F,NIR,XE)
        IF(II(40).EQ.LCDAT)GOTO 220
        IF(II(40).NE.Z4DAT)READ(8,*)NXB,NEND
        IF(II(40).EQ.Z4DAT)READ(8,*)NXB,NEND,NMULT
        WRITE(*,1923)NMULT
1923    FORMAT(' NMULT=',I5)
        IF(II(40).NE.'I88D')NXB=1
        IIO=10
        JB=0
        IF(II(40).EQ.Z4DAT)IIO=20
        IF(II(40).EQ.'I88D')THEN
           IIO=8
           JB=NXB-1
           DO 140 I=1,JB
```

```
140         IHD(I)=0
          ENDIF
          DO 160 J=1,818
          IF(II(40).EQ.HDAT)READ(8,156,END=162)(IHD(JB+I),I=1,IIO)
          IF(II(40).EQ.IDAT)READ(8,157,END=162)(IHD(JB+I),I=1,IIO)
          IF(II(40).EQ.Z4DAT)READ(8,158,END=162)(IHD(JB+I),I=1,IIO)
          IF(II(40).EQ.'I88D')READ(8,159,END=162)(IHD(JB+I),I=1,IIO)
          JB=JB+IIO
156       FORMAT(Z7,9Z8)
157       FORMAT(10I8)
158       FORMAT(20Z4)
159       FORMAT(8I8)
160       CONTINUE
162       NIR=JB+1
          DO 190 I=1,NIR
190       F(I)=IHD(I)*NMULT
220       CLOSE(8)
          RETURN
300        REWIND 8
           N=0
310        READ(8,*,END=320)XT,F(N+1)
           IF(N.GT.0)GOTO 315
           N=XT-1
           IF(N.LT.1)GOTO 315
           N=MIN0(N,8191)
           F(N+1)=F(1)
           DO 313 I=1,N
313        F(I)=0
315        N=N+1
           IF(N.LT.8192)GOTO 310
320        CONTINUE
           NIR=N
           RETURN
1100       CONTINUE
           OPEN(8,FILE=NA,STATUS='OLD',FORM='UNFORMATTED')
           READ(8)N,XE
           READ(8)(F(I),I=1,N)
           N1=AMAX1(0.,XE)
           NIR=N1+N
           DO 1110 I=1,N
           NT=N+1-I
1110       F(N1+NT)=F(NT)
           DO 1120 I=1,N1
1120       F(I)=.1
           CLOSE(8)
           RETURN
999       WRITE(*,1945)
1945      FORMAT(' ERROR IN OPENING FILE ',A40,' TRY AGAIN')
          READ(*,'(A)')NA
          GOTO 5
195       WRITE(*,1978)NA
1978      FORMAT(1X,A/' NO DATA TYPE SPECIFIED IN LINE 4')
          READ(*,*)ITEST
          GOTO 220
2000      CONTINUE
           OPEN(8,FILE=NA,STATUS='OLD',FORM='UNFORMATTED')
           PRINT*,' IN SALLY PART OF BREAD'
```

```
          READ(8)(IHD(I),I=1,128)
          I1=0
         DO 2010 I=1,300
         READ(8,END=2015)(IHD(I1+J),J=1,128)
         I1=I1+128
2010     CONTINUE
2015     NIR=I1+J-1
         PRINT*,' AFTER READING DATA N IS',N
         DO 2190 I=1,NIR
         F(I)=IHD(I)
2190     CONTINUE
         IF(NA(1:2).NE.'F ')RETURN
         DO 2200 I=1,NIR
         F(I)=EXP(.001*F(I))
2200     CONTINUE
         RETURN
         END
C*********************************************************************
         SUBROUTINE BREAD(F,IIDAT,WORDBYTE,N)
C *** Reads Machintosh files and makes conversions
         DIMENSION F(1),IIDAT(1)
         INTEGER*1 WORDBYTE(1),B1,B2,B3,B4
         PRINT*,' IN BREAD'
         DO 416 III=1,8064,128
         READ(8,END=418)(IIDAT(I),I=III,III+127)
416      N=III+127
418      CONTINUE
         N4=4*N
         DO 400 I = 1,N4,4
             B1= WORDBYTE(I)
             B2 = WORDBYTE(I+1)
             B3 = WORDBYTE(I+2)
             B4 = WORDBYTE(I+3)
             WORDBYTE(I)   = B4
             WORDBYTE(I+1) = B3
             WORDBYTE(I+2) = B2
             WORDBYTE(I+3) = B1
400      CONTINUE
         DO 500 I=1,N
         F(I)=IIDAT(I)
500      CONTINUE
         RETURN
       END
C*********************************************************************
         SUBROUTINE EFIT(CONS,PC,PPCCI,PPCC,P,XC,EXC,EC,EEC,N,JM)
C *** Performs an energy fit using the .CA file.
         IMPLICIT REAL*8 (A-H,O-Z)
         DIMENSION CONS(1),PC(1),PPCCI(1),PPCC(1),P(1),XC(1),EXC(1),
      2 EC(1),EEC(1),SM(40)
C *** INITIAL ENERGY CONSTANT ESTIMATION USES A STRAIGHT LINE BETWEEN
C *** THE MOST ACCURATE TWO POINTS
         XS=1D32
         X2S=1D31
         IS=0
         DO 10 I=1,N
         IF(EXC(I).GT.X2S)GOTO 10
         IF(EXC(I).LT.XS)THEN
```

```
          X2S=XS
          I2S=IS
          XS=EXC(I)
          IS=I
        ELSE
          X2S=EXC(I)
          I2S=I
        ENDIF
10      CONTINUE
        PRINT*,' IN EFIT AFTER DO 10'
        CONS(2)=(EC(I2S)-EC(IS))/(XC(I2S)-XC(IS))
        CONS(1)=EC(IS)-CONS(2)*XC(IS)
        IPP=1
        JM=2
        SM(1)=1
        SM(2)=1
22      CHL=1.D32
        CHB=1.D31
        FR=.95
        JMD=JM*(JM+1)/2
25      CONTINUE
        CHI=0
        DO 30 I=1,JM
30      PC(I)=0
        DO 35 I=1,JMD
35      PPCC(I)=0
        RES2B=0
        DO 200 I=1,N
        CALL EPOLY(XC(I),P,JM,CONS,FA,DFA)
        W=1/(EEC(I)*EEC(I)+(DFA*EXC(I))**2)
        ERR=(EC(I)-FA)
        RES2=ERR*ERR*W
        IF(RES2.GT.RES2B)THEN
          IR2B=I
          RES2B=RES2
        ENDIF
        CHI=CHI+RES2
        K=0
        DO 130 J=1,JM
        PC(J)=PC(J)-2*ERR*W*P(J)
        DO 130 L=1,J
        K=K+1
130     PPCC(K)=PPCC(K)+2*W*P(J)*P(L)
200     CONTINUE
        IF((CHI-CHL.LT.1.D-6*CHL.AND.CHL-CHB.LT.1.D-6*CHB)
     2    .OR.CHI.LT.1.D0)THEN
        PRINT*,' CHI-RES2B=',CHI-RES2B,' WITH ',N,' DATA POINTS'
        PRINT'(A/(4G16.6))',' CONSTANTS ARE',(CONS(I),I=1,JM)
        PRINT*,' ENTER A 1 TO ADD MORE CONSTANTS -1 TO EXIT FIT '
        READ(*,*)ITEST
C *** HAVE FOUND A MINUMUM WITH JM CONS
          IF(ITEST.EQ.-1)THEN
            JMT=JM
            DO 220 I=1,JMD
220         PPCCI(I)=PPCC(I)
C *** DO NOT EXPECT TO BE ABLE TO SEE KNOT PARTS
            K=0
```

```
          DO 222 I=1,JM
          K=K+I
          IF(I.LT.3.OR.I.NE.2*(I/2))GOTO 222
          PPCCI(K)=1.D20
222       CONTINUE
223        CALL SMINV(PPCCI,JMT,IFL)
           IF(IFL.EQ.-1)THEN
             PRINT*,' NON INVERTABLE MATRIX ENTER A 1 TO REMOVE CONS'
             DO 225 I=1,JMD
225          PPCCI(I)=0
             JMT=JMT-1
             JMTD=JMT*(JMT+1)/2
             DO 228 I=1,JMTD
228          PPCCI(I)=PPCC(I)
             READ(*,*)ITEST
             IF(ITEST.EQ.1)GOTO 223
           ENDIF
           PRINT*,' PPCCI'
           PRINT'(4G20.6)',(PPCCI(I),I=1,JMD)
           AMULT=DMAX1(CHI/(N-JM+.1D-7),1.D0)
           DO 230 I=1,JMD
230        PPCCI(I)=AMULT*PPCCI(I)
           RETURN
         ELSE
           PRINT'(A/(4G20.6))',' MINIMUM CONS',(CONS(I),I=1,JM)
           CONS(JM+1)=0
           SM(JM+1)=1
           NU=MIN0(N,IR2B+2)
           JM=JM+2
           SM(JM)=1.D6
           CONS(JM)=XC(NU)
           PRINT*,' CONS(JM)',CONS(JM)
           GOTO 22
         ENDIF
        ENDIF
        CALL SMSQ(CHI,CHB,CHL,PC,PPCC,PPCCI,FR,CONS,SM,JM,IPP)
        GOTO 25
        END
C*********************************************************************
        SUBROUTINE EPOLY(X,P,JM,CONS,FA,DFA)
C *** Calculates the energy and its derivative at channel X
        IMPLICIT REAL*8 (A-H,O-Z)
        DIMENSION P(1),CONS(1)
        P(1)=1
        P(2)=X
        FA=CONS(1)+X*CONS(2)
        DFA=CONS(2)
        DO 20 I=3,JM,2
        P(I)=0
        P(I+1)=0
        XM=CONS(I+1)-X
        IF(XM.GT.0)THEN
          P(I)=XM*XM
          P(I+1)=CONS(I)*2*XM
          FA=FA+CONS(I)*P(I)
          DFA=DFA+2*CONS(I)*XM
        ENDIF
```

```
20      CONTINUE
        RETURN
        END
C*********************************************************************
        SUBROUTINE EFPOLY(X,P,JM,CONS,FA,DFA)
C *** Calculates the efficiency and its derivative at channel X
        IMPLICIT REAL*8 (A-H,O-Z)
C *** STRC*FA=STR
        DIMENSION P(1),CONS(1)
        P(1)=1
        P(2)=X
        FA=CONS(1)+X*CONS(2)
        DFA=CONS(2)
        DO 20 I=3,JM,2
        P(I)=0
        P(I+1)=0
        XM=CONS(I+1)-X
        IF(XM.GT.0)THEN
          P(I)=XM*XM
          P(I+1)=CONS(I)*2*XM
          FA=FA+CONS(I)*P(I)
          DFA=DFA+2*CONS(I)*XM
        ENDIF
20      CONTINUE
        FA=DEXP(DMIN1(FA,60.D0))
        DFA=FA*DFA
        DO 25 I=1,JM
25      P(I)=P(I)*FA
        RETURN
        END
C*********************************************************************
        SUBROUTINE EFFFIT(CONS,PC,PPCCI,PPCC,P,XC,EXC,STR,ESTR,
     2 STRC,ESTRC,N,JM)
C *** Performs an efficiency fit using the .CA file
        IMPLICIT REAL*8 (A-H,O-Z)
        DIMENSION CONS(1),PC(1),PPCCI(1),PPCC(1),P(1),XC(1),EXC(1),
     2 STR(1),ESTR(1),STRC(1),ESTRC(1),SM(40)
C *** INITIAL ENERGY CONSTANT ESTIMATION USES A STRAIGHT LINE BETWEEN
C *** THE MOST ACCURATE TWO POINTS
        ANUM=0
        ADEN=0
        DO 10 I=1,N
        W=(STR(I)/ESTR(I))**2
        ANUM=ANUM+W*STR(I)/STRC(I)
        ADEN=ADEN+W
10      CONTINUE
        CONS(1)=DLOG(ANUM/ADEN)
        CONS(2)=0
        IPP=1
        JM=2
        SM(1)=1
        SM(2)=1
22      CHL=1.D32
        CHB=1.D31
        FR=.95
        JMD=JM*(JM+1)/2
25      CONTINUE
```

```
      CHI=0
      DO 30 I=1,JM
30    PC(I)=0
      DO 35 I=1,JMD
35    PPCC(I)=0
      RES2B=0
      DO 200 I=1,N
      CALL EFPOLY(XC(I),P,JM,CONS,FA,DFA)
     #W=1/((ESTR(I)/STRC(I))**2+
     #(ESTRC(I)*STR(I)/(STRC(I)*STRC(I)))**2
     2 +(DFA*EXC(I))**2)
      ERR=(STR(I)/STRC(I)-FA)
      RES2=ERR*ERR*W
      IF(RES2.GT.RES2B)THEN
        IR2B=I
        RES2B=RES2
      ENDIF
      CHI=CHI+RES2
      K=0
      DO 130 J=1,JM
      PC(J)=PC(J)-2*ERR*W*P(J)
      DO 130 L=1,J
      K=K+1
130   PPCC(K)=PPCC(K)+2*W*P(J)*P(L)
200   CONTINUE
      IF((CHI-CHL.LT.1.D-6*CHL.AND.CHL-CHB.LT.1.D-6*CHB)
     2  .OR.CHI.LT.1.D0)THEN
      PRINT*,' CHI-RES2B=',CHI-RES2B,' WITH ',N,' DATA POINTS'
      PRINT'(A/(4G16.6))',' CONSTANTS ARE',(CONS(I),I=1,JM)
      PRINT*,' ENTER A 1 TO ADD MORE CONSTANTS -1 TO EXIT FIT '
      READ(*,*)ITEST
C *** HAVE FOUND A MINUMUM WITH JM CONS
       IF(ITEST.EQ.-1)THEN
         JMT=JM
         DO 220 I=1,JMD
220      PPCCI(I)=PPCC(I)
C *** DO NOT EXPECT TO BE ABLE TO SEE KNOT PARTS
        K=0
        DO 222 I=1,JM
        K=K+I
        IF(I.LT.3.OR.I.NE.2*(I/2))GOTO 222
        PPCCI(K)=1.D20
222     CONTINUE
223      CALL SMINV(PPCCI,JMT,IFL)
         IF(IFL.EQ.-1)THEN
           PRINT*,' NON INVERTABLE MATRIX ENTER A 1 TO REMOVE CONS'
           DO 225 I=1,JMD
225        PPCCI(I)=0
           JMT=JMT-1
           JMTD=JMT*(JMT+1)/2
           DO 228 I=1,JMTD
228        PPCCI(I)=PPCC(I)
           READ(*,*)ITEST
           IF(ITEST.EQ.1)GOTO 223
         ENDIF
         PRINT*,' PPCCI'
         PRINT'(4G20.6)',(PPCCI(I),I=1,JMD)
```

```
              AMULT=DMAX1(CHI/(N-JM+.1D-7),1.D0)
              DO 230 I=1,JMD
230           PPCCI(I)=AMULT*PPCCI(I)
              RETURN
            ELSE
              PRINT'(A/(4G20.6))',' MINIMUM CONS',(CONS(I),I=1,JM)
              CONS(JM+1)=0
              SM(JM+1)=1
              NU=MIN0(N,IR2B+2)
              JM=JM+2
              SM(JM)=1.D6
              CONS(JM)=XC(NU)
              PRINT*,' CONS(JM)',CONS(JM)
              GOTO 22
            ENDIF
          ENDIF
C         PRINT'(A/(4G20.6))','CONS ',(CONS(I),I=1,JM)
C         PRINT'(A/(4G20.6))','PC ',(PC(I),I=1,JM)
          CALL SMSQ(CHI,CHB,CHL,PC,PPCC,PPCCI,FR,CONS,SM,JM,IPP)
          GOTO 25
          END
C**********************************************************************
          SUBROUTINE FAEFA(X,EX,CON,EMAT,JM,P,POLY,FA,DFA,EFA)
C *** Determines which POLY (EPOLY or EFPOLY) to use. It then uses
C *** the returned value and its derivative to calculate the value
C *** of the function and its error at channel X.
          IMPLICIT REAL*8 (A-H,O-Z)
          DIMENSION CON(1),EMAT(1),P(1)
          CALL POLY(X,P,JM,CON,FA,DFA)
          K=0
          EFA=0
          DO 20 I=1,JM
          K=K+I
 20       EFA=EFA-2*EMAT(K)*P(I)*P(I)
          K=0
          DO 30 I=1,JM
          DO 30 J=1,I
          K=K+1
 30       EFA=EFA+4*EMAT(K)*P(I)*P(J)
          EFA=DSQRT(EFA+(DFA*EX)**2)
          RETURN
          END
C**********************************************************************
```

The following is a listing of the XCALIBER.MNU file; this file is read by the MENURD subroutine and the variables are passed to the XCALIBER program to give it its options.

```
         4
         5
         4
       100
 1 PAGE XCALIBER
 2
 3 CALIBRATION INPUT FILES
 4 PEAK INPUT FILES
 1 PAGE XCALIBER1 (CALIBRATION FILES)
```

```
 2
 3 CHANNEL TO ENERGY DATA FILE              FILENAME.CA
 4 ENERGY CONSTANTS                         MAKE
 5 EFFICIENCY CONSTANTS                     NONE
 1 PAGE XCALIBER2 (PEAK FILES)
 2
 3 NAME OF FSPFIT PEAK FILE                 FILE1.PK
 4 NAME OF FSPFIT FILE                      FILE2.PK
```

```
C*********************************************************************
C     The routines that are common to a number of the above programs
C*********************************************************************
        SUBROUTINE BREAD(XE,F,IIDAT,WX,NP,FA)
C *** The basic read program which reads data and either calculates
C *** initial weight estimates or reads them from a file. used by
C *** FSPFIT and FSPDIS
        DIMENSION F(1),WX(1),IIDAT(1),FA(1)
C*******************************************************
C
C DATA FROM MAIN ROUTINE
C
C*******************************************************
      CHARACTER*17 DATAFN,NADIFF
      COMMON/CHBREA/DATAFN,NADIFF
      COMMON/CBREAD/IBEGC,IENDC,IEXTWT
      COMMON/BATCFL/IBATCH
C****************************************************************
        CHARACTER*4 II(40)
        CHARACTER*64 NA
C *** F(N) IS THE SPECTRUM CURRENTLY BEING FITTED.
13      CONTINUE
C3      WRITE(*,109)
109     FORMAT(' ENTER THE NAME OF THE DATA FILE')
C**********************
C NEW BITS
C**********************
        NA=DATAFN
C**********************
        II(1)=NA(:4)
        II(2)=NA(5:8)
1943    FORMAT(A)
C       WRITE(*,1001)
1001    FORMAT(' ENTER BEG CHANNEL, END CHANNEL!')
C     READ(2,*)N1,N2
C************************
C NEW BITS
C************************
      N1=IBEGC
      N2=IENDC
C************************
        WRITE(*,1178)N1,N2
        N2=MIN0(N2,15999+N1)
1178    FORMAT(' IN BREAD N1,N2',2I5)
C*******TEST FOR BATCH OR CMS FILE
      IF(IBATCH.EQ.0)IT=INDEX(NA,' ')
      IF(IBATCH.EQ.1)IT=INDEX(NA,'.')
```

```
C**********END CMS/BATCH SELECT
      IRF=0
       IF(NA(IT+1 :IT+3).EQ.'UF')GOTO 1100
      IF(NA(IT+1 :IT+5).EQ.'SALLY')GOTO 2000
      IF(NA(IT+1:).EQ.'LEO')IRF=1
      WRITE(*,*)IRF
C     CALL CMS('FILEDEF 8 DISK '//NA,IERR)
      OPEN(UNIT=8,FILE=NA,ERR=13)
C     PRINT*,' AFTER CALL TO CMS IERR=',IERR
      IF(IRF.NE.1)GOTO 420
      NREC=(N2-1)/128+1
      NU=128*NREC
      DO 416 III=1,NU,128
416   READ(8)(IIDAT(I),I=III,III+127)
      N=N2-N1+2
      XE=N2-N
      DO 400 I=1,N
      F(I)=IIDAT(I+N1-1)
      WX(I)=1./AMAX1(1.,F(I))
400   CONTINUE
      NP=N
      CLOSE(8)
      RETURN
420   READ(8,101)II
       WRITE(*,'(A,1X,A)')' II(40)=',II(40)
101   FORMAT(10A4)
      IF(II(40).EQ.'FREE')GOTO 2200
      IF(II(40).NE.'HDAT'.AND.II(40).
     #NE.'IDAT'.AND.II(40).NE.'Z4DA')
     # GOTO 195
      IF(II(40).NE.'Z4DA')READ(8,*)NXB,NEND
      NMULT=1
      IF(II(40).EQ.'Z4DA')READ(8,*)NXB,NEND,NMULT
      NXB=N1
      NEND=N2
      I10=10
      IF(II(40).EQ.'Z4DA')I10=20
      NXBS=(NXB-1)/I10
      IF(NXBS.EQ.0)GOTO 145
      DO 144 I=1,NXBS
144   READ(8,156)
145   NIR=NEND-I10*NXBS
      NIR=MIN0(NIR,8192)
      IF(II(40).EQ.'HDAT')READ(8,156)(IIDAT(I),I=1,NIR)
      IF(II(40).EQ.'IDAT')READ(8,157)(IIDAT(I),I=1,NIR)
      IF(II(40).EQ.'Z4DA')READ(8,158)(IIDAT(I),I=1,NIR)
156   FORMAT(Z7,   9Z8)
157   FORMAT(10I8)
158   FORMAT(20Z4)
      CLOSE(8)
      NP=NEND-NXB+1
      NADD=NXB-1-I10*NXBS
      XE=NEND-NP
      DO 190 I=1,NP
      F(I)=IIDAT(I+NADD)*NMULT
190   WX(I)=1./AMAX1(1.,F(I))
      RETURN
```

```
195       N=0
        NS=N1-1
        DO 197 I=1,NS
        READ(8,*)
197     CONTINUE
        N2=N2-N1+1
200   N=N+1
      IF(N.GT.N2)GOTO 220
      READ(8,*,END=220)XT,F(N),ER
      ER=100
         IF(N.EQ.1)XE=XT
        IF(ER.EQ.0.D0)WX(N)=1./AMAX1(1.,F(N))
        IF(ER.NE.0.D0)WX(N)=1./(ER*ER)
        GOTO 200
220   NP=N-1
        CLOSE(8)
        RETURN
1100     CONTINUE
C        CALL CMS('FILEDEF 8 DISK '//NA,IERR)
         OPEN(UNIT=8,FILE=NA,FORM='UNFORMATTED')
         READ(8)N,XE
         READ(8)(F(I),I=1,N)
         NTIME=30
         PRINT*,' IN BREAD N,XE',N,XE,'TI = ',F(NTIME)
         N1=N1-XE
         NT=N+XE
         NC=MIN0(NT,N2)
         N2=N2-XE
         N1=MAX0(1,N1)
        N=N2-N1+1
        XE=N2-N
C        PRINT*,' IN BREAD N1,N2,N',N1,N2,N
        DO 1120 I=1,N
        F(I)=F(I+N1-1)
1120    CONTINUE
C********************************
C
C  CALCULATED WEIGHTS
C
C*********************************
      PRINT*,'IEXTWT=',IEXTWT
      DELT2=0
      DELFI=0
      DO 505 I=1,N
        IF(IEXTWT.EQ.0) THEN
          WX(I)=1./AMAX1(1.,F(I))
        ENDIF
        IF(IEXTWT.EQ.1) THEN
        FIM1=0.0
        FIP1=0.0
        IF(I.GT.1) THEN
          FIM1=F(I-1)
        ENDIF
        IF(I.LT.N) THEN
          FIP1=F(I+1)
        ENDIF
        DELTA=(2*F(I)-FIM1-FIP1)
```

```
C         DELT2=DELTA*DELTA+DELT2
          FA(I)=DELTA*DELTA
C         DELFI=DELFI+F(I)
        ENDIF
505     CONTINUE
        IF(IEXTWT.EQ.1) THEN
C       ALFA2=DELT2/(6*N*DELFI)
        DO 500 I=1,N
         IF(I.LT.8) THEN
           SUMERS=0.0
           DO 502 J=1,16
             SUMERS=SUMERS+FA(J)
502        CONTINUE
         ENDIF
         IF(N-I.LT.8)THEN
           SUMERS=0.
           DO 5021 J=1,16
             SUMERS=SUMERS+FA(N-J)
5021       CONTINUE
         ENDIF
         IF(I.GE.8.AND.I.LE.N-8)THEN
           SUMERS=0.
           DO 501 J=1,16
             SUMERS=SUMERS+FA(I-8+J)
501        CONTINUE
         ENDIF
         SUMERS=SUMERS/(16.*6.)
C       WX(I)=1.0/AMAX1(.000001,ALFA2*F(I))
         WX(I)=1.0/AMAX1(.000001,SUMERS)
500     CONTINUE
        ENDIF
        DO 666 I=1,N
        FA(I)=0.
666     CONTINUE
           NP=N
C          PRINT*,' BREAD, XE,N,F TO 10'    ,XE,N,(F(I),I=1,10)
C          PRINT*,'LAST F=',F(N)
           CLOSE(8)
C*******READ IN WEIGHTS FROM EXTERNAL FILE
        IF(IEXTWT.EQ.2)THEN
          PRINT*,' IN EXTERNAL WEIGHTS READ',NADIFF
C         CALL CMS('FILEDEF 20 DISK '//NADIFF,IERR)
          OPEN(UNIT=20,FILE=NADIFF,FORM='UNFORMATTED')
          READ(20)NDUM,DUM
          PRINT*,' NPOINTS=',NDUM
          READ(20) (WX(I),I=1,NDUM)
          DO 1220 I=1,N
            WX(I)=AMAX1(1.,WX(I+N1-1))
            WX(I)=1./AMAX1(1.,WX(I+N1-1))
1220      CONTINUE
          CLOSE(20)
        ENDIF
         RETURN
2000    CONTINUE
C          CALL CMS('FILEDEF 8 DISK '//NA,IERR)
           OPEN(UNIT=8,FILE=NA,FORM='UNFORMATTED')
C        PRINT*,' IN SALLY PART OF BREAD'
```

```
        READ(8)(IIDAT(I),I=1,128)
        I1=0
       DO 2010 I=1,300
       READ(8,END=2015)(IIDAT(I1+J),J=1,128)
       I1=I1+128
2010   CONTINUE
2015   N=I1+J-1
       CLOSE(8)
C      PRINT*,' AFTER READING DATA N IS',N
         NXB=N1
         NEND=N2
         NP=NEND-NXB+1
         NADD=NXB-1
         XE=NEND-NP
         DO 2190 I=1,NP
         F(I)=IIDAT(I+NADD)
2190     WX(I)=1./AMAX1(1.,F(I))
       RETURN
2200     CONTINUE
         DO 2210 I=1,15
         READ(8,'(10A4)')II
2210     WRITE(*,'(1X,10A4)')II

         NMULT=1
         NXB=N1
         NEND=N2
         I10=8
         NXBS=(NXB-1)/I10
         IF(NXBS.EQ.0)GOTO 2245
         DO 2244 I=1,NXBS
2244      READ(8,156)
2245      NIR=NEND-I10*NXBS
         NIR=MIN0(NIR,8192)
         READ(8,*)(IIDAT(I),I=1,NIR)
         CLOSE(8)
         PRINT*,'FIRST VALUE',IIDAT(1)
         NP=NEND-NXB+1
         NADD=NXB-1-I10*NXBS
         XE=NEND-NP
         DO 2290 I=1,NP
         F(I)=IIDAT(I+NADD)*NMULT
2290      WX(I)=1./AMAX1(1.,F(I))
         RETURN
         END
C*********************************************************************
         SUBROUTINE ROPKS(C,SC,W,IPT,WF,SW,XP,SXP,N,NPP,FW,
     # XOFF)
C *** reads the input peak file (which could have been user modified)
C *** used by FSPFIT to continue fitting with new information and by
C *** FSPDIS to display the fit.
         REAL*8 C,SC,W,SW,WF,XP,SXP,FW,DSQRT,ARG
C*******************************************************
C
C DATA TO MAIN ROUTINE VROBFIT
C
C********************************************************
      CHARACTER*17 IPEKFN
```

```
      COMMON/CHROPK/IPEKFN
      COMMON/CROPKS/IPEKFL
C*******************************************************************
        CHARACTER*64 NA
        DIMENSION C(1),IPT(1),SC(1),W(1),WF(1),SW(1),
     # XP(1),SXP(1),FW(3,5),ARAT(5),WRAT(5)
        DATA ARAT,WRAT/10*1./
C *** THIS ROUTINE IS A FIRST PASS ATTEMPT TO PLACE PEAKS IN
C *** SPECIFIED LOCATIONS -- PEAKS MUST BE ORDERED BY CHANNEL
C *** NUMBER
        NPP=0
3      CONTINUE
C3       WRITE(*,101)
101     FORMAT(' ENTER THE NAME OF THE FILE WITH THE EXPECTED ',
     # 'PEAK LOCATIONS <NONE> IF THERE IS NONE')
      PRINT*, 'NEW CODE'

102     FORMAT(A)
C***************************
C
C  NEW BITS
C
C***************************
      NA=IPEKFN
C***************************
      IF(NA.EQ.'NONE')RETURN
      PRINT*,NA
C     CALL CMS('FILEDEF 10 DISK '//NA,IRT)
      OPEN(UNIT=10,FILE=NA,ERR=3)
      WRITE(*,103)NA
103     FORMAT(' FILE=',A64/' HAD STARTING GUESSES FOR PEAK
     #LOCATIONS')
       READ(10,102)NA
       WRITE(*,104)NA
       READ(10,'(A4,I5)')NA,IP
       PRINT*,NA,IP
       DO 4 I=1,2
       READ(10,102)NA
4     WRITE(*,104)NA
       IF(IP.NE.0)READ(10,'(10F8.4)')(ARAT(I),WRAT(I),I=1,IP)
104     FORMAT(1X,A64)
5     READ(10,*,END=150)XP(NPP+1),DUMM,W(NPP+1),DUM2,STR,DUM3
     # ,IPT(NPP+1)
       IF(XP(NPP+1)-XOFF.LT.0.)GOTO 5
       IF(XP(NPP+1)-XOFF.GT.N)GOTO 150
       NPP=NPP+1
       K=IPT(NPP)
       ARG=FW(1,K)+XP(NPP)*(FW(2,K)+XP(NPP)*FW(3,K))
       ARG=DMAX1(.5D0,ARG)
       WF(NPP)=DSQRT(ARG)
       IF(W(NPP).NE.0.D0)WF(NPP)=W(NPP)
       SW(NPP)=.99D0
       W(NPP)=WF(NPP)
       IRAT=IPT(NPP)
       STR=STR/ARAT(IRAT)
       W(NPP)=W(NPP)/WRAT(IRAT)
       IF(STR.GT.0)C(NPP)=SQRT(STR/W(NPP))
```

```
      GOTO 5
150     CLOSE(10)
235     RETURN
      END
C*********************************************************************
     SUBROUTINE SMSQ(CHI,CHB,CHL,PC,PPCC,A,FR,CONS,SM,NT,NW)
C *** the extended NEWTON RAPHSON minimization method which makes it
C *** all possible, used by NLFIT, BKGFIT, and FSPFIT.
      IMPLICIT REAL*8 (A-H,O-Z)
C      REAL*4 CHI,CHB,CHL,PC,FR,CONS,SM
C *** PPCC AND A MUST BE REAL*8.  THE OTHER ARGUMENTS CAN BE
C *** DECLARED REAL*4 AND THE ROUTINE WILL STILL WORK
C THE ROUTINE FINDS THE NECESSARY CHANGES IN CONS FOR THE NEW
C VALUE OF CHI TO BE EQUAL TO FR*CHI
C CHI IS THE CURRENT VALUE OF CHI (THE QUANTITIY BEING MINIMIZED)
C CHB IS THE VALUE PREDICTED FOR CHI ON THE LAST CALL TO SMSQ
C CHL IS THE LAST VALUE OF CHI, SM IS A VECTOR CONTAINING THE
C RELATIVE SMOOTHING FOR EACH CONSTANT IN THE VECTOR CONS.
C PC IS THE SET OF FIRST DERIVATIVES OF CHI WITH RESPECT TO THE
C CONSTANT PPCC IS THE SET OF SECOND DERIVS (PPCC(I+J*(J-1)/2)
C FOR I < J) STORED IN PACKED FORM, SEE SMINV
C NOTE THAT WHEN CHB=CHL, THE ROUTINE CAN GET NO LOWER
C NT IS THE NUMBER OF CONSTANTS.  PPCC AND AM ARE UPPER TRIANGLE
C  NW DECIDES WHAT TO WRITE
      SAVE NTO,SPC,SPPCC,SCONS,AS
      DIMENSION CONS(1),SM(1),PC(1),PPCC(1),A(1)
     # ,BB(107),B(107),SPC(107),SPPCC(5050),SCONS(107)
C *** THE ABOVE MUST BE DIMENSIONED NT OR LARGER
      DATA X0,X1,X2,AS/3*0.D0,-1.E0/
      DATA NTO/0/,TOL/1.D-6/
      NDT=NT*(NT+1)/2
      IFL=0
      IRT=0
      IAB=0
      QB=1.D0
      ILEFT=0
      ICYCLE=0
       IAA=0
      IF(NT.GT.107)GO TO 2500
      IF(NT.NE.NTO)THEN
      NTO=NT
      CHB=1.D33
      CHL=1.D34
      ENDIF
      AS=DMAX1(DMIN1(AS+2,75D0),-37D0)
C SAVE THE LAST SET OF COEFFS IF CHI<CHL
      IF(NW.GE.1)WRITE(*,'(A,3G20.7)')' CHI,CHB,CHL',CHI,CHB,CHL
      IF(CHI.GT.CHL)GO TO 10
      IF(1.000001*CHI.LT.CHL)IR=0
      DO 8 I=1,NT
      BB(I)=CONS(I)
      SCONS(I)=CONS(I)
8     SPC(I)=PC(I)
      DO 9 I=1,NDT
9     SPPCC(I)=PPCC(I)
      IF((CHI-CHB)/(CHL-CHI+1.E-6).GT..1)FR=.1*(9*FR+1)
      IF((CHI-CHB)/(CHL-CHI+1.E-6).GT..5)GOTO 20
```

```
      FR=DMAX1(1.D-2,FR*FR)
      GOTO 20
10      CONTINUE
C RESET TO THE BEST COEFFS (BEWARE THE ROUTINE MUST HAVE RUN BEFORE)
      AS=DMIN1(AS+17.D0,75D0)
      FR=.25*(3.+CHB/CHL)
      IR=IR+1
      CHI=CHL
      DO 18 I=1,NT
      CONS(I)=SCONS(I)
      BB(I)=CONS(I)
18    PC(I)=SPC(I)
      DO 19 I=1,NDT
19    PPCC(I)=SPPCC(I)
      IF(NW.GE.1)WRITE(*,191)IR
191     FORMAT(' IR=',I5)
      IF(IR.GT.20)THEN
        CHB=CHI
        RETURN
      ENDIF
20      CONTINUE
      IF(NW.GE.2)WRITE(*,173)FR
173     FORMAT(' FR=',E10.3)
      CHB=CHI
25      DE=DEXP(AS)
      KI=0
      DO 32 I=1,NT
      KI=KI+I
32      PPCC(KI)=PPCC(KI)+SM(I)*DE
38    CONTINUE
      CALL GSOLVE(PPCC,A,PC,B,NT)
      KI=0
      DO 40 I=1,NT
      KI=KI+I
40    PPCC(KI)=PPCC(KI)-SM(I)*DE
      ICYCLE=ICYCLE+1
C *** NOTE THAT B HERE IS THE SAME AS X IN GSOLVE
52      AJP=CHI
      IF(NW.GE.4)WRITE(*,10017)(B(I),I=1,NT)
      KI=0
      DO 60 K=1,NT
      KI=KI+K
      IF(DABS(B(K)).GT.1.D10)B(K)=DSIGN(1.D10,B(K))
      AJP=AJP+B(K)*(PC(K)+0.5*PPCC(KI)*B(K))
      IF(K.EQ.1)GOTO 60
      KM=K-1
      KT=KI-K
      DO 55 M=1,KM
      KT=KT+1
55      AJP=AJP+PPCC(KT)*B(M)*B(K)
60      CONTINUE
      IF(NW.GE.3)WRITE(*,104)AJP
104     FORMAT(' AJP=',E20.12)
C *** AJP WAS CALCULATED ABOVE, WILL BE TESTED BELOW *************
C *** IF WE CAN FIND AJP < FR*CHI, WE CAN INTERPOLATE IN AS TO
C *** FIND A VALUE FOR WHICH AJP=FR*CHI, IFL = 1 INDICATES THIS
C *** CONDITION
```

```
      DT=AJP/DMAX1(1.D-35,CHI)
      IF(NW.GE.3)WRITE(*,105)DT
105     FORMAT(' DT=',E20.12)
      IF(DT.GE.FR)IAB=1
      IF(DT.LT.FR)IFL=1
      IF(DT-1.E0.LT.1.E-7)GOTO 1000
      IF(NW.GE.1)WRITE(*,109)AJP,AS
109     FORMAT(' THE MATRIX INVERSION APPEARS WRONG, AJP,AS',2E15.6)
C *********** THIS INVERSION IS WRONG, BUT A PREVIOUS ONE WAS RIGHT
      IF(NW.GE.3)WRITE(*,3421)IAA,AS
3421    FORMAT(' IAA,AS',I5,E20.6)
      IF(IAA.EQ.1)GOTO 1000
C ********** THIS INVERSION IS WRONG AND WE HAVE NEVER HAD A CORRECT
C *** ONE
      IF(AS.GE.76.)THEN
      IR=I+1
      GOTO 10
      ENDIF
115     AS=AS+3.
      GOTO 25
120     CONTINUE
C FINDING X'S ON EACH SIDE OF FR*CHI
      IF(ICYCLE.GT.50)AS=DMIN1(75.D0,AS+30.E0)
      IF(ICYCLE.GT.30.OR.DABS((AJP-FR*CHI)/CHI).LT.TOL)GOTO 3117
      IF(AJP.GE.FR*CHI)GOTO 130
      IRT=IRT+1
       ILEFT=0
      AR=AS
      QR=AJP/CHI-FR
      GOTO 140
130     AL=AS
      QL=AJP/CHI-FR
      IRT=0
      ILEFT=ILEFT+1
140     X0=X1
      X1=X2
      IF(IFL*IAB.EQ.0)GOTO 1150
      IF(QB.GT.1.-FR)GOTO 115
      IF(DABS((AR-AL)).LT..1E-5)GOTO 3117
      AS=AL-QL*(AR-AL)/(QR-QL)
      X2=AS
      IF(NW.GE.2)WRITE(*,102)AL,QL,AR,QR,AS
102     FORMAT(' AL,QL',2E15.6,' AR,QR',2E15.6,' AS=',E15.6)
      IF(IRT.LT.2.AND.ILEFT.LT.2)GOTO 25
C AITKENS EXTRAPOLATION
      ALPHA=(X1**2-X0*X2)*(2.*X1-X2-X0)/((2.*X1-X2-X0)**2+1.E-37)
      IF(ALPHA.LT.AL.AND.AL.LT.AR)ALPHA=.5*(AL+AR)
      IF(ALPHA.GT.AL.AND.AL.GT.AR)ALPHA=.5*(AL+AR)
      IF(ALPHA.LT.AR.AND.AR.LT.AL)ALPHA=.5*(AL+AR)
      IF(ALPHA.GT.AR.AND.AR.GT.AL)ALPHA=.5*(AL+AR)
      IF(DABS((ALPHA-AR)/(AR-AL)).LT..1E-1.OR.
     #DABS((ALPHA-AL)/(AR-AL))
     # .LT..1E-1)ALPHA=.5*(AL+AR)
      AS=ALPHA
      IF(NW.GE.2)WRITE(*,103)AS
103   FORMAT(' AITKENS EXTRAPOLATION, AS',E15.6)
      ILEFT=0
```

```
      IRT=0
      GOTO 25
C ********* WE HAVE FOUND 0 < AJP < CHI ***********************
C ********* AND WE WANT TO SAVE THE BEST SET OF B'S **********
1000  QC=DABS(AJP/CHI-FR)
      IF(IAA.EQ.0)GOTO 1050
      IF(QC.GT.QB)GOTO 1120
1050  CHB=AJP
      IAA=1
      QB=QC
      ASB=AS
      IF(NW.GE.2)WRITE(*,117)QB,AJP,AS
117     FORMAT(' QB,AJP,AS',3E15.6)
      DO 1115 I=1,NT
1115    BB(I)=B(I)
      IF(QB.LT.TOL)GOTO 3117
1120    GOTO 120
C ********* TRY AGAIN WITH MORE OR LESS SMOOTHING **********
1150    AS=AS-4.0
      IF(IR.GT.1)AS=.5*(AS+ASB)
      IF(IAB.EQ.0)AS=AS+8.
      IF(AS.LE.-75.)GOTO 3117
      IF(AS.GE.76.)GOTO 3117
      GOTO 25
C ****************** EXITING WITH THE BEST COEFF'S FOUND ***********
3117    CONTINUE
      CHL=CHI
      DO 3119 I=1,NT
3119    CONS(I)=CONS(I)+BB(I)
      IF(NW.GE.3)WRITE(*,10017)(BB(I),I=1,NT)
10017   FORMAT(1X,6E13.6)
      IF(NW.GE.3)WRITE(*,10018)QB
10018   FORMAT(' THE FINAL QB IS',E20.12)
      IF(NW.GE.1)PRINT'(A,G11.4)',' FINAL AS IS',AS
      RETURN
2500    WRITE(*,4367)NT
4367    FORMAT(' NT OF',I5,' IS TOO LARGE FOR THE'/
     # ' DIMENSION OF B AND BB AND ALSO TERMS IN GSOLVE')
      STOP
      END
C*********************************************************************
      SUBROUTINE SMINV(AP,N,IFL)
C *** THIS ROUTINE BEGINS IN LINPACK CHAPTER 3 - POSITIVE
C *** DEFINITE SYMMETRIC MATRICES
C *** IN OUR CASE (CURVE FITTING) WE CAN ALWAYS WRITE THE APPROXIMATE
C *** FUNCTION AS SUM C*SJ(XI).  IF WE THEN FORM X**2 = SUM ON I
C *** OF SUM ON J (CJ*SJ(XI))**2 WE FIND X**2 = SUM ON K CK *
C *** SUM ON J CJ * AJK WHERE AJK IS SUM ON I SJ(XI)*SK(XJ), THE
C *** POSITIVE DEFINITE SYMMETRIC MATRIX OF INTEREST HERE
C
C *** AP IS A PACKED MATRIX AS IN THE FOLLOWING
C *** K=0
C     DO 20 J=1,N
C     DO 20 I=1,J
C     K=K+1
C     AP(K)=A(I,J)
C 20  CONTINUE
```

```
      IMPLICIT REAL*8 (A-H,O-Z)
      DIMENSION AP(1)
      IFL=0
      JJ=0
      DO 30 J=1,N
      S=0
      KJ=JJ
      KK=0
      IF(J.LT.2)GOTO 20
      JM1=J-1
      DO 10 K=1,JM1
      KJ=KJ+1
      T=AP(KJ)
      KM1=K-1
      IF(KM1.LT.1)GOTO 8
      IF(KM1.GT.32)GOTO 6
      DO 5 I=1,KM1
5     T=T-AP(I+KK)*AP(I+JJ)
      GOTO 8
6     CONTINUE
      CALL TSUM(T,AP,KM1,KK,JJ)
8     KK=KK+K
      T=T/AP(KK)
      AP(KJ)=T
      S=S+T*T
10    CONTINUE
20    CONTINUE
      JJ=JJ+J
      SMT=DMAX1(S,AP(JJ))
      DIFF=AP(JJ)-S
      IF(DIFF.GT.1.D-10*SMT)GOTO 25
C      WRITE(*,1975)J,AP(JJ),S
1975    FORMAT(' POS DEF QUANT LE 0. J=',I5,' AP(JJ)=',E10.3,
     # ' S=',E10.3)
C *** FIX IS MORE SMOOTHING FOR A(J,J)
      IFL=-1
      DIFF=1.D30
25    AP(JJ)=DSQRT(DIFF)
30    CONTINUE
C *** NEXT WE CONSTRUCT THE INVERSE MATRIX
      KK=0
      DO 100 K=1,N
      K1=KK
      KK=KK+K
      AP(KK)=1/AP(KK)
      T=-AP(KK)
      KM=K-1
      IF(KM.LE.0)GOTO 86
      DO 85 I=1,KM
85    AP(I+K1)=T*AP(I+K1)
86    KP1=K+1
      J1=KK
      KJ=KK+K
      IF(N.LT.KP1)GOTO 90
      DO 80 J=KP1,N
      T=AP(KJ)
      AP(KJ)=0
```

```
      IF(K.GT.32)GOTO 72
      DO 70 I=1,K
70    AP(J1+I)=AP(J1+I)+T*AP(K1+I)
       GOTO 77
72    CONTINUE
      CALL VSUM(AP,AP,T,K,K1,J1)
77    J1=J1+J
      KJ=KJ+J
80    CONTINUE
90    CONTINUE
100     CONTINUE
      JJ=0
      DO 130 J=1,N
      J1=JJ
      JJ=JJ+J
      JM1=J-1
      K1=0
      KJ=J1+1
      IF(JM1.LT.1)GOTO 120
      DO 110 K=1,JM1
      T=AP(KJ)
       IF(K.GT.32)GOTO 107
      DO 105 I=1,K
105     AP(I+K1)=AP(I+K1)+T*AP(I+J1)
         GOTO 117
107     CONTINUE
      CALL VSUM(AP,AP,T,K,J1,K1)
117   K1=K1+K
      KJ=KJ+1
110     CONTINUE
120     CONTINUE
      T=AP(JJ)
      DO 125 I=1,J
125     AP(I+J1)=T*AP(I+J1)
C ABOVE LINE ORIGINALLY HAD A CALL TO FUNC ANOV
130     CONTINUE
140     CONTINUE
      RETURN
      END
C *****************************************************************
      SUBROUTINE GSOLVE(A,AM,B,X,NT)
C *** SOLVES AX=-B IN CASES WHERE THE MATRIX CAN BE INVERTED CHEAPLY
C *** AND ACCURATELY WITH NO PROBLEMS
      IMPLICIT REAL*8 (A-H,O-Z)
      DIMENSION A(1),AM(1),B(1),X(1)
      NTD=NT*(NT+1)/2
      DO 10 I=1,NTD
10    AM(I)=A(I)
      CALL SMINV(AM,NT,IFL)
      CALL SMMULT(AM,B,X,NT)
      DO 20 I=1,NT
20    X(I)=-X(I)
      RETURN
      END
C *****************************************************************
      SUBROUTINE SMMULT(A,B,C,N)
C *** MATRIX MULTIPLY
```

```
      IMPLICIT REAL*8 (A-H,O-Z)
      DIMENSION A(1),B(1),C(1)
C *** FOR PACKED MATRICES C(N)=A(N*(N+1)/2)*B(N)
      DO 10 I=1,N
10    C(I)=0
      K=0
      DO 20 I=1,N
      IM=I-1
      DO 15 J=1,IM
      K=K+1
      C(I)=C(I)+A(K)*B(J)
15    C(J)=C(J)+A(K)*B(I)
      K=K+1
20    C(I)=C(I)+A(K)*B(I)
      RETURN
      END
C *******************************************************************
      FUNCTION KIJ(I1,I2)
C *** RETURNS THE LOCATION IN A PACKED MATRIX CORRESPONDING TO I,J IN
A  C *** NORMAL MATRIX
      I=MIN0(I1,I2)
      J=MAX0(I1,I2)
      KIJ=(J*(J-1)/2)+I
      RETURN
      END
C *******************************************************************
     SUBROUTINE VSUM(AP,ATEMP,T,K,K1,J1)
      IMPLICIT REAL*8 (A-H,O-Z)
      DIMENSION AP(1),ATEMP(1)
      DO 75 I=1,K
75    AP(J1+I)=AP(J1+I)+T*ATEMP(K1+I)
      RETURN
      END
C *******************************************************************
      SUBROUTINE TSUM(T,AP,KM1,KK,JJ)
      IMPLICIT REAL*8 (A-H,O-Z)
      DIMENSION AP(1)
      DO 7 I=1,KM1
7     T=T-AP(I+KK)*AP(I+JJ)
      RETURN
      END
C *******************************************************************
      SUBROUTINE BKGFIT(XOFF,F,FA,WX,N,CONS,NV,BKGF,LFLAG,NCALLB,CUT)
C *** Used by FSPFIT and VBKGFI this is the background fitting
C *** routine.
         IMPLICIT REAL*8 (A-H,O-Z)
        DIMENSION XC(256),FC(256),WC(256),FAC(256),
     # CONS(1),P(100),F(1),FA(1),WX(1)
C*********************************************************
C
C DATA TO MAIN ROUTINE
C
C*********************************************************
      CHARACTER*17 IBKGFN
      COMMON/CHBKGF/IBKGFN
      COMMON/CBKGFI/IBKGFL
C*********************************************************
```

```
C
C     KNOT LOCATIONS IN VNODE
C
C***********************************
      DIMENSION VNODE(200)
      COMMON/PNODE/VNODE,NOTS
C***********************************
        CHARACTER*64 FNAME
        CHARACTER*4 BKGF
         CHARACTER*1 ANS
        REAL*4 XOFF,F,FA,WX,CUT,FSMALL
        IF(BKGF.EQ.'NOBF'.AND.NCALLB.GT.0)RETURN
          LFLAG=IBKGFL
      CALL FALPAB(ALP,A,B,CUT,ASHIFT,CHIL)
       PRINT*,' A,B,ALP,ASHIFT ',A,B,ALP,ASHIFT
         NITT=3
         IF(BKGF.EQ.'FITB')NITT=10
         NONEWK=0
      PRINT*,'NCALLB=',NCALLB
         IF(BKGF.EQ.'NONK'.OR.NV.GE.100.OR.BKGF.EQ.'FINK')NONEWK=1
         IF(NCALLB.GT.0)GOTO 60
C *** TRY TO READ THE OLD CONSTANTS AND SET FA TO WHERE IT WAS
15      CONTINUE
        FNAME=IBKGFN
        IF(FNAME.EQ.'NONE')GOTO 40
         NONEWK=1
        OPEN(UNIT=1,FILE=FNAME,ERR=1197)
         READ(1,*)LFLAG
         IBKGFL=LFLAG
        NV=0
20      READ(1,*,END=25)CONS(NV+1)
        NV=NV+1
        GOTO 20
25      CLOSE(1)
         IF(NV.LT.4)GOTO 45
         IF(CONS(NV-1).EQ.0.D0)GOTO 45
        DO 30 J=1,N
         XDP=XOFF+J
        CALL POLY(XDP,P,NV,CONS,FDP)
         FA(J)=FDP
30      CONTINUE
        IF(BKGF.NE.'NOBF')GOTO 60
         NCALLB=1
         RETURN
40      CONTINUE
         NONEWK=1
         NV=1
C *** INITIALIZE FA
45        CONTINUE
60     CONTINUE
         NN=((N-1)/16+1)
C *** ABOVE MAKES THE NUMBER OF POINTS A MULTIPLE OF 16
         WAVE=0
         XAVE=0
       FAVE=0
       DO 120 I=1,10
         WAVE=WAVE+WX(N-I+1)
```

```
        XAVE=XAVE+XOFF+N+1-I
120   FAVE=F(N-I+1)+FAVE
        WAVE=.1*WAVE
        XAVE=.1*XAVE
      FAVE=.1*FAVE
      PRINT*,'DATA COMPRESSION NITT,NN',NITT,NN
160     CONTINUE
        DO 500 ITT=1,NITT
C *** DATA COMPRESSION
        JC=1
        DO 200 I=1,NN
        FC(I)=0
        WC(I)=0
        XC(I)=0
        JCP=JC+15
        DO 180 J=JC,JCP
        XAD=XAVE*WAVE
        FAD=FAVE*WAVE
        WAD=WAVE
        WT=1
        IF(J.LT.N)THEN
        Y=(F(J)-FA(J))**2*WX(J)
        WT=(1+ALP*Y)*A
C *** NOTE THAT WX IS 1/ER**2 IN THE ORIGINAL DATA
        IF(F(J).GT.FA(J))WT=B/(1+ALP*Y)
C *** HELPING THE FIT AT THE BEGINNING
         IF(J.LT.10)WT=WT*4
        WAD=WX(J)*WT
        FAD=(F(J))*WAD
        XAD=(XOFF+J)*WAD
        ENDIF
C *** NOTE THAT WC WILL BE 1/ERR**2
        WC(I)=WC(I)+WAD
        XC(I)=XC(I)+XAD
180     FC(I)=FC(I)+FAD
        XC(I)=XC(I)/WC(I)
        FC(I)=FC(I)/WC(I)
         CALL POLY(XC(I),P,NV,CONS,FAC(I))
200     JC=JC+16
         IF(ITT.NE.1)GOTO 300
         IF(NONEWK.EQ.1)GOTO 300
        IF(NV.LT.4)THEN
          NV=NV+1
          GOTO 300
        ENDIF
        NV=NV+2
         CONS(NV-1)=0
         PRINT*,' BEFORE BRESL NN,NV ',NN,NV
        CALL BRESL(FC,FAC,WC,ILR,NN,LFLAG)
      ILR=MAX0(1,ILR)
      IUP=ILR+MAX0(5,NN/(2*NV))
      PRINT'('' NEW KNOTS AT'',G14.6)',XC(ILR),XC(IUP)
      CONS(NV)=XC(ILR)
      NV=NV+2
      CONS(NV-1)=0
      CONS(NV)=XC(IUP)
C***************************************
```

```
      DO 556 KNT=1,200
        VNODE(KNT)=0.0
556   CONTINUE
      NOTS=0
      LFLAC=1
        DO 555 KNT=1,NV
C***************************************
C  FIND ALL KNOTS
C***************************************
      IF(KNT.GT.5)THEN
        IF(LFLAC.EQ.1)THEN
          NOTS=NOTS+1
          VNODE(NOTS)=CONS(KNT)
          LFLAC=LFLAC+1
        ELSE
          LFLAC=1
        ENDIF
      ENDIF
555     CONTINUE
C***************************************
C  ORDER VNODE
C***************************************
      DO 557 JS=2,NOTS
        VNS=VNODE(JS)
        DO 558 JS2=JS-1,1,-1
          IF(VNODE(JS2).LE.VNS) GOTO 559
          VNODE(JS2+1)=VNODE(JS2)
558     CONTINUE
        JS2=0
559     VNODE(JS2+1)=VNS
557   CONTINUE
      DO 560 KNT=1,NOTS
560   CONTINUE
C****************************************
300     CONTINUE
      IF(BKGF.NE.'FIXK'.AND.BKGF.NE.'FINK')THEN
          CALL SPFIT(XC,FC,FAC,WC,NN,NV,CONS)
        ELSE
           CALL SPFIXK(XC,FC,FAC,WC,NN,NV,CONS)
      ENDIF
        WRITE(*,1190)NN,(CONS(I),I=1,NV)
1190    FORMAT('BKG ',I3/(5E14.6))
C *** REMOVING UNUSED CONSTANTS
      DO 350 I=6,NV,2
      IF(CONS(I)-XOFF.GT.0.D0)GOTO 350
      NV=NV-2
      IM=I-1
      DO 340 J=IM,NV
340   CONS(J)=CONS(J+2)
350   CONTINUE
C *** DATA EXPANSION
        DO 330 J=1,N
         XDP=XOFF+J
        CALL POLY(XDP,P,NV,CONS,FDP)
        FA(J)=FDP
330     CONTINUE
500     CONTINUE
```

```
        NCALLB=1
        RETURN
1197  STOP
      END
C *********************************************************************
        SUBROUTINE POLY(X,P,NV,CONS,FA)
C *** THE FIRST FOUR COEFFICIENTS REPRESENT A CUBIC
C *** THE REST ARE IN THE FORM C(I)*(C(I+1)-X)+ **3
        IMPLICIT REAL*8 (A-H,O-Z)
         COMMON/CBKGFI/IBKGFL
        DIMENSION P(1),CONS(1)
        NVS=NV-4
        P(1)=1.D0
        FA=CONS(1)
        DO 5 I=2,4
        P(I)=X*P(I-1)
5       FA=FA+CONS(I)*P(I)
        DO 20 I=5,NV,2
        P(I)=0.D0
        P(I+1)=0.D0
        IF(CONS(I+1).LE.X)GOTO 20
        DIFF=CONS(I+1)-X
        DIFF2=DIFF*DIFF
        P(I+1)=3*CONS(I)*DIFF2
        P(I)=DIFF2*DIFF
        FA=FA+CONS(I)*P(I)
20      CONTINUE
      IF(IBKGFL.EQ.0)RETURN
        IF(FA.GT.-10.D0.AND.FA.LT.30.D0)THEN
          FA=DEXP(FA)
        ELSE
          IF(FA.LT.0.D0)THEN
             FA=0.D0
          ELSE
             FA=DEXP(30.D0)
          ENDIF
        ENDIF
      DO 30 I=1,NV
30    P(I)=FA*P(I)
        RETURN
        END
C *********************************************************************
      SUBROUTINE FALPAB(ALP,A,B,CUT,ASHIFT,CHIL)
C *** A AND B KEEP THE AVERAGE CORRECT FOR A GAUSSIAN DISTRIBUTION NO
C *** MATTER WHAT THE VALUE OF ALP.  AN OVERALL INCREASE OF BOTH A
C *** AND B ALSO KEEPS CHI**2 CORRECT
      IMPLICIT REAL*8 (A-H,O-Z)
      REAL*4 CUT
      DIMENSION XG1(5),WG1(5),XG2(5),WG2(5),XG0(5),WG0(5)
      DATA XG1,WG1/.5133812615D0,.1188866291D1,.189642447D1,
     # .2661918482D1,.355390392D1,.260877802D0,.1993334055D0,
     # .3797122484D-1,.180587934D-2,.1168498619D-4/
      DATA XG2,WG2/.6568095668D0,.1326557084D1,.2025948015D1,
     # .2783290099D1,.3668470847D1,.185225285D0,.2062921868D0,
     # .4888991002D-1,.2686707670D-2,.1937314074D-4/
       DATA XG0,WG0/.3429013272D0,1.036610829D0,1.756683649D0,
     # 2.532731674D0,3.436159118D0,.6108626337D0,.2401386110D0,
```

```
     # .3387439445D-1,.1343645746D-2,.7640432855D-5/
       TSPI=.636619772
       C15=15.D0*DSQRT(TSPI)
       NC=0
       OALP=0
       ALP=.1D-4
       OFTZER=-2*CUT
10     CONTINUE
      NC=NC+1
C *** B AND C ARE DETERMINED BY GAUSSIAN QUADRATURE
       BI=0
       C=0
       DI=0
       DO 20 I=1,5
       X2=XG1(I)*XG1(I)
       T1=1/(1+2*ALP*X2)
       BI=BI+WG1(I)*T1*(1-2*ALP*X2*T1)
       DI=DI+WG0(I)/(1+2*ALP*XG0(I)*XG0(I))
       DI=DI-20*WG2(I)/(1+2*ALP*XG2(I)*XG2(I))**2
       DI=DI+32*WG2(I)*XG2(I)*XG2(I)/(1+2*ALP*XG2(I)*XG2(I))**3
20     C=C+WG2(I)/(1+2*ALP*XG2(I)*XG2(I))
       BI=4*BI
C *** 1.128379 IS 2/PI**.5
       C=C*1.1283792D0
       DI=DI*15*1.1283792D0
C *** 2=A(1+3ALP)+2*B*C
       A=1
       B=(2-A*(1+3*ALP))/(2*C)
       C2=B*DI+30*A*(1+6*ALP)
       TCUT=1/(1+ALP*CUT*CUT)
       C1=2*B*CUT*TCUT*(1-2*CUT*ALP*TCUT)
       ASHIFT=-C1/C2
       CHIL=.5*C1*C1/C2
       FTZER=C15*(A*(2+8*ALP)-B*BI)-C1
       FP=(FTZER-OFTZER)/(ALP-OALP)
       OALP=ALP
       ALP=ALP-FTZER/FP
       OFTZER=FTZER
       IF(NC.GT.250)THEN
         PRINT*,' CANNOT FIND ALP IN FALPAB'
         STOP
       ENDIF
       IF(DABS((ALP-OALP)/ALP).GT.1.D-7)GOTO 10
       RETURN
       END
C ***********************************************************************
        SUBROUTINE BRESL(F,FB,WX,ILR,N,LFLAG)
C *** Background residual locator, determines where to add additional
C *** knots to background spline.
        IMPLICIT REAL*8 (A-H,O-Z)
        DIMENSION F(N),WX(N),FB(N)
20     ALR=0
        NM2=N-2
        FSP2=(F(2)-FB(2))*DSQRT(WX(2))
          FSP1=(F(1)-FB(1))*DSQRT(WX(1))
       FS=0
        FSM1=0
```

```
        FSM2=0
        FSM3=0
        SUM=FSP1+FSP2
      DO 40 I=1,N
        FSM3=FSM2
        FSM2=FSM1
        FSM1=FS
        FS=FSP1
        FSP1=FSP2
        FSP2=0
        IF(I.LT.NM2)THEN
        FSP2=(F(I+2)-FB(I+2))*DSQRT(WX(I+2))
      ENDIF
        SUM=SUM+FSP2-FSM3
      IF(ALR.GT.DABS(SUM))GOTO 40
      ILR=I
      ALR=DABS(SUM)
40    CONTINUE
      RETURN
      END
C*********************************************************************
        SUBROUTINE SPFIXK(XC,FC,FAC,WC,N,NV,CONS)
C *** THIS ROUTINE MINIMIZES A SPLINE FIT TO THE COMPRESSED
C *** BACKGROUND. KNOT LOCATIONS ARE NOT ALLOWED TO VARY
       IMPLICIT REAL *8 (A-H,O-Z)
        CHARACTER*4 CHAR
      COMMON/PTIALD/ PPCC(666),PPCCI(666),P(36),PC(36),SM(100)
        DIMENSION CONS(NV),TCONS(100)
        DIMENSION XC(1  ),FC(1  ),WC(1  ),FAC(1  )
        DATA NE/0/
        ALAM=25.6D0*N**3/NV**3
        IF(NV.GT.4)PRINT*,' ALAM=',ALAM
        NVT=(NV-4)/2+4
        DO 10 I=1,NVT
10      SM(I)=1.D1
        FR=0.5
        CHB=1.E32
        CHL=1.E33
        NVD=NVT*(NVT+1)/2
C       DO 45 NIT=1,10
         NIT=1
1045    CONTINUE
        DO 13 I=1,NVT
13     PC(I)=0.
        DO 15 I=1,NVD
15       PPCC(I)=0.
        DO 18 I=1,4
18      TCONS(I)=CONS(I)
        DO 20 I=5,NVT
20      TCONS(I)=CONS(3+2*(I-4))
         CHI=0
        DO 38 IT=1,N
        CALL POLY(XC(IT),P,NV,CONS,FAC(IT))
        ERR=(FC(IT)-FAC(IT))
        ERRS=ERR*ERR*WC(IT)
        CHI=CHI+ERRS
        K=0
```

```
          DO 25 I=1,NVT
          IARG=I
          IF(I.GT.5)IARG=3+2*(I-4)
25       PC(I)=PC(I)-2*ERR*P(IARG)*WC(IT)
          W3=2*WC(IT)
          DO 30 I=1,NVT
          IARG=I
          IF(I.GT.5)IARG=3+2*(I-4)
          W3T=W3*P(IARG)
          DO 30 J=1,I
          JARG=J
          IF(J.GT.5)JARG=3+2*(J-4)
          K=K+1
30       PPCC(K)=PPCC(K)+W3T*P(JARG)
38        CONTINUE
           PRINT'(A,3G20.6)',' SPFIXK CHI,CHB,CHL',CHI,CHB,CHL
           G44TE=CHI-CHL-.01
          IF(G44TE.GT.0.D0 )GOTO 44
           IF(NIT.GT.100)GOTO 50
           NIT=NIT+1
           G50TE1=CHB-CHL+.01
           IF(G50TE1.GT.0.D0)GOTO 50
           G50TE2=G50TE1/CHB
           G50TE2=DABS(G50TE2)
           IF(G50TE2.LT.1.D-6)GOTO 50
           IF(DABS(CHL-CHB).LT.1.D-2.OR.DABS((CHL-CHB)/CHB).LT.1.D-6)
     # GOTO 50
44        CALL SMSQ(CHI,CHB,CHL,PC,PPCC,PPCCI,FR,TCONS,SM,NVT,0)
45        CONTINUE
          DO 48 I=1,NVT
          IARG=I
          IF(I.GT.5)IARG=3+2*(I-4)
48        CONS(IARG)=TCONS(I)
           GOTO 1045
50      CONTINUE
          CHIT=CHI
          NE=NE+1
           PRINT'(A,G11.4)',' AT END OF SPFIT CHIT=',CHIT
           RETURN
        END
C*********************************************************************
          SUBROUTINE SPFIT(XC,FC,FAC,WC,N,NV,CONS)
C *** THIS ROUTINE MINIMIZES A SPLINE FIT TO THE COMPRESSED
C *** BACKGROUND. KNOTS ARE VARIED
          IMPLICIT REAL *8 (A-H,O-Z)
          CHARACTER*4 CHAR
        COMMON/PTIALD/ PPCC(666),PPCCI(666),P(36),PC(36),SM(100)
          DIMENSION CONS(1)
          DIMENSION XC(1  ),FC(1  ),WC(1  ),FAC(1  )
          DATA NE/0/
        ALAM=25.6D0*N**3/NV**3
        IF(NV.GT.4)PRINT*,' ALAM=',ALAM
          IF(NE.GT.0)GOTO 14
          DO 10 I=1,4
10        SM(I)=1.D1
          DO 12 I=5,100,2
          SM(I)=1.D1
```

```
12        SM(I+1)=1.D4
14        FR=0.5
          CHB=1.E32
          CHL=1.E33
          NVD=NV*(NV+1)/2
           NIT=1
1045      CONTINUE
          DO 13 I=1,NV
13       PC(I)=0.
          DO 15 I=1,NVD
15         PPCC(I)=0.
           DO 18 I=6,NV,2
18         CONS(I)=DABS(CONS(I))
22         CHI=0
          DO 38 IT=1,N
          CALL POLY(XC(IT),P,NV,CONS,FAC(IT))
          ERR=(FC(IT)-FAC(IT))
          ERRS=ERR*ERR*WC(IT)
          CHI=CHI+ERRS
          K=0
          DO 25 I=1,NV
25      PC(I)=PC(I)-2*ERR*P(I)*WC(IT)
          W3=2*WC(IT)
          DO 30 I=1,NV
          W3T=W3*P(I)
          DO 30 J=1,I
          K=K+1
30      PPCC(K)=PPCC(K)+W3T*P(J)
38        CONTINUE
        NKK=(NV-4)/2
        DO 42 I=1,NKK
        NCOEF=4+I*2
        ISDD=KIJ(NCOEF,NCOEF)
        DO 40 J=1,NKK
        IF(I.EQ.J)GOTO 40
        NCNN=4+J*2
        DIFF=(CONS(NCOEF)-CONS(NCNN))
        DIFF2=DIFF*DIFF
        CHI=CHI+.5*ALAM/DIFF2
        DIFF3=DIFF2*DIFF
        PC(NCOEF)=PC(NCOEF)-2*ALAM/DIFF3
        ISD=KIJ(NCOEF,NCNN)
        DIFF4=DIFF*DIFF3
        PPCC(ISD)=PPCC(ISD)-6*ALAM/DIFF4
        PPCC(ISDD)=PPCC(ISDD)+6*ALAM/DIFF4
40      CONTINUE
42      CONTINUE
           PRINT'(A,3G20.6)',' SPFIT CHI,CHB,CHL',CHI,CHB,CHL
           G44TE=CHI-CHL-.01
         IF(G44TE.GT.0.D0 )GOTO 44
           IF(NIT.GT.20)GOTO 50
           NIT=NIT+1
           G50TE1=CHB-CHL+.01
           IF(G50TE1.GT.0.D0)GOTO 50
           G50TE2=G50TE1/CHB
           G50TE2=DABS(G50TE2)
           IF(G50TE2.LT.1.D-6)GOTO 50
```

```
        IF(DABS(CHL-CHB).LT.1.D-2.OR.DABS((CHL-CHB)/CHB).LT.1.D-6)
     # GOTO 50
44     CONTINUE
        CALL SMSQ(CHI,CHB,CHL,PC,PPCC,PPCCI,FR,CONS,SM,NV,0)
45      CONTINUE
         GOTO 1045
50     CONTINUE
        CHIT=CHI
        NE=NE+1
         PRINT'(A,G11.4)',' AT END OF SPFIT CHIT=',CHIT
         RETURN
      END
C*******************************************************************
       FUNCTION POLYB(X,P,C,XP,W,IPT,NS,IXMID)
C THIS ROUTINE CALCULATES POLYB=SUM CI*SI ALONG WITH DPOLYB/DCI
C THE NEXT NS SI'S, WHICH HAVE DERIVATIVES WITH RESPECT TO C,XP,AND W
C ARE BSPLINES
       IMPLICIT REAL*8(A-H,O-Z)
      COMMON/SCONS/DU(300),XPS(5),XPL(5)
       DIMENSION P(1),C(1),XP(1),W(1),IPT(1)
      NC=1
5     NPART=1
      POLYB=0
      DO 200 J=1,NS
        ITEST=IPT(J)
      P(NPART)=0.
      P(NPART+1)=0.
      P(NPART+2)=0.
      XM=X-XP(J)
      IF(XM.LE.W(J)*XPS(ITEST).OR.XM.GE.W(J)*XPL(ITEST))GOTO 200
      XM=XM/W(J)
C  PARTIAL WRT C(J)
       FADD=C(J)*POLYG(XM,ITEST,SP)
      P(NPART)=2.*FADD
      POLYB=POLYB+C(J)*FADD
C PARTIAL WRT XP(J)
      P(NPART+1)=-C(J)*C(J)*SP/W(J)
C PARTIAL WRT W(J)
      P(NPART+2)=P(NPART+1)*XM
200   NPART=NPART+3
C*******************************
C EXTRA BACKGROUND CONTRIBUTION
C*******************************
      IF(IXMID.EQ.0) GOTO 181
      P(NPART)=1
      NPART=NPART+1
      P(NPART)=X-IXMID
      NPART=NPART+1
      P(NPART)=P(NPART-1)**2
      NPART=NPART+1
      P(NPART)=P(NPART-1)*P(NPART-2)
      NPART=NPART+1
      P(NPART)=P(NPART-2)*P(NPART-2)
      NPART=NPART+1
      P(NPART)=P(NPART-2)*P(NPART-3)
C*******************************
181   CONTINUE
```

```
      RETURN
      END
C*********************************************************************
      FUNCTION POLYG(X,IPT,SP)
C THIS ROUTINE CALCULATES POLYG=SUM CI*SI ALONG WITH DPOLY/DX
C IPT GIVES THE PEAK TYPE RANGING FROM 1 TO 5
       IMPLICIT REAL*8(A-H,O-Z)
        CHARACTER*64 NA
C****************************************************
C
C  DATA FROM MAIN ROUTINE
C
C****************************************************
      DIMENSION STAN(5)
      CHARACTER*17 STAN
      COMMON/CHPOLY/STAN
      COMMON/CPOLYG/NSTAN
C****************************************************
      COMMON/SCONS/W(20,5),XP(20,5),C(20,5),XPS(5),XPL(5),IP,NC
C *** SET UP GRAPHICS ARRAY
      COMMON/G3PLG/CG3(20,5),XPG3(20,5),WG3(20,5),NST(5)
      COMMON/BATCFL/IBATCH
      IF(NC.EQ.1)GOTO 30
      NC=1
        IP=0
3       CONTINUE
        IF(IP+1.GT.NSTAN)GOTO 30
        NA=STAN(IP+1)
      PRINT*,'IBATCH=',IBATCH
        OPEN(UNIT=11,FILE=NA,ERR=3)
      WRITE(*,102)
102   FORMAT(' THE FOLLOWING CONSTANTS,PEAKS,AND WIDTHS ARE THE',
     X ' STANDARD PEAK')
        IP=IP+1
        NS=1
        IF(IP.GT.5)GOTO 300
5     READ(11,101,END=20)C(NS,IP),XP(NS,IP),W(NS,IP)
      CG3(NS,IP)=C(NS,IP)
      XPG3(NS,IP)=XP(NS,IP)
      WG3(NS,IP)=W(NS,IP)
101   FORMAT(5X,3E20.7)
      WRITE(*,103)NS,C(NS,IP),XP(NS,IP),W(NS,IP)
103   FORMAT(I5,3E20.7)
      NS=NS+1
      GOTO 5
20    NS=NS-1
      IF(IBATCH.EQ.0)THEN
        CLOSE(11)
      ELSE
C          IF(IP.EQ.NSTAN)ENDFILE(UNIT=11)
      ENDIF
      XPS(IP)=1.D32
      XPL(IP)=-1.D32
      DO 25 J=1,NS
      XPS(IP)=DMIN1(XPS(IP),XP(J,IP)-W(J,IP))
25    XPL(IP)=DMAX1(XPL(IP),XP(J,IP)+W(J,IP))
        NST(IP)=NS
```

```
        GOTO 3
30     CONTINUE
        POLYG=0.
       SP=0.
         NS=NST(IPT)
       DO 200 J=1,NS
       XM=X-XP(J,IPT)
       IF(XM.LE.-W(J,IPT).OR.XM.GE.W(J,IPT))GOTO 200
       XM=XM/W(J,IPT)
       FADD=C(J,IPT)*((1.+XM)*(1.-XM))**3
       POLYG=POLYG+FADD
        SP=SP+3.*(FADD/(1.+XM)-FADD/(1.-XM))/W(J,IPT)
200    CONTINUE
       RETURN
300      PRINT*,' ATTEMPT TO DEFINE MORE THAN FIVE STANDARDS'
         STOP
       END
C*****************************************************************
       SUBROUTINE CTON(CI,NMISS,ANUM)
C ***  CONVERTS CHARACTERS TO NUMBERS
       CHARACTER*64 CI
       CHARACTER*1 KNUM(10),C
       DATA KNUM/'1','2','3','4','5','6','7','8','9','0'/
       POIFLG=0.
       SFLAG=0.
       NDP=0
       NUM=0
       NMISS=0
       DO 10 I=1,80
       NONUM=0
       C=CI(I:I+1)
       IF(C.EQ.' ')GOTO 100
         IF(C.EQ.'.')POIFLG=1.
         IF(C.EQ.'-')SFLAG=1.
         IF(C.EQ.'.')GOTO 10
         IF(C.EQ.'-')GOTO 10
         IF(POIFLG.EQ.1.)NDP=NDP+1
         DO 20 J=1,10
         IF(C.EQ.KNUM(J))THEN
           IF(J.NE.10)THEN
             NUM=NUM*10+J
           ELSE
             NUM=NUM*10
           ENDIF
           NONUM=1
         ENDIF
 20      CONTINUE
         IF(NONUM.NE.1)THEN
           NMISS=NMISS+1
         ENDIF
 10    CONTINUE
100    CONTINUE
       ANUM=NUM
       IF(SFLAG.EQ.1.)ANUM=-ANUM
       DO 200 J=1,NDP
       ANUM=ANUM/10
200    CONTINUE
```

```
      IF(NMISS.NE.0)THEN
        PRINT*,'THIS IS NOT AN INTEGER NMISS=',NMISS
      ENDIF
      RETURN
      END
C*********************************************************************
        SUBROUTINE LCHEX(F,NF,XE)
C *** This converts character variables to hexadecimal and in
C *** addition corrects for the strange way Lecroy multichannel
C *** analyzers store bytes on tape.
        DIMENSION F(1),IH(6)
        CHARACTER*1 HC(16),HI(64)
        DATA HC/'0','1','2','3','4','5','6','7','8',
     # '9','A','B','C','D','E','F'/
        READ(8,*)NMIN,NMAX
C       WRITE(*,109)NMIN,NMAX
109     FORMAT(' NMIN, NMAX',2I6)
        NOPE=0
        NF=1
        N1=1
        N2=32
        DO 185 ISKIP=1,8
185     READ(8,1994)
10      READ(8,1994,END=50)(HI(I),I=N1,N2)
        N1=1
1994    FORMAT(9X,32A1)
11      IF(NOPE.LT.NMIN)GOTO 17
        DO 15 L=1,6
        J=L+N1-1
        DO 12 K=1,16
        IF(HI(J).EQ.HC(K))GOTO 15
12      CONTINUE
15      IH(L)=K-1
        IF(NF.GE.1)F(NF)=IH(2)+16*IH(1)+256*(IH(4)+16*IH(3)+256*
     # (IH(6)+16*IH(5)))
        NF=NF+1
17      NOPE=NOPE+1
        IF(NOPE.GT.NMAX)GOTO 50
        N1=N1+6
        IF(N1+5.LT.N2)GOTO 11
        NT=N2-N1+1
        DO 20 I=1,NT
20      HI(I)=HI(I+N1-1)
        N1=NT+1
        N2=N1+31
        GOTO 10
50      CONTINUE
        XE=NOPE-1
60      WRITE(*,112)XE,NOPE
112     FORMAT(' XE, NOPE',F7.0,I6)
        IF(F(NF-1).NE.0)RETURN
        NOPE=NOPE-1
        NF=NF-1
        XE=XE-1
        GOTO 60
        END
```

The following routines are non-machine dependent graphics or graphics related routines used by the three display modes. They remain essentially unchanged on all operating systems.

```
      SUBROUTINE NUMOUT(IXE,IYE,RI,NDP,NADJ)
C *** THIS ROUTINE CONVERTS A NUMBER TO TEXT AND TELLS BCHART WHERE
C *** TO PUT IT
C *** NADJ = 0 FOR RIGHT OF IX, 1 FOR LEFT OF IX, 2 TO CENTER ON IX
      CHARACTER*64 CNUMOU,TEMP
      CHARACTER*1 CNUM(10)
      DATA CNUM/'1','2','3','4','5','6','7','8','9','0'/
      ITR=ABS(RI)*10**NDP
      IC=1
      CNUMOU='\'
      DO 10 I=1,11
      INEXT=ITR/10
      ICHAR=ITR-10*INEXT
      IF(ICHAR.EQ.0)ICHAR=10
      TEMP=CNUM(ICHAR)//CNUMOU
      IF(IC.EQ.NDP)THEN
         IC=IC+1
         TEMP='.'//CNUMOU
      ENDIF
      CNUMOU=TEMP
      ITR=INEXT
      IF(INEXT.EQ.0)GOTO 20
10    IC=IC+1
20    CONTINUE
      IF(IC.EQ.0)CNUMOU='0\'
      IF(RI.LT.0.)THEN
        IC=IC+1
        TEMP='-'//CNUMOU
        CNUMOU=TEMP
      ENDIF
      IF(NADJ.EQ.0)IXC=IXE
      IF(NADJ.EQ.1)IXC=IXE-6*IC
      IF(NADJ.EQ.2)IXC=IXE-3*IC
      CALL BCHART(IXC,IYE,CNUMOU)
         RETURN
         END
C*********************************************************************
      SUBROUTINE AXIS(VMIN,VMAX,LFLAG,NRES,HMIN,HMAX,II,CHI)
C *** draws the axis on the screen.
        CHARACTER*8 II
        CHARACTER*64 TITLE,temp
         COMMON/AXISP/BHORI,BVERT,SHORI,SVERT,EHORI,EVERT
C      CALL ANSI
C        READ(*,*)T1,T2
C        IF(T1.EQ.0.)GOTO 15
C *** FOLLOWING VALUES ARE SELANAR SPECIFIC
         EVERT=480
         EHORI=640
C        READ(*,*)T1,T2
C        IF(T1.EQ.0.)GOTO 15
         BVERT=40
         BHORI=80
```

```
15        CONTINUE
          SVERT=(EVERT-79-BVERT)/(VMAX-VMIN)
          SHORI=(EHORI-BHORI-9)/(HMAX-HMIN)
         TITLE=II//' CHIS=\'
          I2=EVERT-34
          I1=EHORI-349
         ADEC=VMAX-VMIN
C        NPY=AMAX1(0.,3.-ALOG10(ADEC))
         NPY=4.-ALOG10(ADEC)
         ADEC=HMAX-HMIN
         NPX=AMAX1(0.,3.-ALOG10(ADEC))
         CALL BCHART(I1,I2,TITLE)
         CALL NUMOUT(I1+150,I2,CHI,1,0)
          I1=BHORI
          I2=EVERT
         CALL STPL(I1,I2)
          I2=BVERT
         CALL PLOT(I1,I2)
          I1=EHORI
         CALL PLOT( I1,I2)
          T1=.25*(EVERT-79-BVERT)
          I1=BHORI
          I2=I1+10
         DO 10 I=1,5
         IVERT=BVERT+T1*(I-1)
          IF(I.EQ.1)GOTO 8
          IF(NRES.EQ.1.AND.I.EQ.5)GOTO 8
         CALL STPL(I1,IVERT)
         CALL PLOT(I2,IVERT)
8        RP=VMIN+(VMAX-VMIN)*.25*(I-1)
         IF(LFLAG.EQ.1)RP=EXP(VMIN+.25*(I-1)*(VMAX-VMIN))+.5
10        CALL NUMOUT(I1-10,IVERT-5 ,RP,NPY,1)
          I1=BHORI-70
          I2=.5*(BVERT+EVERT)
         temp='counts\'
         CALL BCHART(I1,I2,temp)
         IF(NRES.NE.1)GOTO 16
          I1=BHORI
          I2=EVERT-19
          CALL STPL(I1,I2)
          I3=EHORI
        CALL PLOT(I3,I2)
          I2=I2-30
         CALL STPL(I1,I2)
         CALL PLOT(I3,I2)
          I2=I2-30
         CALL STPL(I1,I2)
         CALL PLOT(I3,I2)
16        T1=(EHORI-BHORI-9)*.1
          I1=BVERT
          I2=I1+10
         DO 20 J=1,11
         IXM=BHORI+T1*(J-1)
          IF(J.EQ.1)GOTO 18
         CALL STPL(IXM,I1)
         CALL PLOT(IXM,I2)
18       RPX=HMIN+(J-1)*(HMAX-HMIN)/10.
```

```
20       CALL NUMOUT(IXM,I1-20,RPX,NPX,2)
          I1=.5*(BHORI+EHORI-70)
          I2=BVERT-40
         temp='channel number\'
         CALL BCHART(I1,I2,temp)
         RETURN
      END
C*****************************************************************
         SUBROUTINE PLINTERP(FORIG,XORIG,FINT,XINT,IB,IEX)
C *** interpolates between points using 4 point LAGRANGE
C *** interpolation to provide intermediate values for display
      DIMENSION FORIG(4),XORIG(4),FINT(4),XINT(4)
      DX=(XORIG(IB+1)-XORIG(IB))/(IEX+1)
        DO 100 IP=1,IEX
        XINT(IP)=XORIG(IB)+IP*DX
        FINT(IP)=0
        DO 90 IA=1,4
        A1=1
        DO 80 J1=1,4
        IF(IA.EQ.J1)GOTO 80
        A1=A1*(XINT(IP)-XORIG(J1))/(XORIG(IA)-XORIG(J1))
80      CONTINUE
90      FINT(IP)=FINT(IP)+A1*FORIG(IA)
100   CONTINUE
        RETURN
        END
```

The following routines are machine dependent graphics routines. They are called from all three display routines. Different versions are needed for tectronix graphics(IBM main frame version 1 from EBCIDC and VAX main frame version 2 from ASCII), GDDM(IBM main frame to 3270 terminal and laser plotter), the macintosh (which also requires modifications to the display codes), and the IBM PC and compatibles capable of running GKS graphics (implemented here in its WATFOR77 version). The GKS version which we sincerely hope will one day become a standard is shown here. We also have the others.

```
        SUBROUTINE STPL(IXE,IYE)
C CALL STPL(IX,IY) AT THE FIRST POINT TO START PLOTTING.  CALL STPL
C WITH ANY IX,IY TO EITHER END A PLOT OR SET UP FOR POINT PLOTTING
C THE VECTOR IC CONTAINS THE ENDS OF LINES DRAWN FROM IBP TO IC(1)
C TO IC(2) TO IC(3) TO ETC.
C THE VECTOR ICP CONTAINS ILP LOCATIONS OF POINTS TO BE PLOTTED
C NOTE THAT THE VAX REQUIRES ALL HEX VALUES IN THE CONSTANTS TO BE IN
C  REVERSE ORDER, DOES NOT NEED 0003 IN IVG, BUT DOES NEED <CR> <LF>
C IN IBL.
        real x,y
        COMMON/OR/IL,IBP,IC(16),ILP,ICP(16),ILINE,ixl,iyl
      DATA NC/0/,IVG/Z0003001D/,IBL/Z40404040/
      IF(NC.NE.0)GOTO 5
      IL=0
      ILP=0
      ILINE=0
      NC=1
 5    IX=MAX0(0,MIN0(640,IXE))
      IY=MAX0(0,MIN0(480,IYE))
      IF(ILINE.EQ.0)GOTO 20
```

```
      IF(IY.EQ.420)IY=419
      IF(IY.EQ.421)IY=422
      IF(IY.EQ.443.OR.IY.EQ.444)IY=442
      IF(IY.EQ.445)IY=446
20    CONTINUE
c
c GKS graphics commands
c
        x=float(ix)/640.
        y=float(iy)/480.
        ixl=ix
        iyl=iy
        call gpm(1,x,y)
c
      RETURN
        END
C*******************************************************************
        SUBROUTINE PLOT(IXE,IYE)
C CALL PLOT(IX,IY) TO DRAW A LINE TO IX,IY FROM THE IXL,IYL IN THE
C LAST CALL TO EITHER PLOT OR STPL
      real xp,yp
      dimension xp(2),yp(2)
      COMMON/OR/IL,IBP,IC(16),IDUM(17),ILINE,ixl,iyl
        DATA IVG/Z0003001D/,IBL/Z40404040/
       IX=MAX0(0,MIN0(640,IXE))
       IY=MAX0(0,MIN0(480,IYE))
c
c GKS graphics routines
c
        xp(1)=float(ixl)/640.
        yp(1)=float(iyl)/480.
        xp(2)=float(ix)/640.
        yp(2)=float(iy)/480.
        call gpl(2,xp,yp)
        ixl=ix
        iyl=iy
c
      RETURN
        END
C*******************************************************************
        SUBROUTINE PONT(IXE,IYE)
C TO PLOT A POINT AT IXE,IYE AFTER A CALL HAS AT SOMETIME BEEN MADE
C TO STPL SIMPLY CALL PONT(IXE,IYE).  THE POINT MAY NOT BE ACTUALLY
C SEEN UNTIL THE SET OF PLOT CALLS ARE ENDED WITH A LAST CALL TO STPL
        real x,y
        COMMON/OR/IDUM(18),ILP,ICP(16)
        DATA IVG/Z0003001D/,IBL/Z40404040/
        IX=MAX0(0,MIN0(1023,IXE))
        IY=MAX0(0,MIN0(767,IYE))
c
c GKS graphics routines
c
        x=float(ix)/640.
        y=float(iy)/480.
        call gpm(1,x,y)
c
        RETURN
```

```
        END
C**********************************************************************
      SUBROUTINE BCHART(IX,IY,CHAR)
C *** places a character on the screen at Ix,Iy
      real xf,yf
      COMMON/OR/IL,IBP
      CHARACTER*64 CHAR,VALU
C
C GKS STUFF
C
        XF=float(IX)/640.
        YF=float(IY)/480.
        IT=INDEX(CHAR,'\')
        VALU=CHAR(1:IT-1)
        CALL GTX(XF,YF,VALU)
C
      RETURN
      END
C**********************************************************************
      SUBROUTINE CLEARS
C *** clears the screen
        INTEGER WKID,SCRES
        COMMON/RG1/WKID,SCRES
        call rquitg
        call rsetg
      RETURN
      END
C**********************************************************************
      SUBROUTINE COLOR(CHAR)
C *** changes the color
      CHARACTER*64 CHAR
      IF(CHAR(1:3).EQ.'RED')CALL GSPLCI(2)
      IF(CHAR(1:5).EQ.'WHITE')CALL GSPLCI(1)
         IF(CHAR(1:4).EQ.'BLUE')CALL GSPLCI(3)
         IF(CHAR(1:5).EQ.'GREEN')CALL GSPLCI(4)
         RETURN
         END
C**********************************************************************
        SUBROUTINE ANSI
C *** returns to normal text screen
        INTEGER*2 REGS(10)
        INTEGER WKID/1/
        REGS(1)=3
        CALL GDAWK(WKID)
        CALL GCLWK(WKID)
        CALL GCLKS
        CALL INTR(16,REGS)
        RETURN
        END
C**********************************************************************
       SUBROUTINE GPHMOD
C *** initializes GKS graphics
        INTEGER ERRFID/0/,SCRRES/3/,WKID/1/
        CALL GIVGA18(SCRRES)
        CALL GOPKS(ERRFID)
        CALL GOPWK(WKID,ERRFID,SCRRES)
        CALL GACWK(WKID)
```

```
        CALL GSCHH(0.02)
c Set colour
c
        CALL GSCR(WKID,1,1.0,1.0,1.0)
        CALL GSCR(WKID,2,1.0,0.0,0.0)
        CALL GSCR(WKID,3,0.0,0.0,1.0)
        CALL GSCR(WKID,4,0.0,1.0,0.0)
        CALL GSTXCI(1)
c set marker
c
        CALL GSMK(1)
c
C *** OUT OF CHARLIES
        CALL GSPMCI(15)
        CALL GSPLCI(15)
        CALL GSTXCI(15)
        RETURN
        END
C*******************************************************************
        SUBROUTINE PRTSCN
C *** With DOS 4.0 OR HIGHER PRINTS SCREEN
        INTEGER*2 REGS(10)
        CALL INTR(5,REGS)
        RETURN
        END
C*******************************************************************
      SUBROUTINE RSETG
C ROBFIT graphics initialization
        INTEGER WKID,SCRES
        COMMON/RG1/WKID,SCRES
c
c set screen type
c
        WKID=1
        SCRES=18
c
c init GKS
c
        CALL GECLKS
        CALL GOPKS(0)
        CALL GIVGA18(SCRES)
        CALL GOPWK(WKID,0,SCRES)
        CALL GACWK(WKID)
c
c set marker
c
        CALL GSMK(1)
c
c Set character height
c
        CALL GSCHH(10./500.)
c
c Set colour
c
        CALL GSCR(WKID,1,1.0,1.0,1.0)
        CALL GSCR(WKID,2,1.0,0.0,0.0)
         CALL GSCR(WKID,3,0.0,0.0,1.0)
```

```
        CALL GSCR(WKID,4,0.0,1.0,0.0)
        CALL GSTXCI(2)
c
      RETURN
      END
C
**********************************************************************
      SUBROUTINE RQUITG
C ROBFIT end graphics routine
        INTEGER WKID,SCRES
        COMMON/RG1/WKID,SCRES
c
c quit graphics
c
        CALL GDAWK(WKID)
        CALL GCLWK(WKID)
        CALL GCLKS
      RETURN
      END
```

APPPENDIX C

NLFIT code listing

```
      IMPLICIT REAL *8 (A-H,O-Z)
C *** THIS ROUTINE DOES A WEIGHTED NON-LINEAR LEAST SQUARES FIT TO
C *** USER SUPPLIED DATA IN L DIMENSIONS.  THE ROUTINE POLY MUST
C *** RETURN THE PARTIALS OF THE FITTING FUNCTIONS WITH RESPECT TO
C *** THE CONSTANTS BEING VARIED.
C *** DATA IS SUPPLIED IN FREE FORMAT AS X1,X2,X3,...,FXY,EP WHERE EP
C *** IS THE EXPECTED ERROR IN THE VALUE OF FXY.
      DIMENSION PPCC(5050),PPCCI(5050),P(100),PC(100),CONS(100)
C *** 5050  IS 100*101/2 THE SIZE NEEDED TO FIT 100 CONSTANTS
      DIMENSION X(100),SM(100),XS(100),XB(100),Y(100),YS(100),YB(100)
      CHARACTER*64 NA,NADAT,NACOM
      DATA XS,XB/100*1.D32,100*-1.D32/,YS,YB/100*1.D32,100*-1.D32/
      DATA CONS/100*0.D0/
      PRINT*,' ENTER THE NAME OF THE DIRECTION FILE'
      READ(*,'(A)')NA
      OPEN(13,FILE=NA,STATUS='OLD')
C *** ND IS THE DIMENSION OF X, NV IS THE NUMBER OF CONSTANTS BEING
C *** FITTED CONS ARE THE INITIAL ESTIMATES OF EACH CONSTANT
C *** SM IS THE RELATIVE STIFFNESS OF EACH CONSTANT, 1 FOR LINEAR OR
C *** NEAR LINEAR 10**6 FOR KNOT LIKE IN SPLINES
      READ(13,'(A)')NADAT
        IT=INDEX(NADAT,',')
      IF(IT.GT.0)NADAT=NADAT(1:IT-1)
      READ(13,*)ND
      READ(13,*)NV
      READ(13,'(A)')NACOM
      DO 5 I=1,NV
5     READ(13,*)CONS(I),SM(I)
      PRINT'(A/(2G20.12))',' CONS, SM',(CONS(I),SM(I),I=1,NV)
        PRINT'('' FIT IS BEING MADE TO DATA IN FILE''/1X,A)',NADAT
      CLOSE(13)
      FR=0
      CHB=1.E32
      CHL=1.E33
12    CONTINUE
      DO 13 I=1,NV
13     PC(I)=0.
      NVD=NV*(NV+1)/2
      DO 15 I=1,NVD
   15 PPCC(I)=0.
      CHI=0.D0
      OPEN(8,FILE=NADAT,STATUS='OLD')
      N=0
20    READ(8,*,END=40)(X(I),I=1,ND),FXY,EP
      N=N+1
      DO 22 I=1,ND
```

```
      XS(I)=DMIN1(XS(I),X(I))
22    XB(I)=DMAX1(XB(I),X(I))
      CALL POLY(X,P,NV,ND,CONS,FA)
      CHI=CHI+((FXY-FA)/EP)**2
      W=1./EP**2
      K=0
      DO 25 I=1,NV
25    PC(I)=PC(I)-2*(FXY-FA)*P(I)*W
      DO 30 I=1,NV
      DO 30 J=1,I
      K=K+1
30    PPCC(K)=PPCC(K)+2*W*P(I)*P(J)
       GOTO 20
40    WRITE(*,101)NV,N,ND
      CLOSE(8)
101   FORMAT(' A FIT IS BEING ATTEMPTED USING',I5,
     #' CONSTANTS'/' TO',I5,
     C' OR FEWER DATA POINTS IN',I5,' DIMENSIONS')
       IF(DABS((CHI-CHL)/CHL).LT.1.D-7) GOTO 50
      CALL SMSQ(CHI,CHB,CHL,PC,PPCC,PPCCI,FR,CONS,SM,NV,1)
      WRITE(*,1332)CHI,CHB,CHL,FR
1332  FORMAT(' CHI=',E13.6,' CHB=',E13.6,' CHL=',E13.6,' FR=',E13.6)
      WRITE(*,1235) (CONS(I),I=1,NV)
1235  FORMAT(' CONS'/(4G16.7))
      GOTO 12
50    K=0
      CALL GSOLVE(PPCC,PPCCI,PC,P,NV)
      OPEN(13,FILE=NA,STATUS='OLD')
      DO 55 I=1,3
55    READ(13,'(A)')
      NACOM=' CONS      RELATIVE NONLINEARITY    ERROR IN CONS'
      WRITE(13,'(A)')NA
      PRINT*,' CHI AT END OF FIT',CHI
      CHIT=CHI
      CHI=DMAX1(CHI,1.D0*(N-NV))
      DO 18 I=1,NV
      IYB=10
      IF(ND.LT.2)IYB=1
      K=K+I
      ER=DSQRT(PPCCI(K)*CHI/(N-NV))
      WRITE(13,'(2G20.12,3X,''+-'',G10.3)')CONS(I),SM(I),ER
18    PRINT*,CONS(I),' +-',ER
      WRITE(13,'('' CHI = '',G20.6)')CHIT
      READ(*,*)ITEST
      OPEN(12,FILE='FITMERR.DAT')
      OPEN(10,FILE='FIT.DAT')
      OPEN(11,FILE='FITPERR.DAT')
      DO 80 IY=1,IYB
      DELY=XB(2)-XS(2)/9
      X(2)=XS(2)+(IY-1)*DELY
      DELX=(XB(1)-XS(1))/99
      DO 80  I=1,99
      X(1)=XS(1)+(I-1)*DELX
      CALL POLY(X,P,NV,ND,CONS,FA)
      K=0
      DX=0
      DO 27 J=1,NV
```

```
      K=K+J
27    DX=DX-PPCCI(K)*P(J)*P(J)
      K=0
      DO 67 J=1,NV
      DO 57 L=1,J
      K=K+1
57    DX=DX+2*PPCCI(K)*P(L)*P(J)
67    CONTINUE
      FMDX=0
      FPDX=0
      DX=DSQRT(DX*CHI/(N-NV))
      FMDX=FA-DX
      FPDX=FA+DX
      WRITE(10,*) X(1),FA
      WRITE(11,*) X(1),FPDX
      WRITE(12,*) X(1),FMDX
80    CONTINUE
      STOP
      END
      INCLUDE POLY.RKSPLINE
      SUBROUTINE SMSQ(CHI,CHB,CHL,PC,PPCC,A,FR,CONS,SM,NT,NW)
        IMPLICIT REAL*8 (A-H,O-Z)
C *** PPCC AND A MUST BE REAL*8.  THE OTHER ARGUMENTS CAN BE
C *** DECLARED REAL*4 AND THE ROUTINE WILL STILL WORK
C THE ROUTINE FINDS THE NECESSARY CHANGES IN CONS FOR THE NEW
C VALUE OF CHI TO BE EQUAL TO FR*CHI
C CHI IS THE CURRENT VALUE OF CHI (THE QUANTITIY BEING MINIMIZED)
C CHB IS THE VALUE PREDICTED FOR CHI ON THE LAST CALL TO SMSQ
C CHL IS THE LAST VALUE OF CHI, SM IS A VECTOR CONTAINING THE
C RELATIVE SMOOTHING FOR EACH CONSTANT IN THE VECTOR CONS.
C PC IS THE SET OF FIRST DERIVATIVES OF CHI WITH RESPECT TO THE
C CONSTANT PPCC IS THE SET OF SECOND DERIVS (PPCC(I+J*(J-1)/2)
C FOR I < J) STORED IN PACKED FORM, SEE SMINV
C NOTE THAT WHEN CHB=CHL, THE ROUTINE CAN GET NO LOWER
C NT IS THE NUMBER OF CONSTANTS.  PPCC AND AM ARE UPPER TRIANGLE
C  NW DECIDES WHAT TO WRITE
        SAVE NTO,SPC,SPPCC,SCONS,AS
      DIMENSION CONS(1),SM(1),PC(1),PPCC(1),A(1)
     # ,BB(107),B(107),SPC(107),SPPCC(5778),SCONS(107),PPCCD(107)
C *** THE ABOVE MUST BE DIMENSIONED NT OR LARGER
      DATA X0,X1,X2,AS/3*0.D0,-1.E0/
      DATA NTO/0/,TOL/1.D-6/
        NDT=NT*(NT+1)/2
      IFL=0
      IRT=0
      IAB=0
      QB=1.D0
      ILEFT=0
      ICYCLE=0
       IAA=0
        IF(NT.GT.107)GO TO 2500
      IF(NT.NE.NTO)THEN
      NTO=NT
        CHB=1.D33
        CHL=1.D34
        ENDIF
      AS=DMAX1(DMIN1(AS+2,75D0),-37D0)
```

```
C SAVE THE LAST SET OF COEFFS IF CHI<CHL
      IF(NW.GE.1)WRITE(*,'(A,3G20.7)')' CHI,CHB,CHL',CHI,CHB,CHL
      IF(CHI.GT.CHL)GO TO 10
      IF(1.000001*CHI.LT.CHL)IR=0
        DO 8 I=1,NT
        BB(I)=CONS(I)
        SCONS(I)=CONS(I)
8       SPC(I)=PC(I)
        DO 9 I=1,NDT
9       SPPCC(I)=PPCC(I)
      IF((CHI-CHB)/(CHL-CHI+1.E-6).GT..1)FR=.1*(9*FR+1)
      IF((CHI-CHB)/(CHL-CHI+1.E-6).GT..5)GOTO 20
      FR=DMAX1(1.D-2,FR*FR)
      GOTO 20
10    CONTINUE
C RESET TO THE BEST COEFFS (BEWARE THE ROUTINE MUST HAVE RUN BEFORE)
      AS=DMIN1(AS+17.D0,75D0)
      FR=.25*(3.+CHB/CHL)
       IR=IR+1
       CHI=CHL
         DO 18 I=1,NT
        CONS(I)=SCONS(I)
        BB(I)=CONS(I)
18      PC(I)=SPC(I)
        DO 19 I=1,NDT
19      PPCC(I)=SPPCC(I)
      IF(NW.GE.1)WRITE(*,191)IR
191   FORMAT(' IR=',I5)
      IF(IR.GT.20)THEN
        CHB=CHI
        RETURN
        ENDIF
20    CONTINUE
      IF(NW.GE.2)WRITE(*,173)FR
173   FORMAT(' FR=',E10.3)
      CHB=CHI
25    DE=DEXP(AS)
        KI=0
      DO 32 I=1,NT
        KI=KI+I
      PPCCD(I)=PPCC(KI)
32    PPCC(KI)=PPCC(KI)+SM(I)*DE
38      CONTINUE
        CALL GSOLVE(PPCC,A,PC,B,NT)
        KI=0
        DO 40 I=1,NT
        KI=KI+I
40      PPCC(KI)=PPCCD(I)
      ICYCLE=ICYCLE+1
C *** NOTE THAT B HERE IS THE SAME AS X IN GSOLVE
52    AJP=CHI
      IF(NW.GE.4)WRITE(*,10017)(B(I),I=1,NT)
        KI=0
      DO 60 K=1,NT
        KI=KI+K
      IF(DABS(B(K)).GT.1.D10)B(K)=DSIGN(1.D10,B(K))
      AJP=AJP+B(K)*(PC(K)+0.5*PPCC(KI)*B(K))
```

```
      IF(K.EQ.1)GOTO 60
      KM=K-1
        KT=KI-K
      DO 55 M=1,KM
        KT=KT+1
55    AJP=AJP+PPCC(KT)*B(M)*B(K)
60    CONTINUE
      IF(NW.GE.3)WRITE(*,104)AJP
104   FORMAT(' AJP=',E20.12)
C ****** AJP WAS CALCULATED ABOVE, WILL BE TESTED BELOW *************
C **** IF WE CAN FIND AJP < FR*CHI, WE CAN INTERPOLATE IN AS TO
C **** FIND A VALUE FOR WHICH AJP=FR*CHI, IFL = 1 INDICATES THIS
C *** CONDITION
      DT=AJP/DMAX1(1.D-35,CHI)
      IF(NW.GE.3)WRITE(*,105)DT
105   FORMAT(' DT=',E20.12)
      IF(DT.GE.FR)IAB=1
      IF(DT.LT.FR)IFL=1
      IF(DT-1.E0.LT.1.E-7)GOTO 1000
      IF(NW.GE.1)WRITE(*,109)AJP,AS
109   FORMAT(' THE MATRIX INVERSION APPEARS WRONG, AJP,AS',2E15.6)
C *********** THIS INVERSION IS WRONG, BUT A PREVIOUS ONE WAS RIGHT
        IF(NW.GE.3)WRITE(*,3421)IAA,AS
3421    FORMAT(' IAA,AS',I5,E20.6)
      IF(IAA.EQ.1)GOTO 1000
C *** THIS INVERSION IS WRONG AND WE HAVE NEVER HAD A CORRECT ONE
      IF(AS.GE.76.)THEN
        IR=I+1
        GOTO 10
        ENDIF
115   AS=AS+3.
      GOTO 25
120   CONTINUE
C FINDING X'S ON EACH SIDE OF FR*CHI
      IF(ICYCLE.GT.50)AS=DMIN1(75.D0,AS+30.E0)
      IF(ICYCLE.GT.30.OR.DABS((AJP-FR*CHI)/CHI).LT.TOL)GOTO 3117
      IF(AJP.GT.FR*CHI)GOTO 130
      IRT=IRT+1
       ILEFT=0
      AR=AS
      QR=AJP/CHI-FR
      GOTO 140
130   AL=AS
      QL=AJP/CHI-FR
      IRT=0
      ILEFT=ILEFT+1
140   X0=X1
      X1=X2
      IF(IFL*IAB.EQ.0)GOTO 1150
      IF(QB.GT.1.-FR)GOTO 115
      IF(DABS((AR-AL)).LT..1E-5)GOTO 3117
      AS=AL-QL*(AR-AL)/(QR-QL)
      X2=AS
      IF(NW.GE.2)WRITE(*,102)AL,QL,AR,QR,AS
102   FORMAT(' AL,QL',2E15.6,' AR,QR',2E15.6,' AS=',E15.6)
      IF(IRT.LT.2.AND.ILEFT.LT.2)GOTO 25
C AITKENS EXTRAPOLATION
```

```
      ALPHA=(X1**2-X0*X2)*(2.*X1-X2-X0)/((2.*X1-X2-X0)**2+1.E-37)
      IF(ALPHA.LT.AL.AND.AL.LT.AR)ALPHA=.5*(AL+AR)
      IF(ALPHA.GT.AL.AND.AL.GT.AR)ALPHA=.5*(AL+AR)
      IF(ALPHA.LT.AR.AND.AR.LT.AL)ALPHA=.5*(AL+AR)
      IF(ALPHA.GT.AR.AND.AR.GT.AL)ALPHA=.5*(AL+AR)
      IF(DABS((ALPHA-AR)/(AR-AL)).LT..1E-1.OR.
     #DABS((ALPHA-AL)/(AR-AL))
     # .LT..1E-1)ALPHA=.5*(AL+AR)
      AS=ALPHA
      IF(NW.GE.2)WRITE(*,103)AS
103   FORMAT(' AITKENS EXTRAPOLATION, AS',E15.6)
      ILEFT=0
      IRT=0
      GOTO 25
C ********* WE HAVE FOUND 0 < AJP < CHI ***********************
C ********* AND WE WANT TO SAVE THE BEST SET OF B'S **********
1000  QC=DABS(AJP/CHI-FR)
      IF(IAA.EQ.0)GOTO 1050
      IF(QC.GT.QB)GOTO 1120
1050  CHB=AJP
      IAA=1
      QB=QC
        ASB=AS
      IF(NW.GE.2)WRITE(*,117)QB,AJP,AS
117   FORMAT(' QB,AJP,AS',3E15.6)
      DO 1115 I=1,NT
1115  BB(I)=B(I)
      IF(QB.LT.TOL)GOTO 3117
1120  GOTO 120
C ********* TRY AGAIN WITH MORE OR LESS SMOOTHING **********
1150  AS=AS-4.0
        IF(IR.GT.1)AS=.5*(AS+ASB)
      IF(IAB.EQ.0)AS=AS+8.
      IF(AS.LE.-75.)GOTO 3117
      IF(AS.GE.76.)GOTO 3117
      GOTO 25
C ****************** EXITING WITH THE BEST COEFF'S FOUND ***********
3117  CONTINUE
      CHL=CHI
      DO 3119 I=1,NT
3119  CONS(I)=CONS(I)+BB(I)
      IF(NW.GE.3)WRITE(*,10017)(BB(I),I=1,NT)
10017 FORMAT(1X,6E13.6)
      IF(NW.GE.3)WRITE(*,10018)QB
10018 FORMAT(' THE FINAL QB IS',E20.12)
          IF(NW.GE.1)PRINT'(A,G11.4)',' FINAL AS IS',AS
      RETURN
2500    WRITE(*,4367)NT
4367    FORMAT(' NT OF',I5,' IS TOO LARGE FOR THE'/
     # ' DIMENSION OF B AND BB AND ALSO TERMS IN GSOLVE')
        STOP
      END
        SUBROUTINE SMINV(AP,N,IFL)
        IMPLICIT REAL*8 (A-H,O-Z)
        DIMENSION AP(1)
C *** THIS ROUTINE BEGINS IN LINPACK CHAPTER 3 - POSITIVE
C *** DEFINITE SYMMETRIC MATRICES
```

```
C *** IN OUR CASE (CURVE FITTING) WE CAN ALWAYS WRITE THE APPROXIMATE
C *** FUNCTION AS SUM C*SJ(XI).  IF WE THEN FORM X**2 = SUM ON I
C *** OF SUM ON J (CJ*SJ(XI))**2 WE FIND X**2 = SUM ON K CK *
C *** SUM ON J CJ * AJK WHERE AJK IS SUM ON I SJ(XI)*SK(XJ), THE
C *** POSITIVE DEFINITE SYMMETRIC MATRIX OF INTEREST HERE
C
C *** AP IS A PACKED MATRIX AS IN THE FOLLOWING
C *** K=0
C     DO 20 J=1,N
C     DO 20 I=1,J
C     K=K+1
C     AP(K)=A(I,J)
C 20  CONTINUE
        IFL=0
        JJ=0
        DO 30 J=1,N
        S=0
        KJ=JJ
        KK=0
        IF(J.LT.2)GOTO 20
        JM1=J-1
        DO 10 K=1,JM1
        KJ=KJ+1
        T=AP(KJ)
        KM1=K-1
        IF(KM1.LT.1)GOTO 8
        DO 5 I=1,KM1
5       T=T-AP(I+KK)*AP(I+JJ)
8       KK=KK+K
        T=T/AP(KK)
        AP(KJ)=T
        S=S+T*T
10      CONTINUE
20      CONTINUE
        JJ=JJ+J
        SMT=DMAX1(S,AP(JJ))
        DIFF=AP(JJ)-S
        IF(DIFF.GT.1.D-10*SMT.AND.DIFF.GT.1.D-70)GOTO 25
C        WRITE(*,1975)J,AP(JJ),S
1975    FORMAT(' POS DEF QUANT LE 0. J=',I5,' AP(JJ)=',E10.3,
     # ' S=',E10.3)
C *** FIX IS MORE SMOOTHING FOR A(J,J)
        IFL=-1
        DIFF=1.D30
25      AP(JJ)=DSQRT(DIFF)
30      CONTINUE
C *** NEXT WE CONSTRUCT THE INVERSE MATRIX
        KK=0
        DO 100 K=1,N
        K1=KK
        KK=KK+K
        AP(KK)=1/AP(KK)
        T=-AP(KK)
        KM=K-1
        IF(KM.LE.0)GOTO 86
        DO 85 I=1,KM
85      AP(I+K1)=T*AP(I+K1)
```

```
86        KP1=K+1
          J1=KK
          KJ=KK+K
          IF(N.LT.KP1)GOTO 90
          DO 80 J=KP1,N
          T=AP(KJ)
          AP(KJ)=0
          DO 70 I=1,K
70        AP(J1+I)=AP(J1+I)+T*AP(K1+I)
          J1=J1+J
          KJ=KJ+J
80        CONTINUE
90        CONTINUE
100       CONTINUE
          JJ=0
          DO 130 J=1,N
          J1=JJ
          JJ=JJ+J
          JM1=J-1
          K1=0
          KJ=J1+1
          IF(JM1.LT.1)GOTO 120
          DO 110 K=1,JM1
          T=AP(KJ)
          DO 105 I=1,K
105       AP(I+K1)=AP(I+K1)+T*AP(I+J1)
          K1=K1+K
          KJ=KJ+1
110       CONTINUE
120       CONTINUE
          T=AP(JJ)
          DO 125 I=1,J
125       AP(I+J1)=     T*AP(I+J1)
C ABOVE LINE ORIGINALLY HAD ANOV
130       CONTINUE
140       CONTINUE
          RETURN
          END
          FUNCTION ANOV(X,Y)
          IMPLICIT REAL*8 (A-H,O-Z)
          EQUIVALENCE (IX,XT),(IY,YT)
C         ANOV='FFFF7CFF'X*DSIGN(1.D0,X)*DSIGN(1.D0,Y)
          XT=X
          YT=Y
C         IF((IX.AND.'7F80'X)+(IY.AND.'7F80'X).LT.'BC80'X)
C     #   ANOV=XT*YT
           ANOV=XT*YT
          RETURN
          END
          SUBROUTINE GSOLVE(A,AM,B,X,NT)
          IMPLICIT REAL*8 (A-H,O-Z)
C *** SOLVES AX=-B IN CASES WHERE THE MATRIX CAN BE INVERTED CHEAPLY
C *** AND ACCURATELY WITH NO PROBLEMS
          DIMENSION A(1),AM(1),B(1),X(1)
          NTD=NT*(NT+1)/2
          DO 10 I=1,NTD
10        AM(I)=A(I)
```

```
      CALL SMINV(AM,NT,IFL)
      CALL SMMULT(AM,B,X,NT)
      DO 20 I=1,NT
20    X(I)=-X(I)
      RETURN
      END
      SUBROUTINE SMMULT(A,B,C,N)
      IMPLICIT REAL*8 (A-H,O-Z)
      DIMENSION A(1),B(1),C(1)
C *** FOR PACKED MATRICES C(N)=A(N*(N+1)/2)*B(N)
      DO 10 I=1,N
10    C(I)=0
      K=0
      DO 20 I=1,N
      IM=I-1
      DO 15 J=1,IM
      K=K+1
      C(I)=C(I)+A(K)*B(J)
15    C(J)=C(J)+A(K)*B(I)
      K=K+1
20    C(I)=C(I)+A(K)*B(I)
      RETURN
      END
      FUNCTION KIJ(I1,I2)
C      WRITE(*,'(A,2I5)')' IN KIJ I1,I2',I1,I2
      I=MIN0(I1,I2)
      J=MAX0(I1,I2)
      KIJ=(J*(J-1)/2)+I
      RETURN
      END

C *** Then substitute one of the following poly routines as required.

C *** For Randomly placed splines use
      SUBROUTINE POLY(X,P,NV,ND,CONS,FA)
      IMPLICIT REAL*8 (A-H,O-Z)
      DIMENSION P(1),X(1),CONS(1)
      DIMENSION AK(40)
      SAVE AK,NC
      DATA NC/0/
      NKN=NV-4
      IF(NC.GE.1)GOTO 15
      NC=1
      PRINT*,' ENTER ',NKN,' KNOT LOCATIONS'
      READ(*,*)(AK(J),J=1,NKN)
15    P(1)=1
       DO 20 I=2,4
20     P(I)=X(1)*P(I-1)
      DO 30 I=5,NV
      XM=AK(I-4)-X(1)
      P(I)=0
      IF(XM.LT.0.D0)GOTO 30
      P(I)=XM**3
30    CONTINUE
      FA=0
      DO 32 I=1,NV
32    FA=FA+CONS(I)*P(I)
```

```
      RETURN
      END

C *** For fixed knot splines use

      SUBROUTINE POLY(X,P,NV,ND,CONS,FA)
      IMPLICIT REAL*8 (A-H,O-Z)
      DIMENSION P(1),X(1),CONS(1)
      DIMENSION AK(40)
      NKN=NV-4
      IF(NKN.GT.0)AKN=40./NKN
      DO 10 I=1,NKN
10    AK(I)=I*AKN
       P(1)=1
       DO 20 I=2,4
20     P(I)=X(1)*P(I-1)
      DO 30 I=5,NV
      XM=AK(I-4)-X(1)
      P(I)=0
      IF(XM.LT.0.D0)GOTO 30
      P(I)=XM**3
30    CONTINUE
      FA=0
      DO 32 I=1,NV
32    FA=FA+CONS(I)*P(I)
      RETURN
      END

C *** For a spline fit use

       SUBROUTINE POLY(X,P,NV,ND,CONS,FA)
      IMPLICIT REAL*8 (A-H,O-Z)
      DIMENSION P(1),X(1),CONS(1)
      DIMENSION AK(40)
      NKN=NV-4
       P(1)=1
       DO 20 I=2,4
20     P(I)=X(1)*P(I-1)
      DO 30 I=5,NV,2
      XM=CONS(I+1)-X(1)
      P(I)=0
      P(I+1)=0
      IF(XM.LT.0.D0)GOTO 30
      P(I)=XM**3
      P(I+1)=3*CONS(I)*XM*XM
30    CONTINUE
      FA=0
      DO 32 I=1,4
32    FA=FA+CONS(I)*P(I)
      DO 34 I=5,NV,2
34    FA=FA+CONS(I)*P(I)
      RETURN
      END

C *** For a polynomial fit

      SUBROUTINE POLY(X,P,NV,ND,CONS,FA)
```

```
      IMPLICIT REAL*8 (A-H,O-Z)
      DIMENSION P(1),X(1),CONS(1)
       P(1)=1
       DO 10 I=2,NV
10     P(I)=X(1)*P(I-1)
      FA=0
      DO 32 I=1,NV
32    FA=FA+CONS(I)*P(I)
      RETURN
      END
```

References

1. A. C. Rester, R. L. Coldwell, F. E. Dunnam, G. Eichhorn, J. I. Trombka, R. Starr, and G. P. Lasche, "Gamma-ray observations of SN 1987A from Antarctica," Astrophys. J., **342**, L71-3 (July 15, 1989).
2. G. F. Knoll, *Radiation Detection and Measurement*, New York, John Wiley & Sons (1989).
3. P. Quittner, *Gamma-ray Spectroscopy*, London, Adam Hilger (1973), p. 67.
4. G. W. Phillips and K. W. Marlow, "Automatic analysis of gamma-ray spectra from germanium detectors," Nucl. Instrum. Methods **137**, 525–536 (1976).
5. M. Hillman, "Computer analysis of gamma and x-ray spectra," Nucl. Instrum. Methods **135**, 363–368 (1976).
6. L. Kokta, "Determination of peak area," Nucl. Instrum. Methods **112**, 245–251 (1973).
7. D. D. Burgess, "A comparison of methods for baseline estimation in gamma-ray spectometry," Nucl. Instrum. Methods **221**, 593–599 (1984).
8. M. A. Mariscotti, "A method for automatic identification of peaks in the presence of background and its application to spectrum analysis," Nucl. Instrum. Methods **50**, 309–320 (1967).
9. A. Robertson, W. V. Prestwich and T. J. Kennett, "An automatic peak-extraction technique," Nucl. Instrum. Methods **100**, 317–324 (1972).
10. R. G. Helmer and M. A. Lee, "Analytical functions for fitting peaks from Ge semiconductor detectors," Nucl. Instrum. Methods **178**, 499–512 (1980).
11. P. R. Bevington, *Data Reduction and Error Analysis for the Physical Sciences*, New York, McGraw-Hill (1969).
12. R. L. Coldwell, "Iterative codes for fitting complete spectra," Nucl. Instrum. Methods **A242**, 455–461 (1986).
13. R. L. Coldwell, *Radiative Properties of Hot Dense Matter*, Singapore, World Scientific (1983), pp. 315–349.
14. I. J. Schoenberg, "Contributions to the problem of approximation of equidistant data by analytic functions," Quant. Appl. Math. **4**, 67 (1946).
15. J. C. Holladay, "A Smoothest Curve Approximation," Math. Tables and Aids to Computation, **11**, 233–243 (1957).
16. J. H. Ahlberg, E. N. Nilson, and J. L. Walsh, *The Theory of Splines and Their Applications*, New York, Academic Press (1967), p. 3.
17. C. de Boor, *A Practical Guide to Splines*, New York, Springer-Verlag (1978), p. 180.
18. S. Baker and R. D. Cousins, "Clarification of the use of chi-square and likelihood functions in fits to histograms," Nucl. Instrum. Methods **221**, 437–442 (1984).
19. G. W. Phillips, "Fitting peaks with very low statistics," Nucl. Instrum. Methods **153**, 449–455 (1978).
20. W. H. Press, B. P. Flannery, S. A. Teukolsky, and W. T. Vetterling, *Numerical Recipes*, New York, Cambridge University Press (1987).
21. W. T. Morton, "Knot positions in least-squares fitting of data using cubic splines," Nucl. Instrum. Methods **A272**, 861–865 (1988).

22. D. Mihalas, *Stellar Atmospheres*, New York, Freeman (1979).
23. W. R. Falk, "Data reductions from experimental histograms," Nucl. Instrum. Methods **220**, 473–478 (1984).
24. J. Christensen-Dalsgaard, "An Overview of Helio- and Asteroseismology," *Advances in Helio- and Asterioseismology*, I.A.U. Symposium (123d, 1986, Arhus, Denmark), pp 3–18.
25. G. J. Bamford, H. B. van der Raay, P. L. Palle, and T. Roca Cortes, "Solar Oscillations from Two Widely Separated Stations," Proc. of the Fourth European Meeting on Solar Physics, 1–3 Oct. 1984, Noordwijkerhout, the Netherlands, *The Hydromagnetics of the Sun.*
26. T. L. Duvall, J. W. Harvey, and M. A. Pomerantz, "Intermediate Degree Solar Oscillations," *Advances in Helio- and Asterioseismology*, I.A.U. Symposium (123d, 1986, Arhus, Denmark), pp. 37–40.
27. E. R. Anderson, T. L. Duvall, and S. M. Jefferies, *Modeling of Solar Oscillation Power Spectra*, Tucson, Ariz., National Optical Astronomy Observatories (to be published).
28. R. B. Welch, F. Gyger, D. T. Jost, H. R. von Gunten, and U. Krähenbühl, "Newfit, a computer program for the analysis of alpha, x-ray and gamma-ray spectra," Nucl. Instrum. Methods **A269**, 615–622 (1988).
29. H. Machner, "Automatic analysis of γ-ray spectra including multiplets," Nucl. Instrum. Methods **A258**, 246–249 (1987).
30. T. J. Kennett, W. V. Prestwich, and R. J. Tervo, "An automated background estimation procedure for gamma-ray spectra," Nucl. Instrum. Methods **216**, 205–218 (1983).
31. D. D. Burgess and R. T. Tervo, "Background estimation for gamma-ray spectrometry," Nucl. Instrum. Methods **214**, 431–434 (1983).
32. G. Winter, "Continuum estimation and peak analysis for in-beam gamma ray spectra," Nucl. Instrum. Methods **A258**, 119–126 (1987).
33. W. Westmeier, "Background subtraction in Ge(Li) gamma-ray spectra," Nucl. Instrum. Methods **180**, 205-210 (1981).
34. K. Debertin and R. G. Helmer, *Gamma- and X-Ray Spectrometry with Semiconductor Detectors*, Amsterdam, North-Holland (1988).

INDEX

Printed in the United States
By Bookmasters